BREEDING FOR DISEASE RESISTANCE

Developments in Plant Pathology

VOLUME 1

Breeding for Disease Resistance

Edited for the British Society for Plant Pathology by:

R. JOHNSON

Cambridge Laboratory
Norwich, U.K.

and

G. J. JELLIS

Plant Breeding International Cambridge
Cambridge, U.K.

Reprinted from Euphytica 63(1–2)

KLUWER ACADEMIC PUBLISHERS

DORDRECHT / BOSTON / LONDON

Library of Congress Cataloging-in-Publication Data

Breeding for disease resistance / edited by R. Johnson and G.J. Jellis.
p. cm. -- (Developments in plant pathology ; v. 1)
Includes index.
ISBN 0-7923-1607-X (HB : acid free)
1. Plants--Disease and pest resistance--Genetic aspects--Congresses. 2. Plant breeding--Congresses. I. Johnson, R. II. Jellis, G. J. III. Series.
SB750.B7 1993
632.3--dc20 92 39161

ISBN 0-7923-1607-X

Proceedings of the International Conference on
Breeding for Disease Resistance
held at Newcastle-upon-Tyre, U.K.
on 16–19 December 1991

Published by Kluwer Academic Publishers,
P.O. Box 17, 3300 AA Dordrecht, The Netherlands.

Kluwer Academic Publishers incorporates
the publishing programmes of
D. Reidel, Martinus Nijhoff, Dr W. Junk and MTP Press.

Sold and distributed in the U.S.A. and Canada
by Kluwer Academic Publishers,
101 Philip Drive, Norwell, MA 02061, U.S.A.

In all other countries, sold and distributed
by Kluwer Academic Publishers Group,
P.O. Box 322, 3300 AH Dordrecht, The Netherlands.

Printed on acid-free paper

Printed in the Netherlands

Table of Contents

Euphytica **63**: 1, 1992.
R. Johnson and G.J. Jellis (eds), Breeding for Disease Resistance

Preface

The control of disease in crops is vital in a world with food shortages in many regions and a demand for greater efficiency in food production coupled with environmental protection in others. Breeding for disease resistance can control crop losses cost-effectively, with the minimum of effort for growers and in an environmentally benign manner.

The material presented here is based on the contributions of participants at a conference entitled 'Breeding for Disease Resistance' organised by the British Society for Plant Pathology at the University of Newcastle-upon-Tyne in 1991. It includes both classical and biotechnological developments in resistance breeding, using a number of specific case histories to demonstrate successes and failures in breeders' efforts to control diseases. The problems of pathogen variation and strategies for providing durable resistance are discussed in depth.

There is no attempt to advocate a single method or approach in breeding for resistance. Indeed, by spanning different crops and types of disease, this publication illustrates how breeding for resistance faces a great diversity of challenges requiring a range of different solutions.

R. Johnson
G.J. Jellis
British Society for Plant Pathology

Acknowledgements. The British Society for Plant Pathology gratefully acknowledges financial help towards the costs of organising the meeting on which this publication is based, from the following organisations: *Plant Breeding International Cambridge; British Crop Protection Council; Booker Seeds; ICI Seeds; Elsoms Seeds; Nickerson Seeds; Sogetal.*

Euphytica **63**: 3–22, 1992.
R. Johnson and G.J. Jellis (eds), Breeding for Disease Resistance

Past, present and future opportunities in breeding for disease resistance, with examples from wheat

Roy Johnson
Cambridge Laboratory, AFRC Institute of Plant Science, John Innes Centre, Colney, Norwich NR4 7UJ, UK

Key words: wheat, *Triticum aestivum,* yellow (stripe) rust, *Puccinia striiformis,* septoria, *Septoria tritici, S. nodorum,* eyespot, *Pseudocercosporella herpotrichoides,* resistance genetics, pathogen variation, durable resistance

Summary

This introductory chapter contains some general comments about plant breeding and breeding for disease resistance. The use of disease resistant crop plants is an environmentally favourable method of controlling disease but the process of breeding for disease resistance is subject to several constraints. Among them is the variability of pathogens in relation to host resistance. Some parts of this variation can be resolved into gene-for-gene interactions, but the boundaries within which such interactions can be detected are not sharp. The discussion of this variation is illustrated by reference to some important diseases of wheat, especially yellow rust, septoria and eyespot. The objective of obtaining durable resistance is discussed and some contributions of new genetical and molecular techniques to breeding for resistance are considered. It is suggested that new technology will enhance breeding for disease resistance but that established techniques of plant breeding will remain relevant and important.

Introduction

Breeding for disease and pest resistance is one method of protecting crops from damage due to biotic factors. Inherited resistance is a valuable attribute because it is easy for the grower to use and reduces the need for other methods of control, including the application of chemicals. However, in common with other methods of control, it is subject to some significant biological and financial constraints. In this chapter aspects of breeding for resistance are illustrated with examples mainly from fungal diseases of wheat. These demonstrate an important feature, which is that depending on the type of pathogen, the challenges are very diverse. Thus there is no single model for breeding for resistance, a point that needs to be made because of the many generalisations that are written about disease resistance, its genetic basis and how to obtain it by breeding. Across a wider range of crops this point is even more relevant and before considering particular diseases some comments are included about general features of plant breeding that have a bearing on breeding for disease resistance.

Aspects of plant breeding

Plant breeding is conducted to improve crops ranging from inbreeding or outbreeding annuals through to biennials and diverse perennials including slow-growing trees. It includes crops propagated by seeds and by various vegetative methods and these differences have significant effects on the conduct of the programmes. Occasionally, breeding efforts may be directed towards changing a single character such as an agronomic trait or perhaps a disease resistance character. More frequently the aim is to improve several characters simultaneously whilst maintaining others at levels no lower than in previous cultivars. To achieve this requires the growing and testing of large numbers of plants obtained from numerous crosses between parents carrying desirable characters. For example the winter wheat breeding programme at the Plant Breeding Institute, Cambridge (PBI) was conducted from about 1,500 crosses annually and the F2 plants grown at 10 cm intervals occupied over 200 km of rows and comprised over 2 million plants each year (J. Bingham, Plant Breeding International Cambridge (PBIC), personal communication). Following the F2 selection, F3 and later generations were grown to obtain data on numerous characters including yield and to re-establish genetic homozygosity to give the required uniformity for identification of cultivars. From all this material perhaps only five individual selections would be submitted in any one year to the national trials system. Thus very few of the original crosses ever reached the end of the programme, though many of them were recycled within the crossing system. This description indicates the dynamic nature of a typical breeding programme. Because the characters under selection are usually under the control of separate genetic systems it is very difficult to combine together in a single individual all the required characters, and in fact each new cultivar is a compromise between the optimal and the achievable. As a result resistance to disease may sometimes be less than optimal in a new cultivar. Despite any such weaknesses, the programme conducted at the PBI and PBIC was and remains highly successful in producing new wheats acceptable to commercial farmers and which predominated in the UK wheat area throughout the 1970 to 1990 period and beyond.

Breeding for disease resistance

Deciding which diseases are important

An important consideration in breeding for resistance is to set the appropriate level of priority. Attempts to breed for resistance may be frustrated if the result is seriously detrimental to some other character, such as yield. Such effects may be due to genetic linkage and are, unfortunately, not uncommon. Also, since selection for resistance, like all other characters, requires resources, greatest efficiency will be achieved by placing it at as low a priority level as is consistent with production of a useful cultivar. This in turn depends on the actual and potential importance of a disease and the possibility of incorporating a useful level of resistance during the breeding process and the options for other methods of control.

Some diseases are so prevalent that they are an obvious target for control by plant breeding. For others the decision is less clear and accurate data on disease incidence in crops may not be available. Thus there may be a subjective element in the choice of diseases to target. One challenge of breeding for resistance is to identify or keep in mind diseases that may not be currently important because existing cultivars carry sufficient resistance. The breeder may thus need to expend energy on selecting for resistance, or against high susceptibility for apparently unimportant diseases. The incidence of yellow rust recorded in UK wheat disease surveys between 1970 and 1988 was usually low (Polley & Thomas, 1991) and this might suggest that yellow rust was not an important disease. In fact the usual low incidence of this disease in the UK wheat crop is because of the great efforts by the main UK wheat breeders to produce, and the National Institute of Agricultural Botany to recommend, only cultivars with resistance above a set minimum level. However, this cannot entirely eliminate the risk of epidemics. Even with modern fungicides it is difficult to control yellow rust in highly susceptible cultivars as was illustrated recently by the development of an epidemic when the cultivar Slejpner became susceptible during its commercial use (Bayles et al., 1989). Incidence of yellow rust on highly susceptible cultivars occurs despite the availability of effective fungicides because, due to weather or human factors, these cannot always be used optimally. Fortunately, the potential threat of yellow rust has been known for many years and resistance breeding has remained a high priority in most programmes in the UK. In other countries this is not so and, for example, occasional epidemics of yellow rust on bread wheats in Italy are attributed to the cultivation of highly susceptible cultivars due to a lack of selection for resistance in breeding programmes, combined with more favourable environmental conditions for yellow rust in some years (Chilosi & Corazza, 1990).

Occasionally a disease that has not even been considered to be important may develop in a crop. For example, as a result of wet weather in the UK in summer 1982, following its commercial release, the wheat cultivar Avalon was discovered to be more susceptible to infection with *Fusarium* ear blight than most other cultivars (Lupton, 1983). No selection for resistance to this disease was practised during the breeding of Avalon, but subsequently a method was developed for assessing susceptibility to *Fusarium* spp. by spraying the emerging ears with spore suspensions and maintaining high humidity by mist irrigation (Jenkins, 1984). Using this technique, sources of resistance were identified and exploited.

Reference is made to possible loss of resistance if breeders forget about a disease because it has been successfully controlled for many years (Fletcher, this volume) and to increases in susceptibility to one disease when selection for resistance to another disease was applied (Williamson, this volume).

Surveys of disease incidence are useful in indicating the relative importance of diseases, thus providing information for breeders deciding on priorities. However, as illustrated above, the incidence of disease in commercial crops does not always indicate the potential importance of diseases for the breeder.

The need for sources of resistance

Another criterion that will influence the breeder in choosing targets for breeding for resistance will be information concerning the availability of sources of resistance. For some diseases resistance is readily available, for others much less so. For wheat there are numerous sources of resistance to rust diseases, although not all are of equal value, whereas for diseases such as eyespot, resistance is more difficult to find and to exploit (see below).

Fortunately, resistance to many diseases has been identified and often exploited by breeding, but there are some examples where breeding for resistance has been virtually impossible due to the failure, despite much work, to identify useful sources. This problem is exemplified by the search for resistance to the take-all disease of wheat which, despite many optimistic reports, has not yet identified useful levels of exploitable resistance (Scott, 1981) though the search still continues (Conner et al., 1988). Lack of resistance to some virus diseases is referred to by Fraser (this volume). There are numerous examples where resistance has not been found in existing crop cultivars but in related species from which successful transfers have been achieved or may be in the future. Some examples are given by Knott & Dvorak (1976) and there are many others including the origins of much resistance in raspberries (Williamson, this volume) and in potatoes (Jellis, this volume).

Selecting for resistance

The feasibility of breeding for disease resistance also depends on an appropriate method of selection and this is much more difficult for some diseases than others. Selection for resistance to foliar diseases is often relatively simple by assessing leaf areas affected and sometimes also the type of lesion produced. For root or stem-base diseases, assessments are often more difficult. Usually the pathogen must be present to select for resistance but sometimes linked characters can be used to avoid the need for a direct test. Opportunities for selecting linked characters are likely to increase with the use of molecular techniques which may permit the production of molecular markers genetically linked to resistance genes. Where the disease must be present there is the possibility of using natural infection, or quite often of assisting nature by promoting infection by manipulation of the environment and by the deliberate introduction of the pathogen either in the field or in protected or controlled environments. Where deliberate introduction of pathogens in the field is contemplated consideration must always be given to potential damage to surrounding or subsequent crops.

Pathogen variation in relation to host resistance

All pathogens are variable with respect to host resistance but the variation is, in itself, a variable quality. At one extreme, notably for pathogens that are biotrophic and grow in intimate contact with living host cells, such as rust and powdery mildew pathogens, highly developed specificity occurs. When several different pathogen isolates are used separately to infect an array of host cultivars they often show a variable ability to infect subsets of the array.

Gene-for-gene interactions

Flor (1946) demonstrated that this variation was due to the operation of a gene-for-gene interaction between the host and the pathogen. In this interaction a host gene for resistance possesses alternative alleles (R resulting in resistance and r not resulting in resistance). This locus interacts with a specific, corresponding gene in the pathogen which also has alternative alleles (Av resulting in avirulence and v resulting in virulence). The four possible combinations of the alleles give resistance/avirulence when R in the host coincides with Av in the pathogen and susceptibility for the three other combinations R/v, r/Av and r/v, a pattern that has been referred to as a quadratic check. With a different resistance gene the two allelic forms interact with a different avirulence/virulence gene in the pathogen. This is the critical criterion of gene-for-gene interactions. It cannot be illustrated by the quadratic check for a single corresponding pair of genes because this does not confirm the specificity. Thus the minimum number of gene pairs to depict a gene-for-gene interaction is two (Table 1) which was the original pattern on which Flor formulated the hypothesis (see Day, 1974). This pattern of interaction has implications for investigations of the biochemical source of specificity involved (Johnson & Knott, 1992) and also for plant breeding for disease resistance.

For breeding, the operation of a gene-for-gene system has the implication that resistance will not remain effective if the pathogen acquires the corresponding virulence by losing the avirulence allele that elicits resistance, either by deletion or by genetic change. It also shows that resistance genes can be combined together and that the pathogen must evade the effect of each gene by change at a specific, corresponding locus. Thus it must accumulate the necessary changes to allow virulence. An important question for plant breeding, if such variation is possible, is how quickly the changes will occur leading to failure of resistance. In practice this is very variable and depends on many factors including the extent of deployment of the resistance, the epidemiology and population size of the pathogen and its potential to vary.

Table 1. A gene-for-gene interaction between two host loci for reaction each with two alleles (R1 or r1, R2 or r2) and the corresponding pathogen loci for pathogenicity (Av1 or v1, Av2 or v2)

Pathogen	Host alleles			
	R1R2	R1r2	r1R2	r1r2
Alleles				
Av1, Av2	I[1]	I	I	C[2]
Av1, v2	I	I	<u>C</u>[3]	C
v1, Av2	I	<u>C</u>	I	C
v1, v2	C	C	C	C

[1] I = incompatible interaction (resistant/avirulent).
[2] C = compatible interaction (susceptible/virulent).
[3] __ underlined cells show that R2 does not give incompatibility with Av1 and R1 does not give incompatibility with Av2. This demonstrates the specificity of the R1r1/Av1v1 and R2r2/Av2v2 interactions as originally observed by Flor (see Day, 1974).

General applicability of the gene-for-gene hypothesis

The existence of gene-for-gene interactions has been demonstrated or implicated for many host-pathogen interactions including those with fungi, bacteria (Crute, 1985) and viruses (Fraser, 1985) and its occurrence has had profound consequences for breeding for disease resistance. These have included many examples where initially resistant cultivars have become susceptible. However, the consequences have not been limited to the occurrence of failures of resistance but have included the generation of many arguments about breeding for disease resistance. Foremost among these is whether all resistance operates on a gene-for-gene basis or whether some resistance acts thus while other resistance does not. Views expressed on these topics have often been influenced by the personal experience of workers with particular types of interactions between host and pathogen. For example, it can be suggested that evolution to resistance in the host would be likely to be followed by evolution to virulence in the pathogen, leading to a step-wise progression to a gene-for-gene system (Person, 1959). Thus it could be argued that resistance would always follow a gene-for-gene system, as some publications have suggested (Parlevliet, 1981; Ellingboe, 1975). Ellingboe (1975) argued that failure to demonstrate gene-for-gene interactions was due to inadequate experimental technique.

An alternative view is that although some resistance can be resolved into gene-for-gene interactions, there is other resistance that shows no such interaction but is independant of variation in the pathogen. There are two conditions in which this could be conjectured to occur: one is that in the interaction of a particular species of pathogen with a host species, no discernable gene-for-gene interaction could be observed; the other is where a gene-for-gene interaction occurs but may not account for all known resistance. In the latter case, two further important questions arise. The first is whether resistance not currently accounted for by a gene-for-gene interaction will do so in the future. The second is whether, if there is resistance that does not interact on a gene-for-gene pattern, it can readily be distinguished from that which does or will in the future. Vanderplank (1963) proposed that there were two distinct classes of resistance, vertical which was effective only against certain races and therefore interacted in a gene-for-gene system, and horizontal which was equally effective against all races and outside a gene-for-gene system. Obviously resistance that is not overcome due to a gene-for-gene interaction is of interest to plant breeders because of the promise it holds for achieving more durable resistance. This question is more fully discussed in relation to the resistance of wheat to yellow rust below. Unfortunately, because of the diversity of host-pathogen interactions there is, in fact, no general answer to any of these questions and each must be studied in considerable depth to understand its special features.

Another important consideration in understanding the genetic basis of host-pathogen interactions is that the simplest ideal model of the gene-for-gene interaction is based on the generality that each corresponding gene pair acts independently of the other corresponding gene pairs and that dominance occurs for the resistance allele in the host and the avirulence allele in the pathogen. This was based on Flor's original observations in flax and the flax rust pathogen which were consistent with these conditions (Flor, 1960). However, as more data were accumulated from other host-pathogen systems examples of recessive resistance that act in gene-for-gene systems were encountered. For example, genes *Yr2* and *Yr6* giving race-specific resistance to yellow rust are recessive in at least some crosses and with some pathogen isolates (Singh & Johnson, 1988; El-Bedewy & Röbbelen, 1982). Determination of dominance or recessiveness can depend on the character by which resistance is estimated (see Fraser, this volume). There are also several examples of resistance genes that do not act independently of each other. While genes promoting extreme resistance are usually expressed independently of other genes that may provide lower levels of resistance, there are examples where genes conditioning intermediate levels of resistance act together to give a higher level of resistance. Such combined effects may be additive or non-additive. Examples of non-additive interaction include a few well-documented examples of complementary action of two genes to produce resistance (e.g. Singh & McIntosh, 1984). Examples that may be additive include some genes for resistance to brown (leaf) rust in wheat (Dyck & Kerber, 1985). There are also inhibitors of resistance genes (Kerber & Green, 1980) and inhibitors of avirulence in pathogens, for example in the flax rust pathogen *Melampsora lini* (Lawrence et al., 1981; Jones, 1988).

In addition, expression of many genes is affected by environmental conditions such as light and temperature. Where this involves obligate pathogens it is often not possible to decide whether the observed effects are primarily on the host, the pathogen or on the interaction between them.

Because of these variations in gene action it is not always simple to interpret data on segregation for resistance in crossed materials or to make a full inventory of the action of resistance genes and avirulence genes in a given host-pathogen system in which a gene-for-gene interaction functions.

Pathogenic variation without clear gene-for-gene interactions

As implied above, the full elucidation of gene-for-gene interactions may not be simple. This contributes to the challenge of identifying resistance that may not interact on a gene-for-gene basis and similarly, variations in pathogenicity also possibly independent of gene-for-gene interactions.

Several terms have been used to describe pathogen variation of this type, such as aggressiveness or non-specific pathogenicity. These are used to indicate that pathogen isolates may differ in their ability to cause disease on all host cultivars rather than selectively or differentially on groups of host cultivars. In practice, it may not be easy to decide whether or not there is any differential pathogenicity, such as might fall within the definition of a gene-for-gene interaction.

This has led to various interpretations of existing data in several host-pathogen systems. Some examples are facultative pathogens causing necrotic symptoms on leaves, where maximum susceptibility occurs with maximum necrosis, in contrast to the obligate pathogens such as rusts, where maximum susceptibility occurs when the pathogen causes the least amount of necrosis and grows vigorously in an apparently compatible interaction with the host. Among the facultative pathogens there are several examples where isolates have been compared for their infections of sets of host cultivars. Significant differences in resistance between cultivars and in pathogenicity between isolates have been observed and frequently also, significant pathogen isolate × host cultivar interactions. The question then arises as to whether distinctive pathogen races can be identified, with specific virulence for particular cultivars and whether the interaction indicates a gene-for-gene system.

An example that illustrates this problem is the *Septoria tritici* leaf blotch disease of wheat. Eyal et al. (1973) tested 14 wheat accessions, including hexaploid and tetraploid types, with five cultures of the pathogen and 10 of the same accessions at the boot stage with two of the five cultures. They used the occurrence of pycnidia as a measure of susceptibility and noted large differences in resistance of the accessions and the overall pathogenicity of the cultures. They also stated that the cultures did not follow the same ranking for virulence on the wheat lines and that this indicated their differential interaction with the hosts. They concluded that, contrary to reports from other countries, they showed differential interactions and therefore physiologic specialization of the pathogen (Table 2). The main source of the differential interaction was that a pathogen culture (culture 12) taken from a durum wheat was specifically pathogenic on a durum accession Etit 38 and only weakly pathogenic on hexaploid wheat whereas a culture isolated from a bread wheat (culture 18) had very low pathogenicity on Etit 38 and higher pathogenicity on hexaploid wheats. Apart from this, other possible interactions between isolates and bread wheats were much smaller. Using more extensive data on resistance of wheat and triticale cultivars to isolates of *S. tritici* Eyal et al. (1985) assumed that there was a gene-for-gene interaction between this pathogen and its hosts. The method used to reach this conclusion was to identify in the matrix of data a 'cutpoint' by which to divide the data into resistant and susceptible classes. The method used was to subject weighted means for level of infection of each cultivar to cluster analysis to group the cultivars, with the purpose of identifying a dividing line between those with moderate means. A standard error was computed for the cutpoint and data were then reformulated into resistant and susceptible classes for each isolate. They then applied computer programmes based on gene-for-gene interactions to estimate numbers of resistance genes in the hosts and corresponding virulence genes in the pathogen.

Van Ginkel & Scharen (1988) studied the reaction of numerous durum wheat cultivars to many *S. tritici* isolates taken from durum and bread wheats. They used a single susceptible bread wheat cultivar as a control. Large differences were observed in the resistance of cultivars and in the pathogenicity of pathogen isolates but interaction between isolates and cultivars was not significant. They pointed out that the mean square value for interaction was less than 0.3% of the sum due to cultivars and isolates and that a similar proportion of the variance was due to interaction in the study of Eyal et al. (1985). They considered that their data did not justify the supposition of a gene-for-gene interaction between durum wheat cultivars and *S. tritici* isolates. They concluded, however, that there was sufficient evidence to justify the hypothesis of specificity towards different wheat species such as *T. durum* and *T. aestivum*.

Table 2. Percent of leaf area occupied by pycnidia of *Septoria tritici* in wheat accessions infected at the boot stage with cultures from bread or durum wheats. Adapted from results of Eyal et al. (1975)

Variety	Wheat	*Septoria tritici* cultures (Wheat source)	
		12 (Durum)	18 (Bread)
Bulgaria 88	Bread	0.5	17.5
Etit 38	Durum	30.0	1.0
Mivhor 1177	Bread	10.5	95.0
Racine	Bread	0.0	12.5
Russian	Bread	0.0	30.0

Similar data were presented by several authors for the interaction of wheat with the glume blotch disease of wheat caused by *Septoria nodorum*. Rufty et al. (1981) tested 9 isolates of *S. nodorum* on four wheat cultivars and showed that the cultivars differed in resistance, and pathogen isolates differed in pathogenicity. In addition, significant isolate × cultivar interactions were recorded and the evidence that this could indicate specific interactions was discussed. A subsection of their data is shown in Table 3. This shows changed ranking of Blueboy and Coker 68-15 comparing isolates 1 and 2 with isolates 7 and 8. Despite this apparently sizeable interaction they concluded that the magnitude of the specificity was not high and did not identify specific pathogen strains. Using the same host-pathogen system, Scharen & Eyal (1983) also reported testing pathogen isolates on a series of host cultivars with similar results. These authors also considered that they should not classify the pathogenic patterns into defined physiologic races because the interactions were relatively small compared with the large differences between cultivars and between isolates. However, in contrast to this Scharen et al. (1985) using essentially similar data but applying the same method of handling the data as described in Eyal et al. (1985) concluded that there was also a gene-for-gene interaction between wheat and *S. nodorum* and calculated that there were many genes controlling race-specific interactions. This description of the two *Septoria* diseases

Table 3. Percent wheat leaf area covered by lesions due to isolates of *S. nodorum* extracted from results of Rufty et al. (1981). Data given as: percent, std. error

Isolate	Cultivar				Mean
	Blueboy	Hadden	Anderson	Coker 68-15	
1	74.2, 6.4	46.2, 5.3	46.1, 8.3	34.6, 5.5	50.3
2	70.2, 6.9	54.2, 5.1	53.4, 7.2	40.6, 5.8	54.7
5	85.8, 8.8	58.1, 9.3	66.7, 7.8	50.8, 9.6	65.4
7	33.1, 5.9	60.7, 6.1	57.1, 9.0	60.8, 6.7	52.9
8	35.6, 5.9	46.2, 6.2	44.7, 6.2	73.4, 8.2	49.9

of wheat shows that similar types of data have been interpreted differently, even by the same authors at different times.

In order to recognise specificity, such that a particular pathogen isolate shows a consistent specific pathogenicity towards a particular host cultivar, it would be necessary to demonstrate stability and repeatability of the interactions and this does not seem to have been reported. In practise, there are no reports that unequivocally demonstrate resistant cultivars succumbing to new pathogen races of *S. nodorum* or *S. tritici,* and for the practical purpose of breeding the main challenge is finding sufficient resistance and selecting to a sufficient level in new cultivars. Progress towards this for *S. tritici* was noted by Van Ginkel & Scharen (1988).

The data in Table 3 also indicate that isolate 5 possessed the highest mean pathogenicity. Although the number of cultivars is not large this observation may illustrate the feature of differences in pathogenicity of a non-specific type. There is little doubt that some isolates of pathogens of this type can be weakly pathogenic while others cause more severe infections. Again, observations from cultivars grown commercially suggest that there is an upper limit to such differences in pathogenicity and therefore, providing selection is carried out using adequately pathogenic isolates, resistance should not be eroded in a non-specific manner. Reduced pathogenicity is often encountered in pathogen isolates maintained for prolonged periods in axenic cultures and care is necessary to avoid use of these in selecting for resistance.

This somewhat detailed account of this type of statistical interaction between pathogen isolates and cultivars is given because it seems to represent a characteristic that can be found in several other host-pathogen interactions and has led to alternative interpretations. An example from another crop is of the northern corn leaf blight of maize caused by *Setosphaeria turcica.* Leonard (1993) reported that 12 of the 20 chromosome arms of maize carried genes for resistance and that some specificity was observed in the interactions between pathogen isolates and cultivars. The main evidence on which this was based was the demonstration of enhanced pathogenicity towards particular cultivars by selection among mixed populations of isolates of the pathogen through several generations of infection on those cultivars. The differences between isolates in pathogenicity generated by this technique were very small and Leonard (1993) suggested that they would ordinarily not be detected or be regarded as a noteworthy erosion of resistance. For practical purposes the important point was that a relatively small number of the polygenes was necessary to give adequate resistance in the field, and that such resistance had not been eroded despite prolonged and widespread exposure in agricultural use. It was speculated that the selection for greater pathogenicity in the field is likely to proceed more slowly than in the laboratory. There might also be an upper limit to increased specific pathogenicity in the field.

In contrast, based on similar data on the same disease, it was suggested that because of the statistically significant interaction between pathogen isolates and maize inbred lines the resistance was vertical (using Vanderplank's terminology) or in other words was race-specific (Nelson et al. 1970). The argument developed from this conclusion was that all resistance genes have the same type of action, with a specific interaction with pathogen genes, and the degree to which race-specificity can be detected depends only on the relative numbers of genes operating. Again this shows that the same type of data can be interpreted in different ways. What is lacking, however, in these examples is the ability to show the particular specific virulence of pathogen isolates for particular host genotypes or the identification of individual genes that produce race-specific resistance. In the absence of such evidence it is difficult to conclude that race-specificity as controlled by gene-for-gene interactions has been demonstrated. The parallel with *S. nodorum* and *S. tritici* on wheat is evident. Another example from within wheat is the eyespot pathogen *Pseudocercosporella herpotrichoides* discussed below.

In these examples it appears that a gene-for-gene interaction cannot be identified, at least with present techniques.

Summarizing comments on pathogen variation and host resistance

The above discussion suggests that there is great variation in the interaction of pathogens with their hosts. For some systems there is clear evidence of gene-for-gene interactions but for several pathogens there is a difficult challenge in deciding whether specific interactions occur or not. Among systems in which gene-for-gene interactions are observed there are various levels to which existing resistance has been resolved into the gene-for-gene interaction, and varying evidence for resistance that may remain outside such a system. For other systems there is little or no evidence of a gene-for-gene interaction but often evidence for variation in pathogenicity for the pathogens involved in these systems.

Finally it is quite likely that there are variations in pathogenicity of a similar type even for those pathogens that are shown to interact on a gene-for-gene basis with their hosts. Although these two types of variation may exist together, separating them experimentally is very difficult. Challenges relating to these questions are addressed in considerations of some important diseases of wheat.

Yellow rust on wheat

Yellow (stripe) rust of wheat is caused by *Puccinia striiformis* f. sp. *tritici* which is an obligate parasite and can only be grown in living host material. This *forma specialis* of the pathogen can infect numerous wheat cultivars but also a few barleys and some other grasses (see Stubbs, 1985). It is well adapted to cool maritime or high altitude climates or where wheat is grown in winter seasons in countries that have warm summers, such as north India and Australia. The best method of control is the use of resistant cultivars and breeding for resistance is a high priority in Britain and some other countries and of increasing importance in the programmes of the International Maize and Wheat Improvement Center, Mexico (CIMMYT). However, breeding for resistance to this disease presents some significant challenges and what may be unique features even among the rust diseases of wheat.

Genetics of resistance to yellow rust

Biffen (1905) provided the first evidence for the Mendelian inheritance of disease resistance by reporting that resistance to yellow rust in Rivet wheat was controlled by a single recessive gene. He based programmes of breeding for resistance on this evidence and produced very successful resistant cultivars such as Little Joss and Yeoman which were grown for many years in Britain and remained resistant. Some later attempts to breed for resistance were less successful and this related to variation in the pathogen.

Allison & Eisenbeck (1930) reported physiologic specialisation of *P. striiformis* of wheat into distinct physiologic races, first demonstrating for this disease that cultivars resistant to one pathogen isolate could be susceptible to other isolates. Following this, there was a gradual expansion of the number of races identified, reaching hundreds in the global surveys conducted at the Institute for Plant Protection (IPO) Wageningen (Stubbs, 1985). This was achieved using only a small number of differential cultivars and with rather little information about the genetics of their resistance. Following the demonstration of Mendelian inheritance by Biffen (1905) there were several publications on the genetics of resistance to yellow rust most of which reported single dominant or recessive genes (see Röbbelen & Sharp, 1978). However no systematic attempt to identify and name genes was achieved until Lupton & Macer (1962) began studies on inheritance of resistance in seedlings of several cultivars and introduced the nomenclature of *Yr* genes. The analysis by Lupton & Macer (1962) and Macer (1966) identified genes effective from the seedling stage and throughout the life of the cultivars in which they occurred. The same genes were concurrently used in the PBI wheat breeding programmes and several of them were already known to be race-specific and assumed to operate on a gene-for-gene basis with the pathogen, but in combination to provide effective resistance to known races of the pathogen. The consequences of introducing these genes into the breeding programmes was the creation of a succession of cultivars in which the resistance was rapidly overcome by new pathogen strains (Table 4). By 1968 the number of recognised resistance loci increased to 8 with the naming of *Yr*8 derived from *Aegilops comosa* (Riley et al., 1968) and two further genes were named by Macer (1975) based on reports in the literature, *Yr*9 derived from rye and *Yr*10 from a wheat collected in Turkey. In contrast to the increasing numbers of genes identified for the other two rusts of wheat (brown or leaf rust and black or stem rust) the identification of genes for resistance to yellow rust has been very slow. At the time of writing only 18 genes have been named and entered into the catalogue of genes of wheat (McIntosh, 1988, 1992).

Among the named genes are four that were designated without real genetic data of their mode of inheritance. This arose due to accumulation of data in race surveys of the pathogen. Although most of the named genes were detected in seedlings, specificity for resistance detected after the seedling stage was recorded in the Netherlands (Zadoks, 1961) and in several cultivars in the United Kingdom Pathogen Virulence Survey. Four different specificities of this type were detected in the UK and for convenience of reference were named R11, R12, R13 and R14 (Priestley, 1978). Later Stubbs (1985) proposed the same numbers for other cultivars and R.A. McIntosh (personal communication) suggested that both these sets of numbers would be ignored if new genes were identified with genetic data. It seemed that this would lead to three overlapping systems and as McIntosh was reluctant to leave the numbers vacant, after discussion, those originally listed by Priestley (1978) were designated as genes *Yr*11 to *Yr*14 (McIntosh, 1988).

The resistance that was later named *Yr*11 was first detected in a cultivar of French origin, Joss Cambier when it became highly susceptible to yellow rust after having initially shown slow rusting resistance (Johnson & Taylor, 1972). It was of special importance because it occurred at a time when the idea had gained popularity that resistance that could slow the apparent infection rate 'r' would be horizontal as defined by Vanderplank (1963) and therefore effective against all races. It showed that slow rusting resistance to yellow rust could be highly race-specific and therefore not horizontal. As further examples of similar type were found they reinforced this message. Although the gene *Yr*11 was not of importance to the PBI breeding programmes, those identified as *Yr*13 (first identified in Maris Huntsman) and *Yr*14 (first identified in Maris Bilbo) had far-reaching consequences. Initially they were detected by pathogen strains with virulence for them. These strains possessed virulence for a restricted set of other resistance genes. They did not include, for example, *Yr*6, *Yr*7 or *Yr*9 all of which entered the PBI programmes. As these genes were recombined with the genes *Yr*13 and *Yr*14 they created a succession of lines which could have complete resistance to some races, due to the effectiveness of the *Yr*6, *Yr*7 or *Yr*9 and incomplete resistance to other races due to the effectiveness of the *Yr*13 or *Yr*14, or both together. In either case it only required a single change in virulence in existing pathogen races to overcome the resistance and these lines also succumbed to new races shortly after or sometimes before their commercial release (Table 5). Sometimes, by good fortune, when cultivars with *Yr*13 and *Yr*14 became more susceptible

Table 4. Some wheat cultivars with effective resistance to yellow rust at the seedling stage used in the UK, the year they became susceptible and the genes overcome

Year	Cultivar	Gene combination overcome
1952	Nord Desprez	*Yr*3a, *Yr*4a
1955	Heines VII	*Yr*2
1966	Rothwell Perdix	*Yr*1, (*Yr*2, *Yr*6)[1]
1968	Maris Templar	*Yr*1, *Yr*3a, *Yr*4a
1969	Maris Beacon	*Yr*3b, *Yr*4b
	Maris Ranger	*Yr*3a, *Yr*4a, *Yr*6
1972	Talent	*Yr*7
1975	Clement	*Yr*2, *Yr*9
1983	Stetson	*Yr*1, *Yr*9
1988	Hornet	*Yr*2, *Yr*6, *Yr*9[2]
1991	Hereward	*Yr*2, *Yr*3a, *Yr*4a, *Yrcv*[3]

[1] ( ) it is not clear whether virulence for these genes was necessary.

[2] It was stated erroneously in Johnson (1992) that Hornet did not possess *Yr*2. The statement should have applied to its sister line Haven which also possesses *Yr*6 and *Yr*9.

[3] *Yrcv* indicates virulence for the differential cultivar Carstens V.

they still had some residual resistance of an adequate level. Such cultivars included Maris Huntsman itself, but not its closely related cultivar Maris Nimrod (Johnson & Taylor, 1980).

Durable resistance to yellow rust

As these events of loss of resistance in slow-rusting cultivars occurred it was noted that some cultivars that also possessed resistance developing after the seedling stage had been widely grown for many years and maintained adequate resistance even when other cultivars became severely infected in epidemics. The resistance of these cultivars was described as durable and it was suggested that there was no phenotypic distinction in the appearance of these to distinguish them from the cultivars in which post-seedling resistance was not durable (Johnson & Law, 1973, 1975). Therefore the critical observation was the durability itself. By examining the history of cultivation of cultivars various degrees of exposure were recorded and those that had remained resistant to yellow rust despite widespread use for several years were identified as possible possessors of durable resistance. It was noted that these ranged widely in the levels of infection

Table 5. Some what cultivars with race-specific adult plant resistance to yellow rust, the year they became more susceptible and the gene combinations overcome

Year	Cultivar	Gene combination overcome
1971	Joss Cambier	(*Yr*2)[1], *Yr*11
1972	Maris Bilbo	(*Yr*3a, *Yr*4a), *Yr*14
	Hobbit	(*Yr*3a, *Yr*4a), *Yr*14
1974	Maris Huntsman	(*Yr*2, *Yr*3a, *Yr*4a), *Yr*13
	Maris Nimrod	(*Yr*2, *Yr*3a, *Yr*4a), *Yr*13[2]
1975	Kinsman	(*Yr*3a, *Yr*4a, *Yr*6), *Yr*13
1981	Moulin	(*Yr*6), *Yr*14
	Brigand	(*Yr*2, *Yr*3a, *Yr*4a), *Yr*13, *Yr*14
1985	Slejpner	(*Yr*9), ?[3]
1988	Brock	(*Yr*7), *Yr*14
1992	Atou	?

[1] ( ) seedling genes. Others detected mainly in adult plants.
[2] Maris Nimrod was more susceptible than Maris Hunstman after virulence for *Yr*13 occurred.
[3] ? Adult plant specificity due to unknown gene.

recorded on them in intensively infected plots (Johnson, 1988). There was also a large range in the degree of exposure of cultivars. The best example in the UK was Cappelle Desprez which occupied up to 80 percent of the UK winter wheat area for more than 10 years and substantial areas before and after this period for several further years (Johnson, 1978). Some older cultivars such as Little Joss, Yeoman and Holdfast probably had similar great exposure (Lupton, 1992), but after Cappelle Desprez no cultivar predominated to the same extent, so the diagnosis of durability was correspondingly weaker.

Johnson (1978) argued that the strongest test for durable resistance occurs when a cultivar is widely grown in an environment favourable to the disease. He also proposed that multilocation testing of breeding material, although essential as part of normal breeding programmes, did not provide a powerful test for durable resistance. A recent event reinforces this view. Among several cultivars to which durable resistance to yellow rust was attributed was Atou (Johnson, 1988, 1992). This cultivar was grown in the UK from 1973 to 1979 on up to 12% of the UK winter wheat area (about 200,000 hectares) (Fig. 1) and remained resistant in the field to yellow rust. It continued to be resistant in field trials at PBI and its resistance transmitted in breeding programmes was also effective until 1992 when virulence was detected (G.M.B. Smith, PBIC, personal communication 1992). It must be admitted that the strength of test for durability for Atou was not of the highest, but nevertheless it could be considered more powerful than multilocation testing might have been for the same period of time. This example reinforces the evidence for the need for widespread and prolonged testing to diagnose durable resistance to yellow rust in wheat. Even with a more powerful test it must be concluded that past performance does not guarantee future performance.

Examples of durable resistance were also identified in the USA and were associated with increased expression of the resistance at higher temperatures (Qayoum & Line, 1985; Milus & Line, 1986a, 1986b). In field observations at PBI it was noted that most cultivars become less infected with yel-

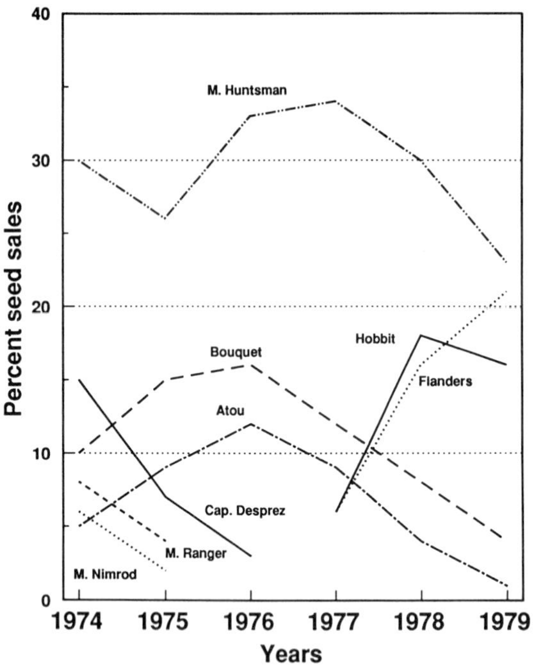

Fig. 1. UK Ministry of Agriculture, Fisheries and Food statistics of seed sales for winter wheat cultivars. (M. = Maris; Cap. = Cappelle)

low rust in warmer summers, even those that are too susceptible for commercial use. Some of the cultivars reputed in the USA to have the high temperature adult plant durable resistance to yellow rust, such as Gaines, Nugaines and Luke were also tested in Cambridge at the PBI (Table 6). Generally they were a little more susceptible in the UK probably due to lower summer temperatures. It is likely that Gaines would have inadequate resistance under UK conditions, compared with Cappelle Desprez which was just adequate. However, one cultivar, Wanser, shown by Qayoum & Line (1985) to have high temperature resistance in the USA proved to be much more susceptible at PBI (Table 6). As it was more susceptible to all the races used at PBI it may have had a greater requirement for high temperature to express its resistance. Although the resistance in these cultivars shows temperature sensitivity it is the maintenance of their resistance during widespread use that is primarily responsible for the recognition of their durable resistance. Temperature sensitivity is extremely common in resistance to rust diseases of wheat and a particular form of temperature sensitivity – more expressed at high temperature – is not necessarily *per se* diagnostic for durable resistance.

Genetic analysis and exploitation of durable resistance

In order to exploit durable resistance in breeding it would be beneficial to understand the genetic basis of the resistance. However, few cultivars with proven durable resistance to yellow rust have been investigated genetically. Among those investigated were those identified in the USA (Milus & Line, 1986a, 1986b). This indicated that each of the cultivars Gaines, Nugaines and Luke possessed a minimum of 2 to 3 genes.Studies were also carried out on some cultivars at PBI, including Hybride de Bersée and Cappelle Desprez using genetically controlled stocks of these cultivars developed by cytogenetical techniques (Johnson & Law, 1975; Law et al., 1978). The contribution of a single chromosome, called 5BS–7BS, to resistance was shown by the increased susceptibility of lines in which this chromosome was either absent or reduced in dosage. (This chromosome is formed by translocation of the short arms of chromosomes 5B

Table 6. Percent leaf area infected with yellow rust in European cultivars with durable resistance and North American cultivars with high temperature adult plant resistance in the USA and in a field trial in the UK at PBI

Cultivar	Place and data of score	
	USA Walla Walla[1] 1975 July 2	UK Cambridge[2] 1980 June 10
Cappelle Desprez	–	35
Hybride de Bersée	–	5
Gaines	40	55
Nugaines	24	33
Luke	17	21
Wanser	28	90
Susceptible controls		
Omar	75	–
Nord Desprez	–	75

[1] Data from Qayoum & Line (1985).
[2] Highest score for any of several races used in the trial.

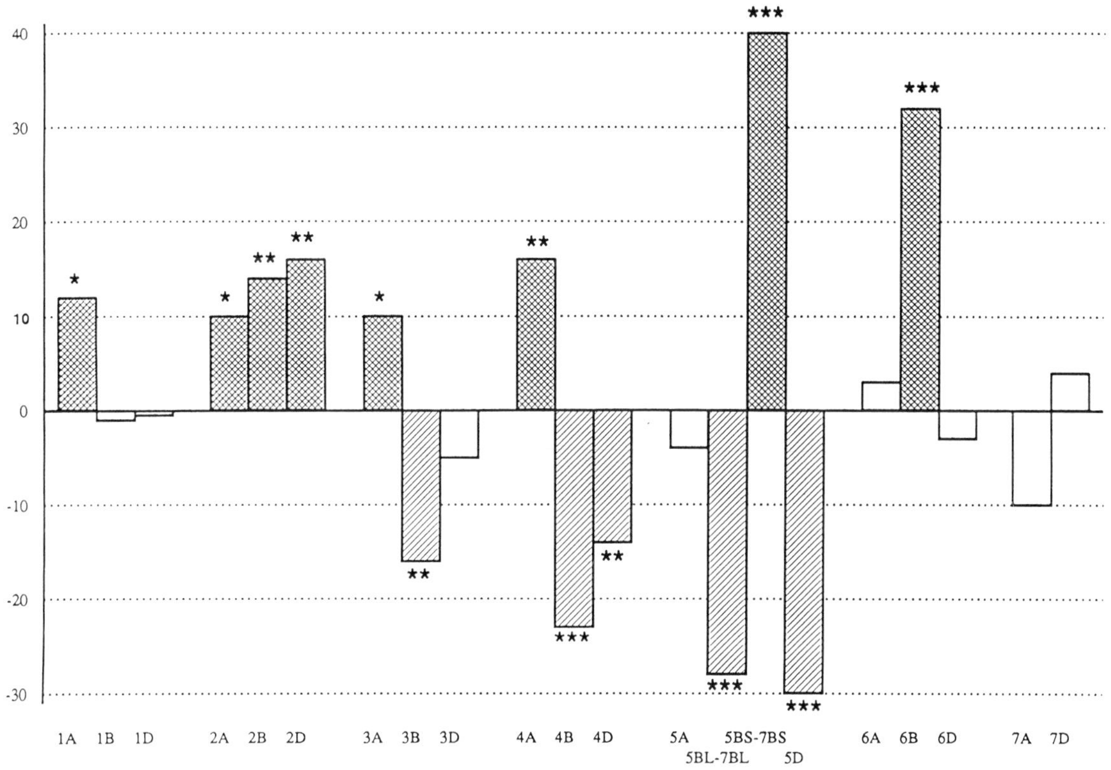

Fig. 2. Deviation from mean of Hobbit sib (34.8%) in leaf area infected with yellow rust in progenies from monosomic plants for 21 monosomics of Hobbit sib. (Significance of deviations is $P < 0.05$ = *, $P < 0.01$ = **, $P < 0.001$ = ***).

and 7B relative to the reference cultivar Chinese Spring). Further investigations of wheat lines with varying levels of resistance indicated that several chromosomes were implicated in the control of resistance. Reduced dose of some of them resulted in greater susceptibility to disease, whereas reduced dose of others resulted in greater resistance. This is illustrated by the observation of monosomic lines of the cultivar Hobbit sib, a yellow rust susceptible line in which monosomic stocks were developed by recurrent backcrossing techniques (Law et al., 1987) (Fig. 2). The observation that the long arms of chromosomes of homoeologous group 5 appeared to carry genes that promote increased susceptibility in many cultivars (Pink et al., 1983) led Law & Worland (1991) to attempt to induce mutations reducing the disease promoting activity of these chromosomes. Seeds of the susceptible cultivar Hobbit sib were treated with fast neutrons by the International Atomic Energy Agency in Vienna. Some of the resulting progeny showed resistance and selection was made for those with normal chromosome number and pairing. Using monosomic analysis and Restriction Fragment Length Polymorphism (RFLP) analysis, it was shown that some of the mutations were on the long arm of chromosomes 5B and 5D that had been shown to promote susceptibility in Hobbit sib. Others were on chromosomes 4B and 4D which also had a similar though less pronounced effect (Fig. 2). It was also observed that there was a correlation in these investigations between resistance to yellow rust, brown rust and powdery mildew. This may indicate an effect on a mechanism common to obligate pathogens, although at present this is speculative. In relation to the potential use of the resistance two further steps are required. The first is to show that grain yields of the mutant lines are equal to those of the parent. The second, more challenging, is to devise a method for in-

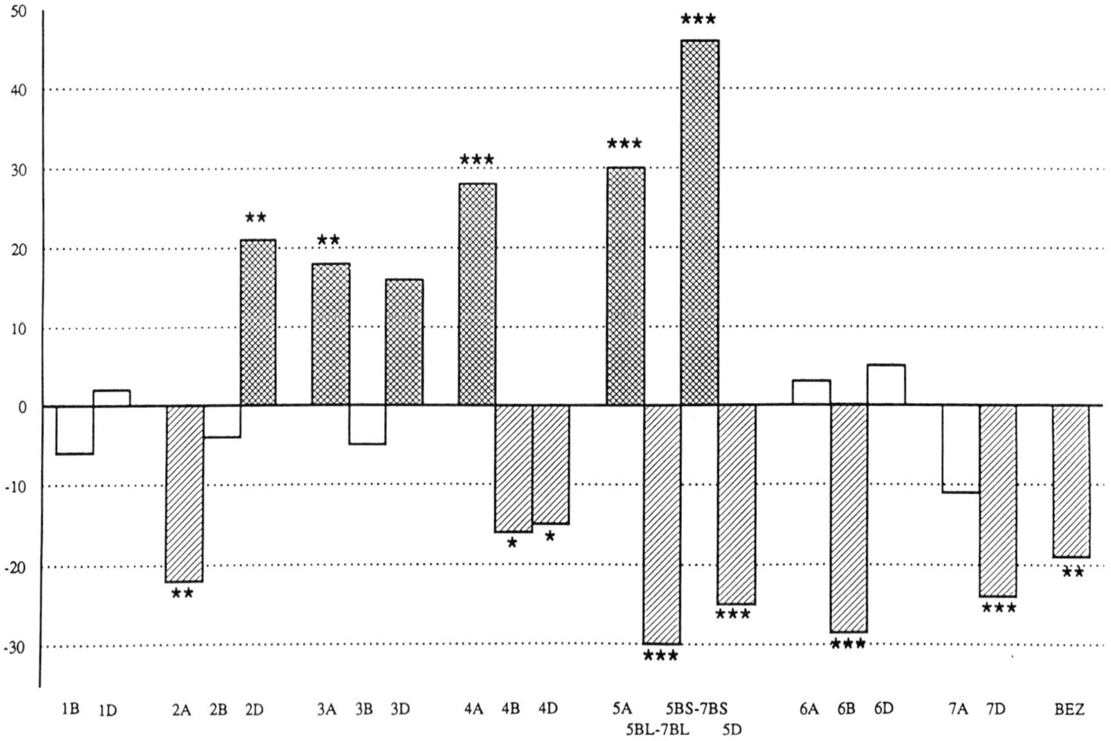

Fig. 3. Deviation from mean of Hobbit sib (30%) in leaf area infected with yellow rust in Bezostaya 1 (BEZ) and 20 chromosome substitution lines of Bezostaya 1 chromosomes into Hobbit sib. (Significance of deviations is $P < 0.05$ = *, $P < 0.01$ = **, $P < 0.001$ = ***).

dicating whether the resistance induced by this technique could be classified as durable. It is not possible to manipulate the pathogen to test for evolution of virulence towards this resistance for two reasons. The first is that it is unlikely that variability occurring within a limited experimental pathogen population would be of the same magnitude as that in a widespread pathogen population during a natural epidemic, which is when resistance is most likely to fail. The second is that it would not be ethical in the main wheat-growing region of the UK to attempt to induce more variation in the pathogen population than occurs already. With present knowledge, therefore, it seems that only widespread and extensive use of the resistance will suffice to test its durability. The results obtained illustrate a novel approach to the use of induced mutations and a contribution from RFLP studies to identify the chromosomal location of the mutations.

Further studies were carried out by chromosome substitution of chromosomes from a Russian wheat, Bezostaya 1, into the same cultivar Hobbit Sib (Law & Worland, 1991). The effects on yellow rust resistance of chromosome substitution into Hobbit Sib were very diverse (Fig. 3). The reduction of disease for substitution of chromosomes of homoeologous group 5 (5BL–7BL and 5D) is notable, suggesting that the Bezostaya 1 chromosomes promote susceptibility much less than the corresponding Hobbit Sib chromosomes. It is evident that intercrossing some of the substitution lines should permit selection of progeny with very high levels of resistance. In contrast, substitution of the Bezostaya 1 5BS–7BS is equivalent to monosomy for Hobbit sib 5BS–7BS suggesting that Bezostaya 1 lacks this component of resistance. It can also be observed that substitution of Bezostaya 1 chromosome 7D into Hobbit Sib increases the resistance. Recent evidence suggests that Bezostaya 1 possesses the gene *Yr*18 on chromosome 7D (McIntosh, 1992) which may be the cause of this effect of chromosome substitution. Bezostaya 1 may possess durable resistance to brown rust (Johnson, 1978) as it was grown extremely extensively for a long period in the USSR and neighbouring countries in which brown rust is endemic. The gene *Yr*18 is tightly linked to the gene *Lr*34 for resistance to brown rust, which is believed to be implicated in durable resistance to brown rust (McIntosh, this volume).

These observations suggest that durable resistance to yellow rust in those cultivars examined so far may be due to genetic complexity in its control. However, it is still not possible to say whether some particular genetic components of the resistance, including those controlling positive or negative effects, are more relevant than others to the durability.

The consequence of these observations is that, until now, the best prospect for breeding for durable resistance to yellow rust in wheat is to start with a cultivar for which there is reasonable evidence of durability, and ensure that the resistance selected is derived from this source. This requires using pathogen isolates that can overcome recognised race-specific components of the resistance (Johnson, 1978, 1992). Unfortunately, the suggestion of Johnson (1992) that Atou would be a simple choice for breeding for durable resistance is rendered obsolete by the discovery of a race with virulence for adult plants of this cultivar. However, other cultivars with stronger evidence of durability still remain resistant in the UK and can be used as proposed.

In the future the possible presence of *Yr*18 in wheat cultivars should be investigated and compared with any evidence for their resistance being of a durable type. If there were evidence that *Yr*18 itself might be implicated in durable resistance, the linkage with *Lr*34, and possible linkage with other markers might produce a significant step forward in exploiting durable resistance to yellow rust as well as brown rust.

Eyespot resistance in wheat and related species

The eyespot disease of wheat, caused by *Pseudocercosporella herpotrichoides,* weakens the stem base and leads to lodging of the crop. The disease presents a very different set of challenges for the breeder from those posed by yellow rust. The pathogen is facultative and can readily be cultured axenically and on straw or grain. It can survive on plant debris in soil for at least three years (Macer, 1961) and is therefore difficult to eliminate by crop rotation. Wheat cultivars differ in resistance as shown by methods measuring depth of penetration of the pathogen into seedlings (Macer, 1966) or frequency and intensity of eyespot lesions on stems (Scott & Hollins, 1974). Both these methods are laborious and subject to large experimental errors and high replication is necessary to achieve significant differences. Breeding for resistance is possible, but obviously selection of individual plants with resistance in segregating material would be subject to error.

The main source of resistance used at PBI was the cultivar Cappelle Desprez. Cytogenetic studies of the resistance of Cappelle Desprez wheat showed that a major component of the resistance was controlled by chromosome 7A (Law et al., 1975). Perhaps partly for this reason breeding for resistance from this source was successful at the PBI where several new cultivars with levels of resistance similar to those of the parent were achieved. Despite this success, the level of resistance in Cappelle Desprez and derived cultivars was not high and, in fact, fungicide sprays were often used on such cultivars after 1970. After intensive use of benomyl fungicides to control this disease in the UK the pathogen population evolved resistance to these fungicides (King & Griffin, 1985).

Variation in the pathogen population was investigated in relation to resistance in wheat and some other species, including rye. Comparing resistance of wheat cultivars infected with a range of pathogen isolates Scott & Hollins (1977) obtained highly significant differences between cultivars, with the French wheat Cappelle Desprez typically representing the resistant end of the range and the British wheat Holdfast the susceptible end. There were also significant differences between pathogen isolates due to overall differences in pathogenicity. In addition, there were significant interactions be-

Fig. 4. Aegilops ventricosa a goat grass related to wheat from which resistance to eyespot was transferred to wheat. (left) whole plant with scale object (T.W. Hollins). (right) two ears.

tween cultivars and pathogen isolates in all tests. Despite this, it was not possible to show that any given isolate possessed a specific virulence for any individual cultivar and no evidence for a gene-for-gene interaction in the resistance. These data from tests of wheat cultivars with isolates of *P. herpotrichoides* show similarities to those from *S. nodorum* referred to above. Again, as with resistance in wheat to *S. nodorum,* such resistance to eyespot, as in Cappelle Desprez, has remained effective during prolonged and widespread use and can be classified as durable. However, as noted above, the level of resistance was not sufficiently high to obviate the possible need for fungicidal control of the disease.

An important advance was achieved by the identification of *Aegilops ventricosa* (Fig. 4) as a source of resistance and its transfer to a wheat line VPM (Maia, 1967). This goat grass is tetraploid and possesses the genome $M^vM^vD^vD^v$. The D genome chromosomes can pair with the D genome of bread wheat and the resistance transferred to VPM was found to be located on chromosome 7D (Worland et al., 1988). The line VPM was used in breeding at PBI and a wheat cultivar Rendezvous was produced with a higher level of resistance than any previous UK wheat cultivar.

Although the resistance of Cappelle Desprez was durable it cannot be assumed that the resistance derived from *A. ventricosa* will necessarily also be durable. This comment arises from the observations of variability within the pathogen population in relation to tests involving wheat and related species. Two major variants in pathogenicity have been noted, with one type attacking wheat but not rye (W-type) and the other attacking both wheat and rye (R-type) (Scott et al. 1975). These two types are different in several respects, with the R-type being slower growing on agar plates and with irregular colonies. They also differ in repetitive, ribosomal and mitochondrial DNA profiles (Nicholson et al., 1991, 1992). In addition to these two types there is also evidence for specificity to other hosts, including some accessions of *Aegilops squarrosa* (Scott et al., 1976).

As already noted, the methods of assessing resistance in wheat to eyespot are laborious and time consuming. However, a linkage was detected between the resistance derived from *A. ventricosa* and an endopeptidase isozyme locus *Ep-D1b* (McMillin et al., 1986; Worland et al., 1987). This linkage provided a rapid and inexpensive method of selection for resistance which has been applied in developing new resistant cultivars. Recently, evidence was presented indicating that another source of the *Ep-D1b* locus obtained by transfer from *A. ventricosa* was susceptible to eyespot (Mena et al., 1992) indicating that the linkage could be broken. The apparently tight linkage of *Ep-D1b* with the eyespot resistance in VPM and derivatives may be due the structure of the translocated segment carrying the two genes, such as an inversion, preventing crossing over between them, rather than close physical proximity on the chromosome segment (A.J. Worland, personal communication). It is therefore possible that derivativation of resistance from *A. ventricosa* could occur without the tight linkage between the endopeptidase and eyespot resistance. For this and other reasons such as the production of toxic chemicals in the endopeptidase test, other linkages might be useful. Chao et al. (1989) reported an RFLP, *Xpsr121,* linked to the resistance locus *Pch-1.* As with the endopeptidase gene, it was suggested that the lack of recombination between the *Xpsr121* and *Pch-1* might have been due to a low rate of recombination between the alien segment of chromosome carrying the loci and the corresponding normal wheat chromosome segment as described above.

The rapid development of the RFLP map and other marker loci in wheat provides opportunities for the identification of closer linkages to resistance genes such as *Pch-1* and will also assist in the separation of such valuable traits of alien origin from other less desirable linked traits such as a gene depressing yield linked to *Pch-1.* It may also be possible to find linkages to the resistance to eyespot identified as being carried by chromosome 7A in Cappelle Desprez.

For eyespot resistance there does not seem to be any prospect of improving resistance from within bread wheat cultivars but there may be alien sources that have not yet been exploited. When testing of alien species is carried out for any character, including disease resistance, there is often only one accession representing a particular species. However, variation is often found in different accessions of wild species. An example for eyespot resistance is that some accessions of *A. squarrosa* are susceptible to some isolates of the eyespot pathogen, but others are resistant and might provide the basis for future improvements in resistance of wheat.

If the linkage between the eyespot resistance and the endopeptidase and RFLP markers is due to a structural feature of the translocated material, such features could, in future be used to create groups of valuable genes that would remain tightly linked during breeding procedures, rather than being separated by crossing over, leading to disruption of the group.

Conclusions

The examples of diseases of wheat and other cereals presented here were chosen to illustrate the diversity of interactions between plants and their pathogens. The diversity is even greater across a wider range or hosts and pathogens as shown in the remainder of this book. This illustrates the importance of the need to understand the particular features of host-pathogen interactions in relation to their manipulation in plant breeding.

Plant breeding for disease resistance is at an important stage of development. Modern molecular techniques hold out the possibility not only of genetic linkage of important characters to easily assessable markers but also of genetic transformation, as indicated in subsequent chapters of this book. Such techniques will also permit more precise investigations into the variation and population genetics of pathogens and the epidemiology of disease.

These are exciting developments which will improve the prospects of successful breeding for disease resistance. However, they will not supplant the continuing need to apply well tried and established techniques of plant breeding for the foreseable future, particularly those concerned with widescale testing of materials before they are released to farmers. For disease resistance introduced by biotechnology its possible durability will remain to be challenged by widespread use of cultivars possessing it.

Acknowledgements

I thank A.J. Taylor, G.M.B. Smith and P.N. Minchin for excellent technical support for many years. Numerous colleagues and visitors have contributed greatly to my understanding of pathology in the context of plant breeding.

References

Allison, C.C. & K. Isenbeck, 1930. Biologische Specialisierung von *Puccinia glumarum tritici* Eriksson and Henning. Phytopathol. 2: 87–98.

Bayles, R.A., M.H. Channell & P.L. Stigwood, 1989. New races of *Puccinia striiformis* in the United Kingdom in 1988. Cereal Rusts Bull. 17: 20–23.

Biffen, R.H., 1905. Mendel's law of inheritance and wheat breeding. J. Agric. Sci. 1: 4–48.

Chao, S., P.J. Sharp, A.J. Worland, E.J. Warham, R.M.D. Koebner & M.D. Gale, 1989. RFLP-based genetic maps of wheat homoeologous group 7 chromosomes. Theor. Appl. Genet. 78: 495–504.

Chilosi, G. & L. Corazza, 1990. Occurrence and epidemics of yellow rust on wheat in Italy. Cereal Rusts and Powdery Mildews Bull. 18: 1–19.

Conner, R.L., M.D. MacDonald & E.D.P. Whelan, 1988. Evaluation of take-all resistance in wheat-alien amphiploid and chromosome substitution lines. Genome 30: 597–602.

Crute, I.R., 1985. The genetic bases of relationships between microbial parasites and their hosts. pp 80–142 In: R.S.S. Fraser (Ed.) 'Mechanisms of Resistance to Plant Diseases'. Martinus Nijhoff/Dr. W. Junk Publishers, Dordrecht.

Day, P.R., 1974. 'Genetics of Host-Parasite Interaction'. W.H. Freeman and Company, San Francisco.

Dyck, P.L. & E.R. Kerber, 1985. Resistance of the race-specific type. pp 469–500 In: A.P. Roelfs & W.R. Bushnell (Eds.) 'The Cereal Rusts, Vol II'. Academic Press, London.

El-Bedewy, R. & G. Röbbelen, 1982. Chromosomal location and change of dominance of a gene for resistance against yellow rust. Z. Pflanzenzücht. 89: 145–157.

Ellingboe, A.H., 1975. Horizontal resistance: an artefact of experimental procedure? Australian Plant Pathology Soc. Newsletter 4: 44–46.

Eyal, Z., Z. Amiri & I. Wahl, 1975. Physiologic specialization of *Septoria tritici*. Phytopathology 63: 1087–1091.

Eyal, Z., A.L. Scharen, M.D. Huffman & J.M. Prescott, 1985. Global insights into virulence frequencies of *Mycosphaerella graminicola*. Phytopathogy 75: 1456–1462.

Flor, H.H., 1946. Genetics of pathogenicity in *Melampsora lini*. J. Agric. Res. 73: 335–357.

Flor, H.H., 1960. The inheritance of X-ray induced mutations to virulence in a urediospore culture of race 1 of *Melampsora lini*. Phytopathology 50: 603–605.

Fraser, R.S.S., 1985. Genetics of host resistance to viruses and of virulence. pp 62–79 In: R.S.S. Fraser (Ed.) 'Mechanisms of Resistance to Plant Diseases.' Martinus Nijhoff/Dr. W. Junk Publishers, Dordrecht.

Jenkins, G., 1984. Winter and spring wheat. pp 23–28, Annual Report of the Plant Breeding Institute 1983.

Johnson, R., 1978. Practical breeding for durable resistance to rust diseases in self-pollinating cereals. Euphytica 27: 529–540.

Johnson, R., 1988. Durable resistance to yellow (stripe) rust in wheat and its implications in plant breeding. pp 63–75 In: N.W. Simmonds & S. Rajaram (Eds.) 'Breeding Strategies for Resistance to the Rusts of Wheat'. CIMMYT, Mexico D.F.

Johnson, R., 1992. Reflections of a plant pathologist on breeding for disease resistance, with emphasis on yellow rust and eyespot of wheat. Plant Pathol. 41: 239–254.

Johnson, R. & D.R. Knott, 1992. Specificity in gene-for-gene interactions between plants and pathogens. Plant. Pathol. 41: 1–4.

Johnson, R. & A.J. Taylor, 1972. Isolates of *Puccinia striiformis* collected in England from wheat varieties Maris Beacon and Joss Cambier. Nature, Lond. 238: 105–106.

Johnson, R. & A.J. Taylor, 1980. Pathogenic variation in *Puccinia striiformis* in relation to the durability of yellow rust resistance in wheat. Ann. Appl. Biol. 94: 283–286.

Johnson, R. & C.N. Law, 1973. Cytogenetic studies on the resistance of the wheat variety Bersée to *Puccinia striiformis*. Cereal Rusts Bull. 1: 38–43.

Johnson, R. & C.N. Law, 1975. Genetic control of durable resistance to yellow rust (*Puccinia striiformis*) in the wheat cultivar Hybride de Bersée. Ann. Appl. Biol. 81: 385–391.

Jones, D.A., 1988. Genetic properties of inhibitor genes in flax rust that alter avirulence to virulence on flax. Phytopathology 78: 342–344.

Kerber, E.R. & J.G. Green, 1980. Suppression of stem rust resistance in the hexaploid wheat cv. Canthatch by chromosome 7DL. Can. J. Bot. 58: 1347–1350.

King, J.E. & M.J. Griffin, 1985. Survey of benomyl resistance in *Pseudocercosporella herpotrichoides* on winter wheat and barley in England and Wales in 1983. Plant Pathol. 34: 272–283.

Knott, D.R. & J. Dvorak, 1976. Alien germ plasm as a source of resistance to disease. Ann. Rev. Phytopathol. 14: 211–235.

Law, C.N., J.W. Snape & A.J. Worland, 1987. Aneuploidy in wheat and its uses in genetic analysis. pp71–108 In: F.G.H.

Lupton, (Ed.) 'Wheat Breeding. Its Scientific Basis'. Chapman & Hall, London.
Law, C.N., R.C. Gaines, R. Johnson & A.J. Worland, 1978. The application of aneuploid techniques to a study of stripe rust resistance in wheat. pp 427–436 In: Proceedings of the 5th International Wheat Genetics Symposium, New Delhi.
Law, C.N. & A.J. Worland, 1991. Improving disease resistance in wheat by inactivating genes promoting disease susceptibility. IAEA Vienna, Mutation Breeding Newsletter 38: 2–5.
Law, C.N., P.R. Scott, A.J. Worland & T.W. Hollins, 1975. The inheritance of resistance to eyespot (*Cercosporella herpotrichoides*) in wheat. Genet. Res. Camb. 25: 73–79.
Lawrence, F.J., G.M.E. Mayo & K.W. Shepherd, 1981. Interactions between genes controlling pathogenicity in the flax rust fungus. Phytopathology 71: 12–19.
Leonard, K.J., 1993. Durable resistance in the pathosystems: maize – northern and southern leaf blights. In: Th. Jacobs & J.E. Parlevliet (Eds.) 'Durability of Disease Resistance'. In press.
Lupton, F.G.H., 1983. Winter wheat. pp 23–26, Annual Report of the Plant Breeding Institute 1982.
Lupton, F.G.H., 1992. Changes in varietal distribution of cereals in central and western Europe. Agro-ecological Atlas of Cereal Growing in Europe. Pudoc, Wageningen.
Lupton, F.G.H. & R.C.F. Macer, 1962. Inheritance of resistance to yellow rust (*Puccinia glumarum* Erikss. & Henn.) in seven varieties of wheat. Trans. Br. Mycol. Soc. 45: 21–45.
Macer, R.C.F., 1961. The survival of *Cercosporella herpotrichoides* Fron in wheat straw. Ann. Appl. Biol. 49: 165–172.
Macer, R.C.F., 1966a. The formal and monosomic analysis of stripe rust (*Puccinia striiformis*) resistance in wheat. pp 137–142 In: Proceedings of the 2nd International Wheat Genetics Symposium, Lund, Sweden, 1963. Hereditas Supplement 2.
Macer, R.C.F., 1966b. Resistance to eyespot disease (*Cercosporella herpotrichoides* Fron) determined by a seedling test in some forms of *Triticum, Aegilops, Secale* and *Hordeum*. J. Agric. Sci. 67: 389–396.
Macer, R.C.F., 1975. Plant pathology in a changing world. Trans. Br. Mycol. Soc. 65: 351–374.
McIntosh, R.A., 1988. Catalogue of gene symbols for wheat. pp 1225–1323 In: T.E. Miller & R.M.D. Koebner (Eds.) Proceedings of the 7th International Wheat Genetics Symposium. IPSR Cambridge.
McIntosh, R.A., 1992. Catalogue of gene symbols for wheat. 1992 supplement. Cer. Res. Commun. 20: In press.
McMillin, D.E., R.E. Allan & D.E. Roberts, 1988. Association of an enzyme locus and strawbreaker foot rot resistance derived from *Aegilops ventricosa*. Theor. Appl. Genet. 72: 743–747.
Mena, M., G. Doussinault, I. Lopez-Brana, S. Arguaded, G. Garcia-Olmedo & A. Delibes, 1992. Eyespot resistance gene *Pch-1* in H-93 wheat lines. Evidence of linkage to markers of chromosome group 7 and resolution from the endopeptidase locus *Ep-D1b*. Theor. Appl. Genet. 83: 1044–1047.
Milus, E.A. & R.F. Line, 1986a. Number of genes controlling high temperature, adult-plant resistance to stripe rust in wheat. Phytopathology 76: 93–96.
Milus, E.A. & R.F. Line, 1986b. Gene action for inheritance of durable, high temperature, adult-plant resistance to stripe rust in wheat. Phytopathology 76: 435–441.
Nelson, R.R., D.R. MacKenzie & G.L. Scheifele, 1970. Interaction of genes for pathogenicity and virulence in *Trichometasphaeria turcica* with different numbers of genes for vertical resistance in *Zea mays*. Phytopathology 60: 1250–1254.
Nicholson, P., T.W. Hollins, H.N. Rezanoor & K. Anamthawat-Jonsson, 1991. A comparison of cultural, morphological and DNA markers for the classification of *Pseudocercosporella herpotrichoides*. Plant Pathol. 40: 584–594.
Nicholson, P., H.N. Rezanoor & T.W. Hollins, 1992. Classification of a world-wide collection of isolates of *Pseudocercosporella herpotrichoides* by RFLP analysis of mitochondrial and ribosomal DNA and host range. Plant Pathol 41 (In press)
Parlevliet, J.E., 1981. Race-non-specific disease resistance. pp 47–54 In: J.F. Jenkyn & R.T. Plumb (Eds.) 'Strategies for the Control of Cereal Disease'. Blackwell Scientific Publications, Oxford.
Person, C.O., 1959. Gene for gene relationships in host : parasite systems. Can. J. Bot. 37: 1101–1130.
Pink, D.A.C., F.G.A. Bennett, C.E. Caten & C.N. Law, 1983. Correlated effects of homoeologous group 5 chromosomes upon infection of wheat by yellow rust and powdery mildew. Z. Pflanzenzücht. 91: 275–294.
Polley, R.W. & M.R. Thomas, 1991. Surveys of diseases of winter wheat in England and Wales, 1976–1988. Ann. Appl. Biol. 119: 1–20.
Priestley, R.H., 1978. Detection of increased virulence in populations of wheat yellow rust. pp 63–70 In: P.R. Scott & A. Bainbridge (Eds.) 'Plant Disease Epidemiology'. Blackwell Scientific Publications, Oxford.
Qayoum, A. & R.F. Line, 1985. High temperature, adult plant resistance to stripe rust of wheat. Phytopathology 75: 1121–1125.
Riley, R., V. Chapman & R. Johnson, 1968. The incorporation of alien disease resistance in wheat by genetic interference with the regulation of meiotic chromosome synapsis. Genet Res. Camb. 12: 199–219.
Röbbelen, G. & E.L. Sharp, 1978. Mode of inheritance, interaction and application of genes conditioning resistance to yellow rust. Advances in Plant Breeding, Suppl. 9. Verlag Paul Parey, Berlin & Hamburg.
Rufty, R.C., T.T. Hebert & C.F. Murphy, 1981. Variation in virulence in isolates of *Septoria nodorum*. Phytopathology 71: 593–596.
Scharen, A.L. & Z. Eyal, 1983. Analysis of symptoms on spring and winter wheat cultivars inoculated with different isolates of *Septoria nodorum*. Phytopathology 73: 143–147.
Scharen, A.L., Z. Eyal, M.D. Huffman & J.M. Prescott, 1985. The distribution of virulence genes in geographically separated populations of *Leptosphaeria nodorum*. Phytopathology 75: 1463–1468.
Scott, P.R., 1981. Variation in host susceptibility. pp 219–236

In: M.J. Asher & P.J. Shipton (Eds.) 'Biology and Control of Take-all'. Academic Press, London.

Scott, P.R. & T.W. Hollins, 1974. Effects of eyespot on yields of winter wheat. Ann. Appl. Biol. 78: 269–279.

Scott, P.R. & T.W. Hollins, 1977. Interactions between cultivars of wheat and isolates of *Cercosporella herpotrichoides*. Trans. Br. Mycol. Soc. 69: 397–403.

Scott, P.R., T.W. Hollins & P. Muir, 1978. Pathogenicity of *Cercosporella herpotrichoides* to wheat, barley, oats and rye. Trans. Br. Mycol. Soc. 65: 529–538.

Scott, P.R., L. Defosse, J. Vandam & G. Doussinault, 1976. Infection of lines of *Triticum, Secale, Aegilops* and *Hordeum* by isolates of *Cercosporella herpotrichoides*. Trans. Br. Mycol. Soc. 66: 205–210.

Singh, H. & R. Johnson, 1988. Genetics of resistance to yellow rust in Heines VII, Soissonais and Kalyansona. pp 885–890 In: T.E. Miller & R.M.D. Koebner (Eds.) Proceedings of the Seventh International Wheat Genetics Symposium, IPSR, Cambridge.

Singh, R.P. & R.A. McIntosh, 1984. Complementary genes for reaction to *Puccinia recondita tritici* in *Triticum aestivum*. I. Genetic and linkage studies. Can. J. Genet. Cytol. 26: 723–735.

Stubbs, R.W., 1985. Stripe rust. pp 61–101 In: A.P. Roelfs & W.R. Bushnell (Eds.) 'The Cereal Rusts. Vol II'. Academic Press, London.

Vanderplank, J.E., 1963. Plant Diseases: Epidemics and Control. Academic Press, New York.

Van Ginkel, M. & A.L. Scharen, 1988. Host-pathogen relationships of wheat and *Septoria tritici*. Phytopathology 78: 762–766.

Worland, A.A. & C.N. Law, 1991. Improving disease resistance in wheat by inactivating genes promoting disease susceptibility. IAEA Vienna. Mutation Breeding Newsletter, 38: 2–5.

Worland, A.A., C.N. Law, T.W. Hollins, R.M.D. Koebner & A. Giura, 1988. Location of a gene for resistance to eyespot (*Pseudocercosporella herpotrichoides*) on chromosome 7D of bread wheat. Plant Breed. 101: 43–51.

Zadoks, J.C., 1961. Yellow rust of wheat: studies in epidemiology and physiologic specialisation. Tijd. Plantenziekt. 67: 69–256.

Euphytica **63**: 23–31, 1992.
R. Johnson and G.J. Jellis (eds), Breeding for Disease Resistance

Gene banks and their contribution to the breeding of disease resistant cultivars

N.L. Innes
Scottish Crop Research Institute, Invergowrie, Dundee, DD2 5DA, Scotland

Key words: genetic resources, gene bank, pearl millet, *Pennisetum glaucum,* potato, *Solanum tuberosum,* rice, *Oryza sativa,* cotton, *Gossypium* spp.

Summary

Genetic variation in crop species and their wild relatives holds the key to the successful breeding of improved crop cultivars with durable resistance to disease. The importance of the conservation, characterization and utilization of plant genetic resources nationally and internationally has been recognised, though much remains to be done. Gene banks have now been established in many countries and at most of the international crop research centres. Cell and tissue culture techniques and biotechnological aids have done much to ensure the creation and safe transfer of healthy germplasm around the world. Multidisciplinary, international research and collaboration are essential to the successful breeding of improved disease resistant cultivars. Examples are given of the effective use of genetic resources in breeding disease resistant cultivars of a number of crops, including cotton, rice, potatoes and pearl millet.

Introduction

Breeding crop cultivars that are resistant to diseases has been part of the plant breeder's armoury for a long time (Roane, 1973; Nelson, 1978). One of the first examples of directed resistance breeding was the transfer of resistance to *Fusarium* wilt from semi-wild, non-edible citron to water melon, which gave rise to the cultivar Conquerer in the USA (Orton, 1911). Since the 1960's reviews on resistance breeding in specific crops have appeared at regular intervals: e.g. vegetables (Walker, 1965), *Phaseolus* beans (Zaumeyer & Meiners, 1975), cotton (Arnold et al., 1976; Brinkerhoff et al., 1984), maize (Sprague & Paliwal, 1984), sorghum (ICRISAT, 1980), rice (Khush & Virmani, 1985), cassava (Jennings & Hershey, 1985), potatoes (Holden, 1977; Davidson, 1980; CIP, 1988), wheat (Simmonds & Rajaram, 1988; Nelson & Marshall, 1990).

Unfortunately, the chemical era in agriculture and horticulture (Bleasdale, 1987), during which pest and disease control was achieved largely by the profligate use of chemicals, served to put a brake on the synthesis of commercial cultivars that are genetically resistant to pests and diseases, especially in the high input farming systems of the western world. However, growing public sensitivity to the excessive use of potentially dangerous chemicals to control diseases and the ephemeral nature of effective control by some chemicals, as well as the escalating costs to develop and bring to the market so-called 'safe' chemicals, have highlighted the attributes of durable host resistance. The advantages of disease resistant cultivars include: no extra seed cost to the farmer; no disruption of the environment; fewer agro-chemicals, and reduction of operating costs. Interestingly, Simmonds (1991a) has pointed out that in sugar cane, pesticides/insecticides are hardly ever used on a crop in which 15–20 diseases are more or less controlled by the use of resistant cultivars, some completely so.

Breeding for resistance is an integrated process that requires development of suitable screening techniques, a search for sources of heritable resistance, and the transfer of resistance – sometimes from related wild species (Knott & Dvorak, 1976; Stalker, 1980) – to advanced breeding lines, which often requires a programme of backcrossing. Such transfers from wild species are often unavoidable, though my own experience leads me to believe that a search for resistance, even at a low level, in locally adapted material should be the plant breeder's first priority.

A major problem in resistance breeding is caused by genetic variation in pathogens 'overcoming' genetic resistance in the hosts (van der Plank, 1984) especially when mobile pathogens are involved. Moreover, environment, especially temperature, often modifies the interaction between host and pathogen. An understanding of the often complex relationships between host, pathogen and environment therefore holds the key to successful resistance breeding (Arnold et al., 1976; Williams, 1989). There is no debate here on the issues of major gene vs polygene, vertical vs horizontal, pathotype-specific vs pathotype non-specific, or durable vs non-durable resistance (van der Plank, 1963; 1968; 1982; Parlevliet & Zadoks, 1979; Robinson, 1971; 1973; Nelson, 1978; Johnson, 1984; Simmonds, 1988; 1991b), as breeding strategies differ for different crops, different diseases and different environments. Also there is no discussion of the relationships between genetic resistance in different tissues of the same plant e.g. leaf and tuber resistance of potato blight (Wastie, 1991) or leaf, stem and boll resistance to cotton blight (Innes, 1983). Whatever the semantics of resistance breeding, experience in a very wide range of crops has shown that in providing adequate resistance effective over a period of time the breeder is unable to rest on his or her laurels with the synthesis of a resistant cultivar. Breeders have to plan their programmes – which for conventional plant breeding are inevitably long term – in an insurance-like fashion so that new and different genes for host resistance are constantly being incorporated into adapted germplasm. In short, the wider the gene base – especially a gene base that incorporates polygenes – the more likely is continued success in containing the ravages of diseases. Conservation of genetic resources in gene banks (where seeds or plant parts are preserved outside their area of growth) and *in situ* (where plants, including wild relatives, are maintained in natural reserves), and the characterisation and utilization of such genetic resources in the breeding of improved cultivars, are among the most important strategies for sustainable agricultural development (Plucknett et al., 1987).

In any breeding programme, the breeder must ensure that, in addition to there being durable resistance in a potential cultivar, that cultivar also possesses the appropriate yield and quality traits to enable it to replace its susceptible counterpart.

Genetic resources

Breeders throughout the world attempt to assemble as wide a gene base as possible and most breeders have working collections containing genetic resources that are likely to be of use to them. For example, in the breeding of cotton resistant to bacterial blight (*Xanthomonas campestris* pv *malvacearum*), the late R.L. Knight (1957) screened over 1,000 different wild and cultivated accessions of *Gossypium* in a search for sources of resistance. His first paper on the genetics of host resistance appeared in 1939 (Knight & Clouston, 1939). Innes (1983) listed 19 major genes or polygene complexes from cultivated tetraploid and diploid *Gossypium* species and diploid wild species that have been used in the breeding of cotton cultivars resistant to blight. Thirteen genes were identified in tetraploid *Gossypium hirsutum,* one in tetraploid *G. barbadense,* two in diploid *G. arboreum,* two in diploid *G. herbaceum* and one in diploid *G. anomalum* (Table 1). An additional major, dominant gene (B_{12}) was identified by Wallace and El-Zik (1989) in a *G. hirsutum* variety S295 (Girardot et al., 1986) bred in Chad that is resistant to the virulent African isolate HV1 of *X. campestris* pv *malvacearum* (Follin, 1983). Long lasting field immunity to bacterial blight appears to have been obtained by cotton breeders in different parts of the world by combining major genes and polygenic complexes (Brinkerhoff et al., 1984).

It is one thing to assemble useful germplasm but another to ensure that such material is properly preserved, described and made available to breeders everywhere. Fortunately methods are now available to store seed in a viable condition for long periods (Roberts, 1975). In addition, tissue culture techniques have been developed to ensure long-term storage of vegetatively propagated species (Withers, 1989).

The conservation of global genetic resources is, however, a task beyond the budget of most national governments, though increasingly individual countries in both the developed and developing parts of the world are making more strenuous efforts to provide the finance and facilities for genetic conservation (Hawkes, 1991).

The International Board for Plant Genetic Resources (IBPGR) was formed in 1974 with a mandate to promote and co-ordinate work on the international collection and preservation of germplasm, and on the data information systems that are central to the efficiency of gene banks (Hawkes, 1985; Plucknett et al., 1987; Brown et al., 1989). Most of the world's preserved genetic resources are now in the public domain and, as a consequence of international agreements, are readily and freely available to *bona fide* breeders

Table 1. Major genes and polygene complexes conferring resistance to bacterial blight of cotton

Gene symbol	Description and source	References
B_1	Weak dominant gene obtained from Uganda B31 (*Gossypium hirsutum*)	Knight & Clouston (1939)
B_2	Strong, dominant gene from Uganda B31; also recorded in Albar from west Africa, UKBR from Tanzania, cultivars in the USA. (All *G. hirsutum*)	Knight & Clouston (1939), Innes (1965b), Innes (1969), Brinkerhoff (1970)
B_3	Partially dominant gene from Schroeder 1306 (an off-type *G. hirsutum* var. *punctatum*)	Knight (1944)
B_4	Partially dominant gene from Multani strain. NT 12/30 (*G. arboreum*)	Knight (1948)
B_5	Partially dominant gene from Grenadine White Pollen (a perennial *G. barbadense*)	Knight (1950)
B_6	Recessive gene from Multani strain NT 12/30 (*G. arboreum*); and possibly from Tanzania, UKBR 61/12 (*G. hirsutum*)	Knight (1953a), Saunders & Innes (1963), Innes (1969)
B_7	Gene from Stoneville 20 and other stocks from the USA (*G. hirsutum*). (Dominance of this gene is dependent upon the genetic background)	Knight (1953b), Green & Brinkerhoff (1956), Innes & Brown (1969)
B_8	Recessive gene from *G. anomalum,* an uncultivated diploid species from Africa	Knight (1954)
B_{9K}	Strong, dominant gene from Wagad 8, an Indian commercial cultivar (*G. herbaceum*)	Knight (1963), Innes (1965a)
B_{9L}	Strong, dominant gene from Allen 51-296 from west Africa (*G. hirsutum*)	Lagière (1960), Innes (1965a)
B_{10K}	Weak, partially dominant gene from Kufra Oasis in Libya (*G. hirsutum* var. *punctatum*)	Knight (1957), Innes (1965a)
B_{10L}	Weak gene from same source as B_{9L}	Lagière (1960), Innes (1965a)
B_{11}	Weak gene from same source as B_{9K}	Innes (1966)
B_{12}	Strong, dominant gene from S295 (*G. hirsutum*) bred in Chad	Girardot et al. (1986), Wallace & El-Zik (1989)
B_{In}	Dominant gene from an unknown cultivar from the USA (*G. hirsutum*)	Green & Brinkerhoff (1956), Brinkerhoff (1963)
B_N	Dominant gene from Northern Star, a cultivar from the USA (*G. hirsutum*)	Green & Brinkerhoff (1956)
B_S	Dominant gene from Stormproof 1, a cultivar from the USA (*G. hirsutum*)	Green & Brinkerhoff (1956)
B_{Sm}	Polygene complex found in Stoneville 2B and Empire, cultivars from the USA (*G. hirsutum*)	Bird & Hadley (1958)
B_{Dm}	Polygene complex found in Deltapine, a cultivar from the USA (*G. hirsutum*)	Bird & Hadley (1958)
$B_?$	Dominant gene from irradiated mutant in Westburn 70 (*G. hirsutum*)	Brinkerhoff et al. (1978)

everywhere. However, the Keystone Center (1991) has recently emphasized the need for a global initiative for the security and sustainable use of plant genetic resources, requiring the joint efforts and involvement of all affected parties and institutions from all parts of the world, including those who are contributors of germplasm, information, technology, funds and systems of innovation.

The increasing importance attached by both developed and developing countries to the health status of seeds and plant material is reflected by the strict quarantine procedures imposed by many governments. Fortunately, biotechnological aids in the form of cell and tissue culture techniques, and kits for the detection of even low levels of pathogens, have served to minimize the risks attached to the international flow of plant material. A top priority at all national and international gene banks is the creation, maintenance and distribution of disease and pest-free material, whether of seed or vegetative propagules.

Utilization of genetic resources

Successes in the use of genetic resources to provide germplasm for improved disease-resistant cultivars are many and varied. I propose to cite examples from programmes of genetic conservation and breeding with which I have been associated, either directly or indirectly. These are the potato (*Solanum tuberosum*) breeding programmes at the Scottish Crop Research Institute (SCRI) and the International Potato Centre (CIP), Peru, and the pearl millet (*Pennisetum glaucum*) breeding programme of the International Crops Research Institute for the Semi-Arid Tropics (ICRISAT), India. First, however, I should like to highlight success in the breeding of one of the world's most important grain crops, rice (*Oryza sativa*).

Rice

Undoubtedly one of the best examples of the utilization of genetic resources is provided by the most widely planted rice cultivar in history, IR36 (Plucknett et al., 1987). This cultivar also serves to illustrate what can be achieved by international cooperation, as well as stressing the effectiveness of collaboration between plant pathologists, breeders, entomologists, physiologists and agronomists. Thirteen rice cultivars from six different countries, and a wild species of rice, *Oryza nivara,* were utilized in the breeding of IR36, and four different countries were involved in screening for resistance to a wide range of pests and diseases after a final cross to combine a number of resistances. IR36 resists many pests and diseases, including green leaf hopper (*Nephotettix virescens*), brown plant hopper (*Nilaparvata lugens*), stem borer (*Chilo* sp), blast (*Pyricularia oryzae*), bacterial blight (*Xanthomonas campestris* pv *oryzae*), tungro and grassy stunt viruses. In addition, it is tolerant of a number of soil toxicities and survives moderate drought. IR36, despite its multiple disease and pest resistance will probably, like its predecessors, ultimately succumb to variant forms of pests and diseases, but its genes will continue to contribute to new improved cultivars which will also contain new genes for resistance from the highly variable genetic pool assembled in the gene bank at the International Rice Research Institute (IRRI), Manila, Philippines (Chang et al., 1982) and in national collections throughout the world.

Potato

The Commonwealth Potato Collection (CPC), which is maintained by the Scottish Crop Research Institute (SCRI), was originally known as the Empire Potato Collection and was founded on material collected in Mexico and South America by expeditions led by E.K. Balls and J.G. Hawkes (Wilkinson, 1991a). The CPC currently has about 2,000 accessions of 63 wild species and cultivated forms of *Solanum.* It has been a source of germplasm for the breeding of improved cultivars at the former Scottish Plant Breeding Station (SPBS) (Holden, 1977; Davidson, 1980; Mackay, 1982; Glendinning, 1983) now part of the SCRI, and the Plant Breeding Institute, Cambridge among others. Clones from SPBS outwith the CPC were also made available to East and Central African and Indian breeders (Black, 1971) and provided improved cultivars with the prefixes Roslin in Kenya (e.g. Roslin Eburu) and Kufri in India (e.g. Kufri Jyoti).

A wide range of material is freely available on request from the CPC, in the form of pollen, true seed, tubers and plantlets in culture. Wilkinson (1991b) has recorded that, despite a lack of publicity, since 1967 over 4,000 requests have been made for accessions in the CPC. Of these, 70% derive from UK institutions, 21% from the rest of Europe, 4% from the Americas, 2% each from Oceania and Asia and 1% from Africa. Thirty-nine requests have been specifically for screening for resistance to late blight, 71 requests for virus resistance studies and 133 for potato genetics/undisclosed screening. Currently the J.G. Hawkes collection (over 1,000 accessions), which was based at the University of Birmingham, is being added to the CPC after passing through quarantine facilities in Scotland.

The most recent cultivar of potato bred at SCRI and released commercially is Brodick, which illustrates the value of a broad genetic base. It has special properties for processing and does not sweeten during low temperature storage (Mackay & McNicol, 1991) as well as an excellent spectrum of resistance to most fungal diseases, especially late blight (*Phytophthora infestans*), and the important virus diseases caused by potato virus Y (PVY) and potato leafroll virus (PLRV). It has within its complex pedigree *S. tuberosum* spp. *andigena, S. demissum, S. phureja, S. simplicifolium* and *S. vernei* as well as other *Solanum* species that undoubtedly feature in the parentage of some of the established cultivars that are part of Brodick's pedigree.

The potato gene bank at the International Potato Centre (CIP), Lima, Peru (which has nearly five and a half thousand accessions – Table 2) has served well the needs of potato breeders both at CIP and in the National Agricultural Research Systems (NARS) of developing countries and CIP's advanced breeding material is to be found in farmers' fields throughout the world (Figs. 1 and 2). Some of these cultivars have been released but not named by NARS, some named but not officially released and some are simply being grown by farmers with no official governmental action (J.E. Bryan & Z. Huaman, personal communication).

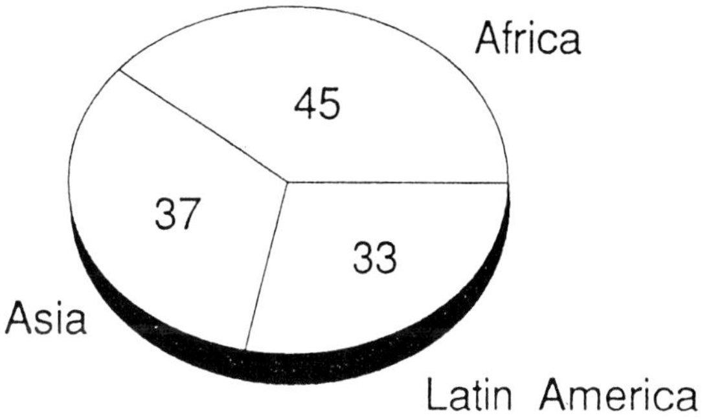

Fig. 1. Potato cultivars in farmer fields from germplasm distributed by the International Potato Centre to National Agricultural Research Systems (July, 1991).

Among CIP clones with good disease resistance are P-3 (or CIP 374080.5) and Serrana Inta (or CIP 720087). P-3, which is well adapted to cool, wet, tropical highlands, was bred by CIP in 1974 from the cross I-1085 × 700764 and has excellent resistance to late blight as well as tolerance of frost and wide adaptability to different ecoregions in Peru, where it is known as Perricholi. In Bolivia it has the name Runa Toralapa, in Guatemala it is Icta Paquix and in Burundi it is known as Lupita. It is also grown by farmers in Vietnam and the Philippines. P-3's female parent, I-1085, was bred in India, using germplasm from Scotland which in turn was derived from Mexican germplasm with the wild species *Solanum demissum* in its pedigree, providing P-3 with both race-specific and non-specific resistance to late blight. The male parent, 700764, was a native Peruvian cultivar of *S. tuberosum* spp. *andigena,* known as Casa Blanca, which donated non race-specific resistance to blight and frost tolerance.

Serrana Inta was derived from B70.178.2, an

Table 2. Number of accessions in the potato genebank at the International Potato Centre (July, 1991)

Wild species	1500
Native cultivars	3465
Improved cultivars	294
Breeding lines	196
Total	5455

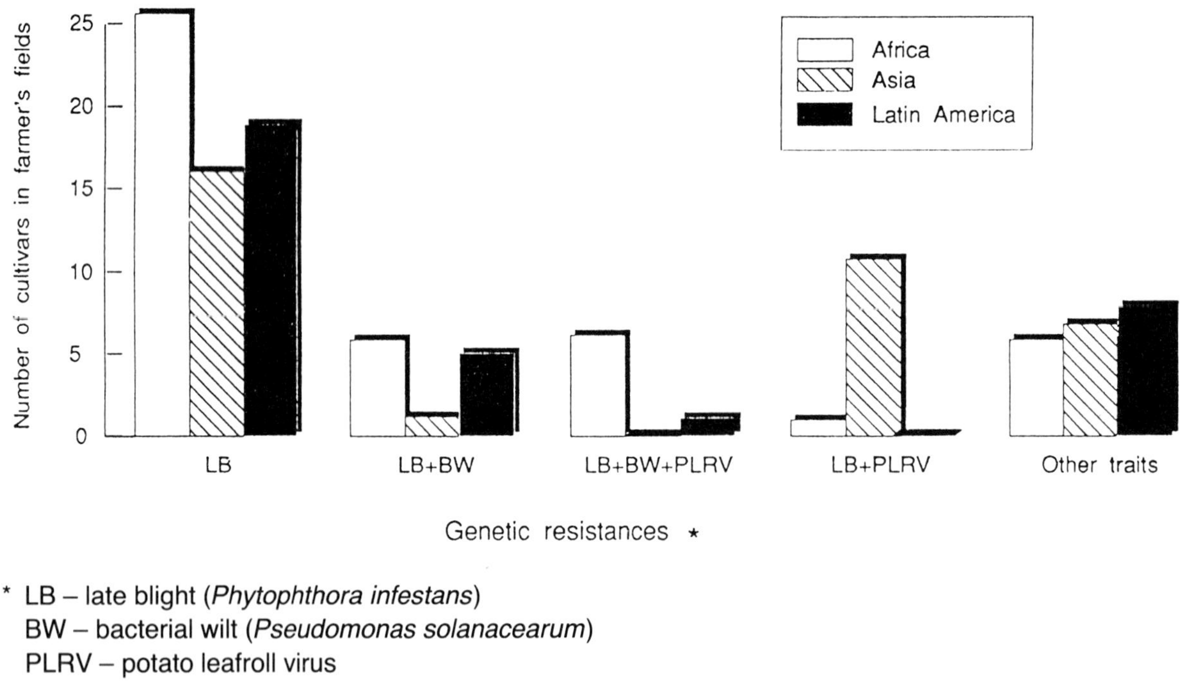

Fig. 2. Genetic resistances identified in potato germplasm distributed by the International Potato Centre to National Agricultural Research Systems (July, 1991).

Argentinian bred clone which was selected from the cross MPI 59703/21 × B2.63 and was received by CIP in 1976. Its female parent, MPI 59703/21, came from an accession of *S. stoloniferum* resistant to PVY supplied by Dr Hans Ross of the Max Planck Institute, Germany. Potato virus X resistance had also been incorporated from another parent. The male parent, B2.63, was derived from the cross B13 × Huinkul MAG (an Argentinian cultivar known for its good storage and excellent culinary qualities). In its screening work, CIP confirmed the PLRV resistance and storage quality of B70.178.2, found the clone was immune to PVX and had hypersensitivity to PVY. It was also well adapted to diverse ecoregions, especially to the warm tropics.

CIP distributed the clone as B70.178.2 until 1978 when Argentina named it Serrana Inta and subsequent distributions by CIP were under that name. Serrana Inta is now grown in China, Papua New Guinea, Tonga, Fiji, Vanuata and Tahiti, where its heat tolerance, long term storage attribute and virus resistances are needed. Serrana Inta is also widely used as a parent by both CIP and NARS breeders, especially as a female parent in CIP's true potato seed (TPS) programmes. Serrana × DTO-28 and Serrana × LT-7 are two of CIP's four most widely grown TPS progeny of major importance in Egypt, Bangladesh, Indonesia and Paraguay.

Pearl millet

The International Crop Research Institute for the Semi-arid Tropics (ICRISAT) has a global mandate for research on five major crops of the semi-arid tropics: the cereals sorghum (*Sorghum bicolor*) and pearl millet (*Pennisetum glaucum*) and the legumes groundnut (or peanut) (*Arachis hypogea*), chickpea (*Cicer arietinum*) and pigeonpea (*Cajanus cajan*), and also maintains a germplasm collection of minor millets (ICRISAT, 1990a). Within its gene bank are about 100,000 accessions made up as follows (June 1991):

Sorghum	Pearl millet	Chickpea	Pigeonpea	Groundnut	Minor millets
32,890	21,919	15,995	11,482	12,841	7,082

M.H. Mengesha, Programme Leader of Genetic Resources at ICRISAT, provided the data in Table 3, which shows the number of accessions with resistance/tolerance to diseases in one of ICRISAT's five mandate crops, pearl millet. Some of these resistances may, of course, depend on the same genes and further work is needed to resolve this issue.

A successful pearl millet release serves to reemphasize the value of an international approach to breeding for resistances to diseases (Harinarayana & Rai, 1989). ICMV-1 or WC-C75, an open-pollinated cultivar was released in 1982 by the Government of India and by 1990 occupied about 1.5 million hectares in India and has been introduced to Zambia (ICRISAT, 1990b). One of the WC-C75's main strengths is a broadly based genetic resistance to downy mildew (*Sclerospora graminicola*). It was bred at ICRISAT from a 'World Composite' originally constituted in 1971 at the Institute for Agricultural Research, Ahmadu Bello University, Nigeria from numerous crosses between world-wide sources of pearl millet and Nigerian early-maturing landraces. Full-sib recurrent selection was practised at ICRISAT with testing at Coimbatore (southern India), Hisar (northern India) and Patancheru, near Hyderabad (central India). From a total of 441 full-sib families seven were selected, and disease-free plants in these families in a downy mildew screening nursery at Patancheru were selfed. The resulting S_1 bulk was subsequently sown in the next season's downy mildew nursery and bulk pollen used to force intermating. The cultivar so produced became WC-C75, which was subjected to five generations of relatively light selection pressure for agronomic characters before the production of breeder's seed.

Table 3. ICRISAT pearl millet accessions resistant/tolerant to diseases (June, 1991)

Disease	Accessions
Downy mildew (*Sclerospora graminicola*)	1220
Ergot (*Claviceps fusiformis*)	151
Smut (*Tolyposporum penicillariae*)	161
Rust (*Puccinia penniseti*)	392

Conclusions

Some of the examples that I have given to illustrate the contributions made by breeders and pathologists to the synthesis of disease-resistant cultivars involve genetic resource material from breeders' collections made prior to the establishment of gene banks. Such collections now form an integral part of national and international gene banks which hold the key to the future well-being of sustainable agriculture throughout the world.

I have concentrated on breeding products from so-called conventional or traditional plant breeding. The accelerated and targeted transfer of genes of both plant and non-plant origin by the use of recombinant DNA technology and cell and tissue culture is now likely to complement the breeder's traditional technology (Innes, 1989). Gene banks will, however, continue to provide the plant genes for genetic engineers, especially useful genes from wild species.

Quarantine problems may delay the movement of parental breeding material but I firmly believe that the real hold up in resistance breeding is as a consequence of an inadequate understanding of the complex relationships involving host, pathogen and environment. In discussing the breeding of potato cultivars resistant to late blight, Wastie (1991) wrote 'Plenty of resistance to blight exists: the problem is to define more efficient and cost-effective methods of screening for and using it.' That statement is applicable to quite a number of diseases. For success, better interdisciplinary collaboration on a wide front is needed, between breeders and pathologists, tissue culture specialists and molecular biologists, national and international programmes.

Acknowledgements

I am grateful to colleagues at SCRI for comments on the text and to M.H. Mengesha (ICRISAT) and

J.E. Bryan and Z. Huaman (CIP) for providing information.

References

Arnold, M.H., N.L. Innes & S.J. Brown, 1976. Resistance breeding. In: M.H. Arnold, (Ed.) Agricultural Research for Development. The Namulonge Contribution. pp. 175–195. Cambridge University Press, Cambridge.

Bird, L.S. & H.H. Hadley, 1958. A statistical study of the inheritance of Stoneville 20 resistance to the bacterial blight disease of cotton in the presence of *Xanthomonas malvacearum* races 1 and 2. Genetics 43: 750–767.

Black, W., 1971. Researches on potatoes at the Scottish Plant Breeding Station. 50th A. Rep. (1970–71) Scottish Plant Breeding Station, 52–60.

Bleasdale, J.K.A., 1987. Future research and development. Prof. Hort. 1: 35–38.

Brinkerhoff, L.A., 1963. Variability of *Xanthomonas malvacearum:* the cotton bacterial blight pathogen. Oklahoma State University Technical Bulletin, T-98.

Brinkerhoff, L.A., 1970. Variation in *Xanthomonas malvacearum* and its relation to control. A. Rev. Phytopath. 8: 85–110.

Brinkerhoff, L.A., L.M. Verhalen, R. Mamaghani & W.M. Johnson, 1978. Inheritance of an induced mutation for bacterial blight resistance in cotton. Crop. Sci. 18: 901–903.

Brinkerhoff, L.A., L.M. Verhalen, W.M. Johnson, M. Essenberg & P.E. Richardson, 1984. Development of immunity to bacterial blight of cotton and its implications for other diseases. Plant Dis. 68: 168–173.

Brown, A.H.D., O.H. Frankel, D.R. Marshall & J.T. Williams, 1989. The Use of Plant Genetic Resources. Cambridge University Press, Cambridge.

Chang, T.T., C.R. Adair & T.H. Johnston, 1982. The conservation and use of rice genetic resources. Adv. in Agron. 35: 38–91.

CIP, 1988. Bacterial diseases of the potato. Rep. of the Planning Conference on Bacterial Diseases of the Potato, 1987. International Potato Centre, Lima, Peru.

Davidson, T.M.W., 1980. Breeding for resistance to virus disease of the potato (*Solanum tuberosum*) at the Scottish Plant Breeding Station. 59th A. Rep. (1979–80) of Scottish Plant Breeding Station, 100–108.

Duncan, J.M. & L. Torrance, 1992. Techniques for the Rapid Detection of Plant Pathogens. Blackwell Scientific Publications.

Follin, J.C., 1983. Races of *Xanthomonas campestris* pv *malvacearum* (Smith) Dye in Western and Central Africa. Coton Fibr. trop. 38: 277–280.

Girardot, B., E. Hequet, M.T. Yehouessi & P. Guibordeau, 1986. Finding a variety of *Gossypium hirsutum* L. resistant to strains of *Xanthomonas campestris* pv *malvacearum* (Smith) Dye virulent on associations of major genes (B_2B_3 or B_9–B_{10}). Coton Fibr. trop. 41: 67–69.

Glendinning, D.R., 1983. Potato introductions and breeding up to the early 20th Century. New Phytol. 94: 479–505.

Green, J.M. & L.A. Brinkerhoff, 1956. Inheritance of three genes for bacterial blight resistance in Upland cotton. Agron. J. 48: 481–485.

Harinarayana, G. & Kn Rai, 1989. Use of pearl millet germplasm and its impact on crop improvement in India. Collaboration on genetic resources, pp 89. ICRISAT (International Crops Research Institute for the Semi-Arid Tropics), Patancheru, India.

Hawkes, J.G., 1985. Plant genetic resources. The impact of the International Agriculture Research Centres. Consultative Group on International Agricultural Research, Washington DC.

Hawkes, J.G., 1991. Genetic conservation of world crop plants. Biol. J. Linn. Soc. 43.

Holden, J.H.W., 1977. Potato breeding at Pentlandfield. 56th A. Rep. (1976–77) of Scottish Plant Breeding Station, 66–97.

ICRISAT, 1980. Sorghum diseases. A world review. Proceedings of the International Workshop on Sorghum Diseases, 1978. International Crops Research Institute for the Semi-Arid Tropics, Patancheru, India.

ICRISAT, 1990a. A. Rep. (1989) International Crops Research Institute for the Semi-Arid Tropics. Patancheru, India.

ICRISAT, 1990b. ICRISAT's contribution to pearl millet production. International Crops Research Institute for the Semi-Arid Tropics, Patancheru, India.

Innes, N.L., 1965a. Resistance to bacterial blight of cotton: the genes B_9 and B_{10}. Exp. Agric. 1: 189–191.

Innes, N.L., 1965b. Inheritance of resistance to bacterial blight of cotton. 1. Allen (*Gossypium hirsutum*) derivatives. J. Agric. Sci. 64: 257–271.

Innes, N.L., 1966. Inheritance of resistance to bacterial blight of cotton. 3. *Herbaceum* resistance transferred to tetraploid cotton. J. Agric. Sci. 66: 433–439.

Innes, N.L., 1969. Inheritance of resistance to bacterial blight of cotton. 4. Tanzania selections. J. Agric. Sci. 72: 41–57.

Innes, N.L. & S.J. Brown, 1969. A quantitative study of the inheritance of resistance to bacterial blight in Upland cotton. J. Agric. Sci. 73: 15–23.

Innes, N.L., 1983. Bacterial blight of cotton. Biol. Rev. 58: 157–176.

Innes, N.L., 1989. Biotechnology and plant breeding. AgBiotech News and Inf. 1: 27–32.

Jennings, D.L. & C.H. Hershey, 1985. Cassava breeding: a decade of progress from international programs. In: G.E. Russell (Ed.) Progress in Plant Breeding – 1. pp. 89–116. Butterworths, London.

Johnson, R., 1984. A critical analysis of durable resistance. A. Rev. Phytopath. 22: 309–330.

Keystone Center, 1991. Global initiative for the security and sustainable use of plant genetic resources. Keystone International Dialogue Series on Plant Genetic Resources. 3rd Plenary Session 31 May – 4 June 1991, Oslo, Norway, p. 42.

Knight, R.L., 1944. The genetics of blackarm resistance. 4.

Gossypium punctatum (Sch. & Thon.) crosses. J. Genet. 46: 1–27.
Knight, R.L., 1948. The genetics of blackarm resistance. 6. Transference of resistance from *Gossypium arboreum* L. to *G. barbadense*. J. Genet. 48: 359–369.
Knight, R.L., 1950. The genetics of blackarm resistance. 8. *Gossypium barbadense*. J. Genet. 50: 67–76.
Knight, R.L., 1953a. The genetics of blackarm resistance. 9. The gene B_{6m} from *Gossypium arboreum*. J. Genet. 51: 270–275.
Knight, R.L., 1953b. The genetics of blackarm resistance. 10. The gene B_7 from Stoneville 20. J. Genet. 51: 515–519.
Knight, R.L., 1954. The genetics of blackarm resistance. 11. *Gossypium anomalum*. J. Genet. 52: 466–472.
Knight, R.L., 1957. Blackarm disease of cotton and its control. Plant Protection Conference, 1956. p53. Butterworth, London.
Knight, R.L., 1963. The genetics of blackarm resistance. 12. Transference of resistance from *Gossypium herbaceum* to *G. barbadense*. J. Genet. 58: 328–346.
Knight, R.L. & T.W. Clouston, 1939. The genetics of blackarm resistance. 1. Factors B_1 and B_2. J. Genet. 38: 133–159.
Khush, G.S. & S.S. Virmani, 1985. Breeding rice for disease resistance. In: G.E. Russell (Ed.) Progress in Plant Breeding – 1. Butterworths, London, pp. 239–279.
Knott, D.R. & J. Dvorak, 1976. Alien germplasm as a source of resistance to disease. A. Rev. Phytopath. 14: 211–235.
Lagière, R., 1960. La bactériose du cotonnier (*Xanthomonas malvacearum*) (E.F. Smith) Dowson dans le monde et en République Centrafricaine (Oubangui-Chari) Paris, I.R.C.T., p. 252.
Mackay, G.R., 1982. Breeding for resistance to pests and diseases. Producing quality seed potatoes in Scotland. Proc. Scottish Society for Crop Research. Bulletin No. 1: 27–35.
Mackay, G.R. & R.J. McNicol, 1991. Plant genetics. A. Rep. Scottish Crop Research Institute (1990): 8–9.
Nelson, R.R., 1978. Genetics of horizontal resistance to plant diseases. A. Rev. Phytopath. 16: 359–378.
Nelson, L.R. & D. Marshall, 1990. Breeding wheat for resistance to *Septoria nodorum* and *Septoria tritici*. Adv. Agron. 44: 257–277.
Orton, W.A., 1911. The development of disease resistant varieties of plants. 4th Conf. Internat. Genet., Paris, 247–265.
Parlevliet, J.E. & J.C. Zadoks, 1977. The integrated concept of disease resistance; a new view including horizontal and vertical resistance in plants. Euphytica 26: 5–21.
Plucknett, D.L., N.J.H. Smith, J.T. Williams & N.M. Anishetty, 1987. Gene Banks and the World's Food. Princeton University Press, New Jersey, 247p.
Roane, C.W., 1973. Trends in breeding for disease resistance in crops. A. Rev. Phytopath. 11: 463–486.
Robinson, R.A., 1971. Vertical resistance. Rev. Plant Path. 50: 233–239.
Robinson, R.A., 1973. Horizontal resistance. Rev. Plant Path. 52: 483–501.
Roberts, E.H., 1975. Problems of long-term storage of seed and pollen for genetic resources conservation. In: O.H. Frankel & J.G. Hawkes (Eds.) Crop Genetic Resources for Today and Tomorrow, pp. 269–295. Cambridge University Press.
Saunders, J.H. & N.L. Innes, 1963. The genetics of bacterial blight resistance in cotton. Further evidence on the gene B_{6m}. Genet. Res. 4: 382–388.
Simmonds, N.W., 1988. Synthesis: the strategy of rust resistance breeding. In: N.W. Simmonds & S. Rajaram (Eds.) Breeding Strategies for Resistance to the Rusts of Wheat, pp. 119–136. CIMMYT, Mexico D.F.
Simmonds, N.W., 1991a. Reflections on sugar cane. Tropic. Agric. Assoc. Newsletter June, 1991, 30–31.
Simmonds, N.W., 1991b. Genetics of horizontal resistance to diseases of crops. Biol. Rev. 66: 189–241.
Simmonds, N.W. & S. Rajaram, 1988. Breeding strategies for resistance to the rusts of wheat. CIMMYT, Mexico D.F..
Sprague, E.W. & R.L. Paliwal, 1984. CIMMYT's maize improvement programme. Outlook on Agric. 13: 24–31.
Stalker, H.T., 1980. Utilization of wild species for crop improvement. Adv. Agron. 33: 112–147.
van der Plank, J.E., 1963. Plant diseases: Epidemics and Control. Academic Press, London.
van der Plank, J.E., 1968. Disease resistance in plants. Academic Press, New York.
van der Plank, J.E., 1982. Host-pathogen interactions in Plant Disease. Academic Press, New York.
van der Plank, J.E., 1984. Disease Resistance in Plants. Academic Press, New York.
Walker, J.C., 1965. Disease resistance in the vegetable crops. III Bot. Rev. 31: 331–380.
Wallace, T.P. & K.M. El-Zik, 1989. Inheritance of resistance in three cotton cultivars to the HV1 isolate of bacterial blight. Crop. Sci. 29: 1114–1119.
Wastie, R.L., 1991. Breeding for resistance. *Phytophthora infestans,* the cause of late blight of potato. Adv. Plant Path. 7: 193–224.
Wilkinson, M.J., 1991a. The revitalisation of the Commonwealth Potato Collection. *Solanum Newsletter.* In press.
Wilkinson, M.J., 1991b. Systematics research at SCRI. Scottish Crop Research Institute.
Williams, P.H., 1989. Screening for resistance to diseases. In: A.H.D. Brown, O.H. Frankel, D.R. Marshall, J.T. Williams (Eds.) The use of plant genetic resources, pp. 335–352. Cambridge University Press.
Withers, L.A., 1989. In vitro conservation and germplasm utilisation. In: A.H.D. Brown, O.H. Frankel, D.R. Marshall & J.T. Williams (Eds.) The use of plant genetic resources, pp. 309–334. Cambridge University Press.
Zaumeyer, W.J. & J.P. Meiners, 1975. Disease resistance in beans. A. Rev. Phytopath. 13: 313–334.

Euphytica **63**: 33–49, 1992.
R. Johnson and G.J. Jellis (eds), Breeding for Disease Resistance

Disease resistance in protected crops and mushrooms

J.T. Fletcher
ADAS Boxworth, Cambridge CB3 8NN, UK; present address, ADAS, Olantigh Road, Wye, Ashford, Kent TN25 5EL, UK*

Key words: tomato, *Lycopersicon esculentum,* cucumber, *Cucumis sativus,* pepper, *Capsicum annuum,* lettuce, *Lactuca sativa,* mushroom, *Agaricus* spp., carnation, *Dianthus caryophyllus,* chrysanthemum, *Dendranthema grandiflorum,* disease resistance

Summary

Cultivars of tomatoes, cucumbers, lettuce and peppers have been bred for resistance to one or more pathogens. Some tomato and cucumber cultivars have resistance to a wide range of diseases. Resistance has been transient in many cases and a succession of cultivars with new genes or new combinations of resistance genes has been necessary to maintain control. There has been a number of notable exceptions and these have included durable resistance to such pathogens as *Fulvia fulva* and tomato mosaic virus. With lettuce the resistance situation is complicated by the occurrence of fungicide resistant pathotypes. There are no strains of *Agaricus bisporus* purposely bred for disease resistance.

In protected flower crops only resistance to Fusarium wilt in carnations has been purposely bred but differences in disease resistance are apparent in cultivars of many ornamental crops. This is particularly so in chrysanthemums where there are cultivars with resistance to many of the major pathogens. Similar situations occur with other flower crops and pot plants. Cultivars of some species have not been systematically investigated for resistance.

The need for genetic resistance will increase with the further reduction, in the limits on pesticide use and an increasing public awareness and importance of pesticide pollution.

Introduction

Production of crops in glasshouses and in mushroom sheds was well established in the UK at the beginning of the century. A great diversity of crops were grown including some vegetables, flowers and even fruits such as peaches and grapes. During the two world wars there was an emphasis on food production and many flower growers were compelled to grow tomatoes. In the 1960's plastic covered structures were developed and are now a substantial part of protected cropping in England and Wales. Throughout this entire period the tomato crop has been the most widely grown. The total area of commercial glasshouses and protected structures has remained more or less constant for some time, at around 2,200 ha. The total gross value of all protected crops at present is about £556 m which, put in the perspective of the more major crops grown, is about a third of the value of the wheat crop. Relative values and areas of the type of protection and of the individual crops (Tables 1, 2, 3) show that mushrooms have the largest gross value followed by tomatoes. Pot plants are the most valuable ornamental crop overall, but

* ADAS is an executive agency of the Ministry of Agiculture, Fisheries and Food and the Welsh Office.

Table 1. Areas (ha) of protected crops in England and Wales 1991

Tomato		Chrysanthemum AYR+	72
heated	413	Pots	20
unheated	157	Other	150
Cucumber	213	Pinks	20
Lettuce	1455	Carnation	10
Celery	158	Alstroemeria	21
Sweet Pepper	70	Rose	12
Mushroom	549	Pot plants	173

+ All-year-round.

chrysanthemums have the highest value of the cut flowers.

This background gives some indication of the commercial relevance of protected cropping in relation to other crops and explains to a large extent why companies developing disease resistant cultivars or pesticides, cannot afford to embark on expensive research programmes for the UK industry alone. Even taken on a European or global scale it is financially debatable whether large expenditure can be justified on the development of disease resistance in cultivars of many protected crops.

It is therefore not surprising that there are relatively few cultivars that have been bred specifically for disease resistance that are available to growers of protected crops. The exception is the tomato, where there are many cultivars with resistance to a range of pathogens. Lettuce, cucumbers and peppers have been bred with resistance to one or more pathogens. Even where resistance has been introduced, it is not always present in acceptable cultivars. In protected ornamental crops, which include various cut flowers and pot plants, almost no cultivars have been bred especially for disease resistance. The exception is the carnation where resistance to Fusarium wilt (*Fusarium oxysporum* f. sp. *dianthi*) has been bred into cultivars which are now widely used.

Table 2. Area (ha) of types of protected cropping in England and Wales 1991

Glass	heated	1,451.7
	unheated	299.7
Plastic	heated	99.4
	unheated	320.5
Total		2,171.3

During the past twenty years there have also been major changes in methods of crop culture which have influenced the spectrum of diseases which occur in protected crops and therefore the grower's requirements. By far the biggest change has been the move out of the soil for crops such as tomatoes, cucumbers, peppers and pot plants into inert media such as rockwool or perlite, and in the case of pot plants into peat and peat/bark composts. The importance of some soil-borne diseases of protected vegetable crops has decreased as a result of this change and disease resistance requirements changed with it.

Fuel crises and the development of greater precision in the control of the environment have resulted in growers being able to control temperature and relative humidity more precisely. In this way some air-borne fungal pathogens have been successfully controlled without genetic resistance or fungicides. Conversely, other factors have increased the need for resistant cultivars, such as the development of fungicide resistant strains and the increasing restriction on the use of pesticides. Overall the size of the market for disease resistant cultivars will always determine the financial input into disease resistance breeding, which is likely to be concentrated on those crops which are grown world-wide and by the companies who have world wide sales of their propagation material.

Table 3. Gross values of protected crops in England and Wales 1990

Crops	Values in £M
Mushrooms	173.7
Tomato	90.4
Lettuce	36.2
Cucumber	43.7
Celery	5.0
Sweet pepper	2.6
ornamentals	199.9
Total	551.5

Tomato

In 1949 there were 1,380 ha of protected tomatoes in England and Wales, producing some 112,000 t of tomatoes (Hitchins, 1951), equivalent to about 81 t ha^{-1}. Maximum yield at this time would have been about 144 t ha^{-1}.

In 1990 there were 570 ha producing approximately 138,800 t equivalent to about 243 t ha^{-1}. Maximum yields are now near to 450 t ha^{-1}. Yield from the best crops and per hectare has more than trebled over the 43 year period and some of this increase is attributable to better disease control which is partly because of the development of disease resistant cultivars. Resistance is available to a range of pathogens as well as to root know nematodes (*Meloidogyne incognita*) and many cultivars have combined resistance to many of these (Table 4). The use of resistant cultivars has resulted in reduced fungicide use and some of the previously common diseases have become very uncommon, eg. Fusarium wilt (*F. oxysporum* f. sp. *lycopersici*) and Verticillium wilt (*V. albo-atrum, V. dahliae*), leaf mould (*Fulvia fulva*) and tomato mosaic. Root disease caused by *Pyrenochaeta lycopersici* is also less common, not because of resistance but because of the change in cultural practice away from the soil to the use of synthetic substrates or nutrient film. In 1973 it was estimated that 10–21% of the crop was lost annually due to tomato mosaic, leaf mould, Fusarium and Verticillium Wilt (Fletcher, 1973). In 1990 it is likely that losses from these diseases are insignificant and this is almost entirely due to the universal use of resistant cultivars. In addition to savings on lost yield, growers now spend no money on fungicides to control these diseases.

Tomato mosaic (Tomato mosaic virus, ToMV). The majority of tomato cultivars commercially cultivated are resistant to ToMV, once the most common and damaging disease of the crop (Fletcher, 1973). Resistance is largely dependent upon the gene *Tm*-2^2 which is the major source of resistance in the majority of cultivars. The first TMV resistant cultivars depended upon the gene *Tm*-1 alone but these remained resistant for a very short period, in some cases less than 6 months (Pelham, Fletcher & Hawkins, 1970). The use of *Tm*-2^2 either in combination with *Tm*-1 and *Tm*-2 or alone, has given durable resistance for more than 20 years. Although resistance breaking strains have been reported (Hall & Bowes, 1979) they have not become established in crops. There are only two records in the UK of such strains and both were from Sussex (1975 and 1976 respectively). At this time, resistant and susceptible cultivars were grown, often together. Cross protection of susceptible cultivars using an avirulent strain of the virus MII-16, was also practised (Rast, 1972). The predominant cultivar was Sonato (claimed to be homozygous for *Tm*-2^2) and the resistance breaking strain, although able to propagate in this cultivar, did so very poorly and spread within the affected crop was slow. By removing the diseased plants, the disease was totally arrested, a measure which would have had no noticeable effect on the rate of spread of the wild-type strain in susceptible cultivars. Similar resistance-breaking strains occurred at the same time in the Netherlands but since then there have been no further records. ToMV remains a problem and cross protection by mild strain inoculation is used where the susceptible cherry cultivar Gardeners Delight is grown.

Occasionally ToMV resistant cultivars have been grafted onto ToMV susceptible rootstocks which have become infected. The *Tm*-2^2 resistant scion cultivar develops severe distortion of the foliage and extensive necrosis of the fruit. ToMV resistant roodstocks (Hires and Signaal) are now available for graft combinations with *Tm*-2^2 resistant scions.

In early experiments with grafted plants it was shown that inoculation of a *Tm*-0 host with strain O, purified by single lesion transfer, grafted to cultivars with resistance genes *Tm*-1, *Tm*-2, *Tm*-1 + *Tm*-2 and *Tm*-2^2 with each genotype represented in a single graft combination, resulted in the recovery of strain 0 from the *Tm*-0 inoculated component, strain 1 from the *Tm*-1 component, strain 2 from *Tm*-2 and 1 : 2 from *Tm*-1 + *Tm*-2. No new strains were recovered from the *Tm*-2^2 component. The *Tm*-2^2 host showed necrosis and stunting which is considered to be a resistant reaction (McNeil & Fletcher, 1971; Hall & Bowes, 1979).

Table 4. Disease resistance in Tomato cultivars

Cultivar	Diseases						
	M	LM	FW	VW	FC & RR	B & CR	RK
2101	+	C5	F2	+	+	–	–
Abunda	+	C5	F2	+	–	–	–
Account	+	C5	F2	+	–	–	–
Angela	+	C3	F1	+	–	–	–
Arasta	–	C2	–	–	–	–	–
Astrid	+	C5	F2	+	+	–	–
Belcanto	+	C5	F2	+	–	–	+
Blizzard	+	C5	F2	+	–	–	–
Calypso	+	C5	F2	+	–	–	–
Cantatos	+	C5	F2	+	–	–	–
Carusa	+	C5	F2	+	–	–	–
Castel	+	C5	F2	–	–	–	+
Cheresita (FL52)	+	–	–	–	–	–	+
Choice	+	C5	F2	+	+	–	–
Concord	+	C5	F2	–	–	–	–
Concreto	+	C5	F2	–	–	–	–
Cossack	+	C5	F2	–	–	–	–
Counter	+	C5	F2	+	–	–	–
Criterion	+	C5	F2	+	–	–	–
Curabel	+	C5	F1	+	–	–	–
Dombella	+	C5	F2	+	–	–	+
Dombito	+	C2	F2	–	–	–	–
Duranto	+	C5	F2	–	–	–	–
Duro	+	C5	F1	+	–	–	+
Else	+	C5	F1	+	–	–	–
Estafette	+	C5	F1	+	–	–	+
Evita	+	–	–	–	–	–	+
Fianto	+	C5	F2	–	–	–	–
G's D'lit	–	–	–	–	–	–	–
Goldstar	+	C5	F2	+	–	–	–
Kontiki	+	C5	F2	+	–	–	–
Larganto	+	C2	–	–	–	–	–
Larma	+	C2	F2	–	–	–	–
Laura	+	C2	F2	+	–	–	–
Liberto	+	C5	F2	+	–	–	+
Libra	+	C5	F2	+	+	–	–
Lingano	+	C4	F2	–	–	–	–
Locanda	+	C4	F2	–	–	–	–
Lotus	+	–	F2	–	–	–	–

Cultivar	Diseases						
	M	LM	FW	VW	FC & RR	B & CR	RK
Mammoth	+	–	F2	–	–	–	–
Manhattan	+	–	F2	–	–	–	–
Marathon	+	C5	F2	+	–	–	–
Marcanto	+	C5	F2	–	–	–	–
Meltine	+	C2	F2	+	–	–	+
Mercator	+	C5	F2	–		–	–
Monza	+	C3	F1	–	–	–	–
Nomato	+	C5	F1	+	–	–	+
Ostona	+	C5	F1	–	–	–	–
Pannoy	+	C5	F2	+	–	–	–
Perfecto	+	C5	F2	–	–	–	–
Pinto	+	–	F1	+	–	–	–
Piranto	+	C5	F2	–	–	+	–
Portanto	+	C5	F2	–	–	–	+
Primato	+	C5	F2	+	–	–	–
Pronto	+	C5	F2	+	+	–	+
Rainbow	+	C5	F2	–	–	–	–
Red Ensign	–	C5	–	–	–	–	–
Restino	+	C5	F2	+	–	–	+
Rimini	+	C2	F2	+	–	–	+
Rocco	+	C5	F2	–	–	–	–
Samoa	+	C5	F2	+	+	–	–
Santana	+	C4	F2	–	–	–	–
Shirley	+	C3	F2	–	–	–	–
Sierra	+	C3	F1	–	–	–	–
Small Fry	–	–	F1	+	–	–	+
Sonatine	+	C5	F2	–	–	–	–
Sonato	+	C2	F1	–	–	–	–
Spectra	+	C5	F2	+	–	–	–
Supsweet 100	–	–	F1	+	–	–	–
Tahiti	+	C5	F2	+	–	–	–
Tipico 2055	+	C5	F2	+	+	–	–
Trend	+	C5	F2	+	+	–	–
Turbo	+	C5	F2	+	–	–	–
Typhoon	+	C5	F2	+	–	–	–
Vendettos	+	C5	F2	–	–	–	–
Virosa	+	C5	F2	–	–	–	–
Wiranto	+	C5	F2	–	–	–	–

M = Tomato Mosaic Virus; LM = Leaf mould (*Fulvia fulva*); C = Resistant to groups of races of *Fulvia fulva*; FW = Fusarium wilt (*F. oxysporum* f. sp. *lycopersici*); F = Resistance to races 1 or 2; VW = Verticillium wilt (*V. albo-atrum* & V. Dahliae); FC & RR = Fusarium crown and root rot (*F. oxysporum* f. sp. *radicis lycopersici*); B & CR = Brown and corky root (*Pyrenochaeta lycopersici*); RK = Root knot (*Meloidogyne incognita*); – = susceptible; + = resistant; G's D'lit = Gardener's Delight; Supsweet = Supersweet.

Similar necrosis had been reported by Cirulli & Alexander (1969) who found it to occur more commonly at high temperatures. Hall & Bowes (1979) showed that repeated transfer of ToMV (strain 0) through *Tm*-2^2 hosts increased the amount of systemic necrosis and they speculated that ToMV types could evolve which were capable of producing high levels of systemic necrosis in homozygous *Tm*-2^2 plants, at normal temperatures. They concluded that the strains from the two UK outbreaks on Sonato and similar strains from The Netherlands were not strictly strain 2^2 because they induced partial hypersensitive reactions and lacked the high transmissibility of ToMV on non resistant hosts.

The reactions of 52 cultivars, many with breeders numbers, to the two resistance breaking strains from Sussex, were tested by sap inoculation of the cotyledons. Three of the 52 cultivars showed typical ToMV symptoms, with mosaic and leaf narrowing. Two of these were known to be universally susceptible but the third, 1315/72, was claimed to have *Tm*-1 and *Tm*-2^2 resistance genes (Leo van den Berkmortel, Bruinsma Seeds, personal correspondence). Previous tests with strains 0, 1, 2 and 1.2 had shown 1315/72 to be resistant to these. Five cultivars showed no symptoms at all including Pagham and Kirdford Cross and all were claimed to have all three resistance genes. The remaining cultivars showed stunting and mottling of the leaves with pale green and dark green blotches. Leaves developing after inoculation were twisted and puckered but not reduced in width. Occasional necrotic flecks occurred in some of the youngest leaves. The plants grew only slowly but new leaves were less severely distorted, and developed a mild mosaic. Thus, with the exception of 1315/72, the cultivars tested which were claimed to have *Tm*-2^2 behaved in a somewhat resistant way to the two Sussex isolates obtained from a *Tm*-2^2 host. The behaviour of 1315/72 cannot be explained as it appeared to be resistant to strains with virulence for *Tm*-1 and *Tm*-2 but did not show the necrotic reaction when inoculated with the ToMV isolates from Sonato.

When typed on Pelham's isogenic differentials the Sussex isolates behaved as strain 1 (Pelham 1968) with the *Tm*-2^2 line showing some necrosis as it does with many isolates of strains 0 and 1. Fraser (1990) commented that the resistance breaking isolates of ToMV capable of overcoming the *Tm*-2^2 gene appear to be defective in their ability to establish on *Tm*-2^2 resistant plants. It is remarkable that resistance to such a virus as ToMV with its well known variation has remained effective for such a long time. The results of the earlier grafting experiments gave a clear indication of the ease of breakdown of *Tm*-1 and *Tm*-2 host resistance but also of the durability of the *Tm*-2^2 gene even when subjected to the extreme pressure of graft contact with an infected susceptible host.

Leaf mould. Many of the currently grown tomato cultivars have resistance to all five race-groups of *Fulvia fulva* (Hubbeling, 1971). These groups enable cultivars to be classified but may not give a clear indication of the resistance genes present, particularly if modifying genes are involved (Table 5). The early use of leaf mould resistance genes met with limited success and were fairly rapidly overcome. In Europe, cultivars with the combination of two genes (*Cf*-2 and *Cf*-4) became available in the 1960's and in spite of the fact that individually these genes were no longer useful, the combination remained effective for a number of years. From 1967 onwards effectiveness began to decline but the *Cf*-2 and *Cf*-4 gene combination continued to give good resistance in the UK up until the early 1980's. *Cf*-5 was introduced in 1975 and race-5 (group D) overcame this resistance in Belgium in 1976 and soon

Table 5. Race classification for *Fulvia fulva* on tomatoes

	Race groups				
	A	B	C	D	E
Virulence genes	1	4	2.4	5	2.3.4.5
	2	1.4	1.2.4		
	3	3.4	2.3.4		
	1.2	1.3.4	1.2.3.4		
	1.3				
	2.3				
	1.2.3				

(after Hubbeling, 1971).

afterwards a complex race R.2, 3, 4, 5 (group E) appeared in The Netherlands. Surveys of races in the UK indicated that virulence to *Cf*-5 lines was also present. But the resistance of the then commonly cultivated cultivar Sonatine remained effective against all races (Hall, Glasshouse Crops Research Institute, Littlehampton, personal communication).

Leaf mould in commercial tomato crops is now an unusual disease and is restricted to those crops that have no genetic resistance. Only cherry tomato growers and amateurs have a problem. Very few, if any, growers use fungicides to control this disease. The resistance genes in cv Sonatine and its successors, designated C5 by the seed suppliers, have therefore been very durable in spite of strains of the pathogen being recorded that can overcome the resistance genes individually. Perhaps, because of very low disease levels, cultivars have not been subjected to high selection pressure, and the use of fungicides to control other diseases especially in unheated crops may also have affected the development of epidemics of virulent strains of *F. fulva*. Recent analysis of the virulence patterns of strains has not been done and an explanation of the durability of leaf mould resistance has not been reported.

Vascular wilt pathogens. Many tomato cultivars are resistant to *Fusarium oxysporum* f. sp. *lycopersici* races 0 and 1 (American nomenclature races 1 and 2) (Gabe, 1975) to *Verticillium albo-atrum* and *V. dahliae* and more recently to *F. oxysporum* f. sp. *radicis lycopersici.* Since their use, Fusarium and Verticillium wilt diseases have been rare in the UK. There have been a number of reports of apparent breakdown of Fusarium resistance (Hall & Bowes, 1979) but none has been confirmed as due to resistance breaking strains. Infection has been attributed to incomplete resistance or exceptional cultural conditions which have allowed avirulent isolates to colonise the vascular system. Paternotte (1991) was able to infect two Verticillium resistant cultivars with a strain of *V. albo-atrum* found in Holland. This is the first report in Europe of a Verticillium resistance-breaking strain.

In 1988 *Fusarium oxysporum* f. sp. *radicis lycopersici* was found in the UK (Hartman & Fletcher, 1990) and is currently confined to the south of England. The use of new resistant cultivars in 1991 gave complete control and in 1992 growers had a choice of eight such cultivars.

Other tomato pathogens. Resistance is available to *Pyrenochaeta lycopersici* in a limited number of cultivars. But due to the development of soil-less systems the disease has become less important and because of their relatively poor yield these cultivars have never become well established in the UK. Even in soil, they have not yielded as well as susceptible cultivars grafted onto resistant rootstocks or grown in soil treated with dazomet or methyl bromide. However, it is likely that they would outyield susceptible cultivars growing in untreated, infested soil.

The two major diseases of the protected tomato crop where resistance is not available, are powdery mildew (*Erysiphe* sp.) and grey mould (*Botrytis cinerea*). At present powdery mildew is the one disease for which early season growers use fungicides, whereas *Botrytis* tends to be more of a problem in later unheated crops. It appears that there is little prospect at present of the development of cultivars resistant to either of these pathogens.

Disease resistance breeding of protected tomatoes has had considerable success and this has contributed to yield increases, and also to a reduction in the use of fungicides. Unfortunately, the occurrence of powdery mildew means that some growers must still use fungicides in order to obtain disease control.

Cucumber

Like tomatoes, cucumbers have been bred with resistance to a number of pathogens.

Leaf spot and gummosis. The earliest record of a disease resistant cultivar of any protected crop in the UK is that of Butcher's Disease Resister (BDR), a cucumber resistant to Cercospora Leaf Spot (*Corynespora cassiicola* syn. *Cercospora melonis*). This cultivar was selected by a grower, Mr

Butcher, from a crop which was affected by Cercospora Leaf Spot, a disease which threatened to eliminate the cucumber crop in the Lea Valley between 1896–1907. Following the introduction of BDR in 1907 Cercospora Leaf Spot declined and has been an insignificant problem since. There have been isolated outbreaks on susceptible cultivars (Green, 1932) but the disease has not been seen in commercial cucumber crops for many years. Most, if not all, modern cultivars are resistant to this disease and many may have the original BDR resistance. Little appears to be known about the genetics of resistance although it has been suggested that a single dominant gene is involved (Abdul-Hayja et al., 1978).

Although most modern cultivars are thought to be resistant to Cercospora Leaf Spot, only a relatively small number make claim for this resistance (Table 6). The same is also true for gummosis or scab (*Cladosporium cucumerinum*). Gummosis was occasionally a devastating disease in the 1960's but is now rare (van Steekelenburg, 1986). These two diseases, because of effective genetic resistance which is controlled by single dominant genes (Walker, 1950; Abdul-Hayja et al., 1978) have been uncommon in commercial crops for at least 25 years. There is the danger that they may be forgotten not only by the growers but also by the seed trade. It is to be hoped that plant breeders continue to include them in their screening trials.

Fusarium wilt. Fusarium wilt (*Fusarium oxysporum* f. sp. *cucumerinum*) has never been a serious problem in the UK in spite of the susceptibility of Butcher's Disease Resister. Fortunately the first occurrence of the disease in 1967 coincided with the

Table 6. Disease resistance in cucumber cultivars

Cultivar	Diseases							Cultivar	Diseases						
	LS	GorS	PM	DM	GM	BSR	V		LS	GorS	PM	DM	GM	BSR	V
468 Hazera	–	–	R	R	–	–	CMV	Femspot	R	R	–	–	–	–	–
472 Hazera	–	–	–	–	–	–	CMV	Fidelio	–	–	T	–	–	–	–
							MMV	Flamingo	–	–	T	–	–	–	–
492 Hazera	–	–	R	R	–	–	–								
								Girola	R	R	–	–	–	–	–
Aidas	R	R	–	–	–	–	–								
Allure	–	–	–	–	T	T	–	Jessica	–	–	–	–	–	–	–
Andora	R	R	–	–	–	–	–								
Aramon	–	–	T	–	T	–	–	Marana	–	R	–	–	–	–	–
								Marello	–	–	T	–	–	–	–
Bastion	R	R	–	–	–	–	–	Midistar	–	–	T	–	–	–	–
Bella	–	–	R	P	–	–	–	Mildana	–	–	P	P	–	–	–
Brucona	R	R	–	–	–	–	–								
Brudania	R	R	–	–	–	–	–	Pepita	–	–	R	–	–	–	CMV
Brustar	R	R	–	–	–	–	–	Prestige	–	–	–	–	–	–	CMV
								Pyralis	R	R	R	–	–	–	–
Carmen	–	R	P	P	–	–	–								
Cordito	–	–	T	–	–	–	–	Rebella	–	–	–	–	–	–	–
Corona	–	–	–	–	–	–	–								
								Separator	R	R	–	–	–	–	–
Euphya	–	–	R	–	–	–	–								
								Telstar	–	–	T	–	–	–	–
Farbiola	–	–	–	–	–	–	–	Type 7	–	–	R	–	–	–	–
Femdan	R	R	–	–	–	–	–								
								Vitalis	R	R	–	–	–	–	–

LS = Leaf Spot (*Corynespora cassiicola*); GorS = Gummosis or Scab (*Cladosporium cucumerinum*); PM = Powdery mildew (*Sphaerotheca fuliginea*); DM = Downy mildew (*Pseudoperonospora cubensis*); GM = Grey mould (*Botrytis cinerea*); BSR = Black stem rot (*Didymella bryoniae*); V = Virus; – = susceptible; T = tolerant; CMV = Cucumber mosaic resistance; R = resistant; P = Partial resistance; MMV = Melon mosaic resistance.

introduction of new Dutch cultivars which, by chance, were resistant. Wilt resistance is believed to be governed by a single dominant gene and quantitative differences in resistance have been recorded between homozygous and heterozygous lines. The same is true for Verticillium wilt (*V. albo-atrum* and *V. dahliae*). Resistance may be linked as cultivars resistant to one appear, in practice, to be resistant to both. But there have been no reports of systematic checks of cultivars for susceptibility to these pathogens apart from the initial work in the UK when a number were found to have marked resistance to Fusarium wilt (Fletcher & Kingham, 1966).

Root and stem diseases. The main root and stem pathogens of cucumbers in the UK are *Botrytis cinerea, Didymella bryoniae, Penicillium oxalicum* and *Phomopsis sclerotioides.* There are claims of reduced susceptibility in some cultivars to *Botrytis* and *Didymella* but these have not been substantiated in experimental work. Breeder's lines have been shown to have marked resistance to *Didymella* (van der Mear et al., 1978; Wyszogrodska et al., 1986). van Steekelenburg (1986) considers that breeding for resistance to gummosis and Cercospora Leaf Spot as well as bitter free fruits and all female cultivars, has resulted in greater susceptibility to stem and fruit rot diseases. *Phomopsis* can be controlled by steam treatment of the soil, by good hygiene and the use of rockwool or by grafting onto *Cucurbita ficifolia* rootstocks which are not only less susceptible to *Phomopsis* but also to Fusarium wilt and Fusarium basal rot (*F. solani* f. sp. *cucurbitae*). *Penicillium oxalicum* was first recorded in the UK in 1989 (O'Neill et al., 1991). Differences in the resistance of cultivars was demonstrated but nothing is known about the genetics of resistance.

Powdery mildew. Cucumber mildew (*Sphaerotheca fuliginea*) is a commonly occurring disease and is generally well controlled with fungicides although there are reports of resistance to dimethirimol and the triazoles (Bent et al., 1971; Schepers, 1985). A number of cultivars have resistance to this disease but have not found favour with the industry because of their poor yields and tendency to show leaf necrosis under low light conditions (van Steekelenburg, 1986). Breeders have variously described cultivars as tolerant, partially resistant or resistant but in a recent experiment at HRI Stockbridge House, little difference in resistance was seen between the cultivars compared (Table 7). Growers frequently make second crop plantings in July and it is these crops which are most likely to be affected by powdery mildew. A resistant cultivar is commonly chosen for such later planted crops. Cultivars resistant to powdery mildew also have some resistance to downy mildew (*Pseudoperonospora cubensis*), a disease which has occurred in the UK on a number of occasions during the past 5 years, but has rarely been epidemic. Although a wide spectrum of disease resistance is available in cucumber cultivars, most crops are sprayed, usually to control powdery mildew, *Botrytis* or *Didymella.*

Pepper

There are only 70 hectares of protected sweet peppers in the UK. Consumption considerably exceeds home production and the bulk of the retailed crop is imported. Pests and diseases are generally not important; pests are more common than diseases and are usually effectively controlled biologically. However, there are a number of pepper cultivars on the market with resistance to virus diseases. Various tobamoviruses have been isolated from sweet peppers. Some have been recognised as distinct viruses whilst others have been found to be identical to or related to tobamoviruses from other hosts (Boukema et al., 1980).

In order to convey the results of their resistance breeding programmes in a way which pepper growers can understand, some breeders have adopted a system of virus classification based upon the virus-host interactions in which pathogenicity of the tobamoviruses is expressed as a number which relates to the L-genes in the host *Capsicum* (Table 8). Others have not adopted this system. Rast (1988) suggests that this may lead to confusion because of the conflicting interests of breeders and virologists and has suggested a different system (Table 9). His classification involves representative strains or isolates of each tobamovirus which are known pathotypes on sweet peppers. Rast's concern is well illustrated in the seed catalogues where cultivar resistance to the tobamoviruses is expressed in various ways and is confusing (Table 10). It appears that the pathotypes P0, P1, P1.2 and P1.2.3 of Boukema et al. (1980) are referred to by the seedhouses as TM0, 1, 2 & 3 respectively.

Lettuce

Lettuce is both a field and protected crop in the UK which is significant in terms of disease inoculum. The protected crop is normally affected by *Botrytis cinerea, Bremia lactucae, Rhizoctonia solani* and occasionally lettuce mosaic virus and big vein. Control of these diseases is generally by chemical or environmental means. Lettuce downy mildew (*Bremia lactucae*) is the exception in that its control relies upon a combination of genetic resistance and fungicidal application (Crute, this volume). The situation is further complicated by the existence of phenylamide resistant strains and the control strategy has relied upon various combinations of the available fungicides and the current choice of resistant cultivars.

There has been confusion in the literature over the nomenclature of races of *Bremia*. All the protected lettuce cultivars grown in the UK are of Dutch origin and resistance is cited in terms of the Dutch NL nomenclature for strains of the pathogen. During the past 20 years there has been a succession of cultivars with various genes and combinations of genes for resistance. These have lasted a relatively short period particularly during the time when fungicidal control relied exclusively on the dithiocarbamates. Since the introduction of the phenylamide fungicide metalaxyl and also fosetyl-al, fungicidal control has been better and by reducing pathogen selection pressure, genetic resistance has lasted longer. The development of metalaxyl resistance in some of the virulent strains has further complicated the situation (Crute & Harrison, 1988). For a period the R11 resistance factor gave effective protection but recently a virulent strain, NL15 which is also metalaxyl resistant, has become widespread in the UK. In order to obtain mildew free crops growers are advised to choose cultivars resistant to NL15 but also resistant to those earlier strains which were metalaxyl resistant. In effect this means choosing cultivars with resistance genes R6 + R11 or R16 or R18 (O'Neill, ADAS Cambridge, personal communication). There is a range of such cultivars available to growers (Table 11) and by using these and combinations of fungicides with different modes of action downy mildew control is maximised.

Cultivars are also available that are resistant to lettuce mosaic virus eg., Oriana, Ermosa, Valuta, Vitana, Voluma but none of these are commonly grown under protection.

Table 7. A comparison of various cucumber cultivars with powdery mildew resistance

Cultivars	Resistance seedhouse category	Mean mildew score % leaf cover	Yield kg/m^2
Jessica	S	78.0	10.2
Euphya	R	1.3	11.7
Flamingo	T	2.6	12.5
Carmen	P	0.6	10.4
Aramon	T	4.3	10.1
Millio	T	2.3	10.0

S = susceptible; R = resistant; P = partially resistant; T = tolerant.

Table 8. Pathotype-genotype interaction of tobamoviruses in Capsicum hosts (+ = susceptible; − = resistant)

Host	Genotype	Pathotype			
		P0	P1	P1.2	P1.2.3
C. annuum					
'Early California Wonder'	$L^+ L^+$	+	+	+	+
C. annuum					
'Bruinsma Wonder'	$L^1 L^1$	−	+	+	+
C. frutescens 'Tabasco'	$L^2 L^2$	−	−	+	+
C. chinense P.I. 159236	$L^3 L^3$	−	−	−	+
C. chacoense P.I. 260429	$L^4 L^4$	−	−	−	−

Adapted by Rast (1988) from Boukema et al. (1980).

Mushroom

Agaricus bisporus and *A. bitorquis* are widely cultivated throughout the world. The *A. bisporus* crop is the most valuable horticultural crop in the UK. There are numerous strains available varying from smooth whites to white with rough surface to creams and browns. Most growers now grow hybrid white strains which are not completely smooth, but give high yields of good quality mushrooms.

Very little work has been reported on a comparison of spawn strains and disease incidence and there are no strains marketed that claim to be resistant to any diseases. van Zaayen & van der Pol-Linton (1977) examined various strains of *A. bitorquis* in relation to false truffle disease (*Diehliomyces microsporus*). They found that of the five types examined, K26 and K32, were the least sensitive. Peng (1986) screened 42 strains of *A. bisporus* for resistance to *Verticillium fungicola*. He detected differences in strain susceptibility which he found to be consistent. Similarly, Peng screened eight strains for resistance to bacterial blotch (*Pseudomonas tolaasii*) and found strain variation in response to different levels of inoculum.

The incidence of virus diseases in *A. bisporus* varies with the strain but this is thought to be the result of incompatibility between strains preventing an anastamosis which is a major means of virus transmission (Fletcher et al., 1989).

Challen & Elliott (1987) took the unusual approach of breeding novel strains of *A. bisporus* which were resistant to four fungicides, thereby increasing the potential range of fungicides available for the control fungal diseases of the crop.

Ornamentals

There are a large number of plant species grown as protected ornamentals. They are sold as cut flowers, flowering pot plants, foliage plants and bedding plants.

Derbyshire & Ann (1986) described the most important diseases of 47 different species of pot plants in the UK and Chase (1987) described the diseases of 50 species whilst recognising that nearly 500 are grown as pot plants in Florida. Many are propagated from clonal material and others are produced from seeds. No cultivars have been bred for disease resistance although differences in susceptibility to pathogens have been observed in a number of species.

Internationally, chrysanthemums, carnations and roses are probably the most widely grown greenhouse flowers and in the UK begonias, pelargoniums, cyclamen, poinsettias and pot chrysanthemums are among the most commonly grown flowering pot plants. There are relatively few reports in the literature of screening the commonly grown cultivars for resistance to particular pathogens and even fewer where disease resistance breeding programmes have been developed.

Table 9. List of tobamoviruses and corresponding pepper pathotypes

Tobamovirus	Strain/isolate	Pathotype
Tobacco mosaic virus (TMV)	type or common strain, vulgare strain, U1	P0
Tomato mosaic virus (ToMV)	Dahlmense strain, Y-TAMV	P0
Bell pepper mottle virus (BePMV)	unusual pepper strain, FO, eggplant strain A1	P0
Tobacco mild green mosaic virus (TMGMV)	para-tobacco mosaic virus, T2MV, U2, South Carolina mild mottling strain, G-TAMV	P0 or P1
Unnamed	P11	P1
Tomato mosaic virus (ToMV)	Pepper strain Ob	P1 or P1.2
Pepper mild mottle virus (PMMV)	Samsun latent strain, SL-TMV, P8, P14, Capsicum mosaic virus	P1.2.3

After Rast (1988).

Carnation

There are numerous cultivars of this crop which is grown worldwide. In the UK it is now of minor importance, largely because of the imports of cheaply produced flowers from elsewhere. The crop is prone to a number of major diseases (Fletcher, 1984), in particular Fusarium wilt (*Fusarium oxysporum* f. sp. *dianthi*) which occurs wherever the crop is grown. An examination of various cultivars by Garibaldi (1975) showed that there were differences in susceptibility and although some of the Mediterranean and miniature cultivars showed resistance, the larger flowered American types were completely susceptible. Garibaldi initially recognised two forms of the pathogen which could be differentiated on these different cultivar types. Since then further work in various countries, (Garibaldi & Gullino, 1987; Blanc, 1983; Carrier, 1977; Matthews, 1979; Baayen et al., 1988; Demmink et al., 1989) has resulted in the production of resistant cultivars and some understanding of the host pathogen relationship. In 1983, Garibaldi recognised eight pathotypes. Garibaldi & Rossi (1987) reported pathotypes 1, 2, 4, 5, 6 and 8 to be commonly found in the Liguria area of Italy. Race 2 is believed to predominate in the UK and the Netherlands (Matthews, 1979). Garibaldi (1983) screened 112 cultivars using these pathotypes and although many were resistant to six or seven of them, only one, cv. Duca was resistant to all. Demmink et al. (1989) examined the virulence spectrum of three pathotypes, 1, 2 and 4 on nine carnation cultivars (Table 12). They concluded that resistance to pathotype 1 is monogenically inherited and is complete but resistance to pathotypes 2 and 4 is probably polygenically inherited.

Table 10. Pepper cultivars and tobamovirus resistance as listed by seedhouses

Cultivar	Resistance
Bell Boy	TMV
Belmont	TMV
Bianca	TM 0
Capri	TMV
Carlos	TMV race P0, PVY 0 & 1
Cubico	TM 2
Domina	TMV
Eagle	TM 2
Elea	TM2
Gloria	TMV
Gold Flame	TM 0 & PVY
Herpa	TMV
Jetta	TMV
Lambada	TM 0
Latina	PVY, TMV (tomato strains)
Locas	TM 0
Marraf	TM 2
Martel	TMV
Mazurka	TM 0
Medeo	TM 0
Pantser	TMV
Parma	TMV
Paula	TMV
Propa Rumba	TM 0
Ranger	TMV
Salsa	TM 0
Siraki	TM 0
Tasty	TM 3
Tenno	TMV race P0, PVY 0 & 1
Tequila	TM 0
Tonika	–
Valeta	TM 0
Zerto	TMV race P 0
Zico	TMV race P0, PVY 0 & 1

TMV = Tobacco mosaic virus; TM = Tomato mosaic virus; PVY = Potato virus Y.

Table 11. Lettuce cultivars resistant to all known metalaxyl resistant strains including NL15

Cultivar	Resistance gene(s)
Animo	R11 + R16
Banjo	R11 + R16
Berlo	R11 + R16
Clarisse	R6 + R11
Charlene	R3 + R11 + R16
Desso	R11
Disney	R11 + R16
Impala	R18
Liset	R6 + R11
Luxor	R2 + R16
Mirage	R6 + R11
Pantra	R3 + R11 + R16
Rosana	R16
Ricardo	R11 + R16
Titania	R16
Vicky	R11 + R16
Virginia	R16

Fusarium wilt in carnations is now well controlled by the use of resistant cultivars. A wide range of colours of high yielding cultivars is available and is widely used. Wilt resistance in carnations represents the main success in the breeding of cultivar resistance into protected ornamentals.

There are a few other reports of differences in the resistance of carnations to diseases. Rattink (1972) reported some differences in the susceptibility to *Phialophora cinerescens* in crosses made between cultivars William Sim and Royalette and breeding material. This vascular wilt was the most important disease of the crop until the appearance of Fusarium wilt in the 1960's. Its demise coincided with the increasing importance of Fusarium wilt and it had already become an insignificant disease before Fusarium wilt resistant cultivars were available. Various workers have noted differences in the susceptibility of cultivars to rust (*Uromyces dianthi*). Semina & Shestachenko (1981) reported 12 resistant cultivars in screening tests and Sezgin & Esentepe (1986) examined six cultivars and found one, Minirosa, to be resistant whilst Aliatta and Ernesto were highly and moderately susceptible respectively. In the UK, Spencer (1981) confirmed differences in cultivar reactions to rust.

Similar cultivar variability has been recorded to *Alternaria dianthi* by Strider (1978a). He inoculated a wide range of commercial cultivars and found them to be susceptible but Dusty, Imp, New Pink Sim, Light Pink Barbi, Maj Britt, Pink Ice and Red Gayety were significantly less susceptible than most.

Table 12. Virulence pattern of three pathotypes of *F. oxysporum* f. sp. *dianthi* on 9 cultivars

Cultivar	Pathotype of *F. oxysporum* f. sp. *dianthi*		
	Race 2	Race 4	Race 1
Sam's Pride	S	S	S
Alice, Sacha	S	S	R
Lena	S	MR	R
Pallas	MR	S	R
Niky, Elsy	MR	R	S
Revada, Novada	R	R	R

(after Demmink et al., 1989).

Chrysanthemum

This is probably the most widely grown protected ornamental. There are a vast number of cultivars and many pathogens have been recorded. In the UK the most serious diseases include Phoma root rot (*Phoma chrysanthemicola*), ray blight (*Didymella chrysanthemi*), Verticillium wilt (*V. dahliae*), petal blight (*Itersonilia perplexans*), grey mould (*Botrytis cinerea*), rusts (*Puccinia horiana* and *P. chrysanthemi*), and various virus diseases – most recently tomato spotted wilt. There are no cultivars marketed that claim resistance to any of these diseases but most growers know that different cultivars vary not only in their susceptibility to most diseases but also to pests. In the latter case, apparent differences in susceptibility to tomato spotted wilt virus reflects in part the preferential feeding of the vector, western flower thrips (*Frankinella occidentalis*), on some cultivars. There are various reports in the literature of differences in susceptibility to various pathogens and one company specialising in the development of new cultivars has recently begun a breeding programme to include disease resistance as one of its major aims (C. Scharfenberg, Yoder Bros., Florida, personal communication).

Englehard (1969) reported observations on 52 cultivars of commercial chrysanthemums in relation to Ascochyta blight (*Didymella chrysanthemi*) rust (*Puccinia chrysanthemi*) and flower spots caused by *Botrytis cinerea* or *Alternaria* sp. The results show a complete range of resistance for each of the diseases with the majority of the cultivars showing some resistance to one or more of the pathogens but with a minority being very susceptible. Englehard reports 'growers often disregard disease resistance or susceptibility when selecting cultivars to grow' and this is still the case in spite of resistance being present in some of the existing stocks. Recently Matteoni & Allen (1989) have reported on the sensitivity of cultivars to tomato spotted wilt virus which has spread internationally at alarming rates with the distribution of the vector, western flower thrips, probably on cuttings. They found considerable variation in the severity of symptom expression in a wide range of cultivars which they inoculated. Only one cultivar was symptomless but about a quarter of the inoculated plants of that cultivar were infected. This underlines the necessity to examine cultivars systematically in resistance tests particularly when screening for resistance to systemic pathogens.

Byrne et al. (1980) reported the results of screening 87 cultivars for resistance to *Verticillium dahliae* and *Puccinia chrysanthemi*. They found a wide range of resistance. Growers in the UK of all year round (AYR) crops recognise a range of susceptibility of cultivars to Verticillium wilt. For instance, cvs Hurricane, Rhino, Jaguar, Nikita, Mecca, Texas Improved, Garland and Princess Anne are known to be very susceptible whereas Rose Swan appears to be resistant. More extensive tests have been done on Fusarium wilt (*F. oxysporum* f. sp. *chrysanthemi* and *F. oxysporum* f. sp. *tracheiphilum*) not yet recorded in the UK. Strider (1985) reported the results of screening 183 cultivars and found a number with resistance to both species. Cultivars Airborne, Royal Trophy, Yellow Delaware were most susceptible and Jamboree, Puritan and Tune-up most resistant to *F. oxysporum* f. sp. *chrysanthemi*. None of the cultivars was highly susceptible to *F. oxysporum* f. sp. *tracheiphilum*.

In 1960 a root disease of chrysanthemums caused by *Phoma chrysanthemicola* was first recorded in England. Hawkins et al. (1963) found considerable variation in the susceptibility of cultivars and, in subsequent tests with a range of isolates of the pathogen, Wilcox (1963) showed that cvs Heyday, Supertop, Snowcap and Princess Anne had a high degree of resistance. Some cultivars varied in their reactions to some isolates suggesting that there could be a degree of specialisation in the pathogen.

Variations in the susceptibility of cultivars to other chrysanthemum pathogens have been recorded. Semina & Babkina (1981) showed considerable differences between cultivars in their resistance to powdery mildew. Water et al. (1984) found that of 10 cultivars tested, three, Carfour Album, Stateman and Coppa were resistant to *Puccinia horiana*. Similar reports were made by Rademaker & Jong (1985). Variation in resistance to *Septoria chrysanthemi* found by Zhang & Li (1986). Two of the 14 cultivars they tested had a degree of resistance to this leaf spot, whereas Strider & Jones (1986) found some resistance to bacterial leaf spot and bud blight (*Pseudomonas cichorii*) in eleven out of 131 cultivars tested.

Miller et al. (1975) examined 237 cultivars for susceptibility to *Agrobacterium tumefaciens*. They used the American culture collection type strain B6 and also an isolate from chrysanthemum which proved to be more virulent. Resistance to both strains was observed in 10% of the cultivars.

Although a large amount of variation in the resistance of chrysanthemum cultivars to various pathogens has been recorded, growers have not yet benefited. Fashions in cultivars change quickly and growers must grow what the market requires. This usually precludes the choice of disease resistant cultivars. There is clearly much scope for the development of disease resistant chrysanthemums but it is unlikely to happen unless the industry changes its policy and uses a more limited range of cultivars into which resistance could be bred.

Other ornamentals

The incidence of diseases and the resistance of cultivars of other protected crops are recorded in the literature. There are a number which refer to roses but these are concerned with field grown rather than protected crops. Differences in cultivar susceptibility were recorded to dieback (*Diplodina rosarum*) by Kove et al. (1977), to black spot (*Diplocarpon rosae*) by Palmer & Salac (1977); Svejola & Bolton (1980); Knight & Wheeler (1978); and to powdery mildew (*Sphaerotheca pannosa* var. *rosae*) by Semina & Timoshenko (1979), Deshpande et al. (1979) Bender & Coyier (1986). The latter authors found evidence for physiological specialisation in the pathogen in greenhouse crops.

Of the various pot plants, there are a number of records which relate to disease incidence in begonias. Strider (1978b) recorded the reactions of Rieger elatior begonias to powdery mildew (*Oidium begoniae*) and found cv. Aphrodite Red to be highly resistant and Stella, Ballerina and Hawaiian Punch to be resistant. O'Riordan (1979) in his tests on similar cultivars found that they were all susceptible with the exception of Aphrodite Red. Strider (1975) had previously reported on the resistance of begonias to bacterial blight (*Xanthomonas begoniae*). In contrast to the mildew reactions, the Aphrodite cultures were found to be susceptible whereas Goldachs and Bernstein Gelbe were only slightly affected. Later work (Strider, 1978) showed that the cultivars Ballerina, Nixie, Elfe and Stella also had some resistance. Differences between the resistance of the species of begonia in cultivation to *X. begoniae* were recorded by Harri et al. (1977). They tested elatior types and also fibrous and tuberous rooted begonias. All proved to be susceptible but all the Rex begonias were resistant when spray inoculated.

Begonias not only vary in their resistance to pathogens but also to damage by ozone. Reinert & Nelson (1979) tested twelve *Begonia* × *hiemalis* cultivars to 25 and 50 ppm ozone. They found distinct differences with Whisper O'Pink and Improved Krefeld Orange the most sensitive and Ballerina, Mikkell, Limelight and Turo the least.

Strider (1978c, 1980) studied the resistance of saintpaulia cultivars to *Phytophthora nicotianae* var. *parasitica* and also to powdery mildew (*Oidium* sp). He found great variation in the resistance of the Ballet and Rhapsodie series to Phytophthora root rot with cultivars Erica, Helga, Inge and Karth of the Ballet series and Barbara, Astrid and Ruby of the Rhapsodie series being the most resistant. Of the 48 cultivars tested for mildew resistance, in the Melodie and Ballet series he found differences between cultivars but also between the flowers and leaves of the same cultivar. Most resistant overall were cultivars Allison, Brilliant, Eva, Dolly, Mitzi, Pearl and Rachel.

Pelargonium rust. (*Puccinia pelargonii-zonalis*) has been investigated by various workers. McCoy (1975) examined the susceptibility of 17 species and cultivars. He found five cultivars of *Pelargonium hortorum* and one of *P. domesticum* to be highly susceptible, a high level of resistance in *P. radula* and *P. limoneum,* whilst all the other species were immune. Harwood & Raabe (1979) examined cultivars of *P. hortorum* and found variation in their resistance but also in the virulence of isolates of the pathogen, suggesting race specialisation. Similar variation in the resistance of pelargonium cultivars has been reported for virus diseases. Albouy et al. (1979) reported resistances to tobacco ring spot, tomato ring spot and tomato black ring in a red cultivar. Three other cultivars were susceptible.

Significant differences in the resistance of cyclamen cultivars to the relatively recently described anthracnose disease (*Crypocline cyclaminis*) were described by Brielmaier-Liebetanz & Buhmer (1988), although none were highly resistant.

Discussion

The range of plant species grown under protection is very wide, particularly of ornamentals. In some cases there are only a few growers in any one country with a particular type. In these circumstances it is not surprising that expensive breeding programmes have not been developed to provide resistant cultivars. The greatest use of resistance has been with tomatoes where breeding programmes can provide cultivars suitable for culture in the field as well as under protection. The estimated annual worldwide production of the tomato crop in 1985 was in the region of 60 million metric tonnes (Jones et al., 1991) making it a worthwhile market for breeders.

Some of the troublesome diseases of greenhouse vegetable crops are caused by pathogens which rot stems and leaves, in particular *Botrytis cinerea.* As a group, these diseases are not easily controlled by genetic resistance although it is interesting that there is now a number of cucumber cultivars which the breeders claim are less susceptible to some of them. Resistance to this type of pathogen appears to be more difficult to find and detection methods are not available for its quantification.

The durability of resistance has been variable, sometimes requiring a constant change of resistance genes or combinations as in the case of lettuce downy mildew and, for some years, tomato leaf mould and tomato mosaic virus diseases. Resistance to the latter two diseases has now been maintained for some years in spite of the plasticity of the pathogens involved. Single gene resistance in the case of tomato mosaic and Cercospora leaf spot of cucumbers has been extremely durable.

The need to include resistance to some of the major root rot pathogens of tomatoes and cucumbers has been largely overcome by the growth of these crops in a soil-free media. Such systems will become more sophisticated and in order to avoid pollution of the underlying soil with run-off fertilisers and pesticides, a change to recirculatory systems is likely. Recirculation could favour a different spectrum of root disease pathogens, in particular those which prefer the wet environment. The need to control such diseases by all available means is likely to increase in the next 5 years.

Flower producers, with the exception of carnation growers, have little to choose from if they are looking for disease resistance, although resistance is known to occur in many crops. Growers and plant raisers know from their observations that some cultivars are more susceptible to certain diseases than others; this information is never given in sales catalogues which usually concentrate on colour, form, season and keeping quality. New cultivars of such crops as chrysanthemums are numerous and fashions change. Although a grower with a particular disease problem, Verticillium wilt for instance, knows that there are some cultivars that will always be affected, he often has to grow some of them, perhaps in sterilised soil, in order to meet market demands. Soil sterilisation and fungicide use are expensive operations but so is plant breeding. Perhaps the time will come when public pressure will make breeding for disease resistance in flower crops worthwhile for the larger plant producers, and growers will benefit from a more economic means of disease control.

Pathologists, with some exceptions, have frequently not identified sources of resistance within existing cultivars. Such work, although mundane, is needed not only for immediate use but also to enable breeders to understand the possible interactions between their cultivars and pathotypes which may exist in various regions. Disease resistance in protected crops has had some notable successes, some at low cost (eg. cucumber, cv Butcher's Disease Resister).

The effect of disease resistance and the use of fungicides on protected crops is difficult to quantify. There can be no doubt that without the resistance available to tomato and cucumber growers, fungicides would be more extensively used. With the increased ability to control the protected environment, main crop tomato growers have little need to use fungicides. Unfortunately, the occurrence of the new powdery mildew disease has changed the situation for some of them. Cucumber growers, likewise, benefit from resistance to leaf spot and gummosis but still spray regularly to control mildew, grey mould and black stem rot. Lettuce growers spray mildew resistant cultivars to protect the resistance genes. Control of pests by biological means restricts the choice of fungicides that can be used on some crops but occasionally growers must resort to the use of fungicides that are disruptive of their biological pest control programmes. This pressure, together with a reduction in the availability of pesticides for horticultural crops and an increasing public awareness of pollution, makes the need to breed or identify good disease resistant cultivars very important for the future of the industry.

Acknowledgements

The author thanks the Ministry of Agriculture, Fisheries and Food for financial support in the preparation of this paper.

References

Albouy, J., J.C. Morand & J.C. Poutier, 1979. Effect of three ring spot-type viruses on growth and flowering of *Pelargonium* × *hortorum* grown from seed. Revue Hortic. No. 193, 29–33.

Abdul-Hayja, Z., P.H. Williams & C.E. Peterson, 1978. Inheritance of resistance to anthracnose and target leaf spot in Cucumbers. Plant Dis. Reptr. 62: 43–45.

Baayen, R.P., D.M. Elgersma, J.F. Demmink & L.D. Sparnaaij, 1988. Differences in pathogenesis observed among susceptible interactions of carnation with four races of *Fusarium oxysporum* f. sp. *dianthi*. Neth. J. Plant Path. 94: 81–94.

Bender, C.L., & D.L. Coyier, 1986. Pathogenic variation in Oregan populations of *Spaerotheca pannosa* var *rosae*. Plant Disease 70: 383–385.

Bent, K.J., A.M. Cole, J.A.W. Turner & M. Woolmer, 1971. Resistance of cucumber mildew to dimethirimol. Proceedings 6th Br. Insectic. Fungic. Conf.: 274–282.

Blanc, H. 1983. Carnation breeding for resistance to *Fusarium oxysporum* f. sp. *dianthi*. Practical achievement of resistant cultivars. Acta Horticulturae 141: 43–47.

Boukema, I.W., K. Jansen & K. Holman, 1980. Strains of TMV and genes for resistance in Capsicum. Eucarpia Capiscum Working Group. Synopses of the IVth Meeting, Wageningen, 14–16 October, p. 44–48.

Brielmaier-Liebetanz, U. & B. Buhmer, 1988. *Crypocline cyclaminis*, studies of the susceptibility of cyclamen cultivars and on the range of host plants. Gesunde Pfl. 40: 253–256.

Byrne, T.G., A.H. McCain & T.M. Kretchun, 1980. Testing chrysanthemums for disease resistance. California Agric. 34: 14–15.

Carrier, L.E., 1977. Breeding carnations for disease resistance in Southern California. Acta. Horticulturae 71: 165–168.

Challen, M.P. & T.J. Elliott, 1987. Production and evaluation of fungicide resistant mutants in the cultivated mushroom *Agaricus bisporus*. Trans. Br. mycol. Soc. 88: 433–439.

Chase, A., 1987. Compendium of Ornamental Foliage Plant Diseases. American Phytopathological Society, St Paul, Minnesota.

Cirulli, M. & L.J. Alexander, 1969. Influence of temperature and strain of tobacco mosaic virus on resistance in a tomato seedling line derived from *Lycopersicon peruvianum*. Phytopathology 59: 1287–1297.

Crute, I.R. & J.H. Harrison, 1988. Studies on the inheritance of resistance to metalaxyl in *Bremia lactucae* and on the stability and fitness of field isolates. Plant Pathol. 37: 231–250.

Demmink, J.F., R.P. Baayen & L.S. Sparnaaij, 1989. Evaluation of the virulence of races 1, 2 & 4 of *Fusarium oxysporum* f. sp. *dianthi* in carnation. Euphytica 42: 55–63.

Derbyshire, D.M. & D. Ann, 1986. Control of Diseases of Protected Crops – Pot Plants. MAFF Publications, Alnwick, Northumberland.

Deshpande, G.D., K.W. Anserwadekar & D.C. Warke, 1979. A note on the varietal reaction of hybrid-T roses to powdery mildew. Research Bulletin of Marathwada Agricultural University 3: 81–83.

Englehard, A.W., 1969. Observations on cultivars of commercial chrysanthemums to Ascochyta blight, rust and three petal spot diseases. Florida State Horticultural Society No. 3377: 340–343.

Fletcher, J.T., 1973. Glasshouse crops disease control – current developments and future prospects. Proceedings of the 7th Br. Insectic. Fungic. Conf.: 857–864.

Fletcher, J.T., 1984. Diseases of Greenhouse Plants. Longman, London.

Fletcher, J.T. & H.G. Kingham, 1966. Fusarium wilt of Cucumbers in England. Plant Pathol. 15: 85–89.

Fletcher, J.T., P.F. White & R.H. Gaze, 1989. Mushrooms; Pest and Disease Control. Second Edition. Intercept, Andover, Hants.

Fraser, R.S.S., 1990. The genetics of resistance to plant viruses. Ann. Rev. of Phytopathol. 28: 179–200.

Gabe, H.L., 1975. Standardisation of nomenclature for pathogenic races of *Fusarium oxysporum* f. sp. *lycopersici*. Trans. Br. mycol. Soc., 64: 156–159.

Garibaldi, A., 1975. Race differentiation in *Fusarium oxysporum* f. sp. *dianthi*. First contribution. Meded. Fac. Landbwet. Rijksuniv., Gent, 40: 531–537.

Garibaldi, A., 1983. Resistance of carnation cultivars to 8 pathotypes of *Fusarium oxysporum* f. sp. *dianthi*. Revista della Ortoflorofrutticoltura Italiana 67: 261–270.

Garibaldi, A. & M.L. Gullino, 1987. Fusarium wilt of carnation: Present situation, problems and perspectives, Acta. Horticulturae 216: 45–54.

Garibaldi, A. & G. Rossi, 1987. Observations on the resistance of carnations to *Fusarium oxysporum* f. sp. *dianthi*. Panorama Floricola 12: 5–9.

Green, D.E., 1932. Note on the disease resistance shown by Butcher's Disease Resister Cucumber to Cercospora leaf spot. Jl. R. Hort. Soc., Lvii: 63–64.

Hall, T.J. & S.A. Bowes, 1979. Disease resistance. Report of the Glasshouse Crops Research Institute: 49–53.

Harri, J.A., P.O. Larsen & C.C. Powell, 1977. Bacterial leaf spot and blight of Rieger elatior begonia: systemic movement of the pathogen, host range and chemical control trials. Plant Dis. Reptr. 61: 649–653.

Hartman, J.R. & J.T. Fletcher, 1990. Fusarium crown and root rot of tomatoes in the UK. Plant Pathol. 40: 85–92.

Harwood, C.A. & R.D. Raabe, 1979. The disease cycle and control of geranium rust. Phytopathology 69: 923–927.

Hawkins, J.H., P. Wiggell & H.J. Wilcox, 1963. A root rot of chrysanthemums. Plant Pathol. 12: 21–22.

Hitchins, P.E.N., 1951. Production of Tomatoes under Glass. Ernest Benn Limited, London.

Hubbeling, N., 1971. Attack of hitherto resistant tomato varieties by a new race of *Cladosporium fulvum* and resistance against it. Meded. Fac. Landbwt. Rijksuniv. Gent 36: 1011–1016.

Jones, J.B., J.P. Jones, R.E. Stall & T.A. Zitter, 1991. Compendium of Tomato Diseases, American Phytopathology Society, St. Paul, Minnesota.

Knight, C. & B.E.J. Wheeler, 1978. Evaluating the resistance of roses to blackspot. Phytopathologische Zeitschrift 91: 218–229.

Kove, S.S., D.O. Nirmal & K.W. Ansarwadekar, 1977. Field screening of rose cultivars for die-back. South India Horticulture 25: 167.

McCoy, R.E., 1975. Susceptibility of Pelargonium species to geranium rust. Plant Dis. Reptr. 59: 618–620.

McNeill, B.H. & J.T. Fletcher, 1971. The influence of tolerant tomato hosts on the pathogenic characteristics of tobacco mosaic virus. Can. J. Bot. 49: 1947–1949.

Matteoni, J.A. & W.R. Allen, 1989. Symptomatology of tomato spotted wilt virus infection in florist's chrysanthemum. Can. J. of Plant Pathol. 11: 379–380.

Matthews, P., 1979. Variation in English isolates of *Fusarium oxysporum* f. sp. *dianthi*. Proceedings of Eucarpia Meeting on Carnation and Gerbera. Alassio, 1978: 115–126.

Miller, H.N., J.W. Miller & G.L. Crone, 1975. Relative susceptibility of chrysanthemum cultivars to *Agrobacterium tumfaciens*. Plant Dis. Reptr. 59: 576–587.

O'Neill, T.M., M. Bagabe & D.M. Ann, 1991. Aspects of biology and control of stem rot of cucumber caused by *Penicillium oxalicum*. Plant Pathol. 40: 78–84.

O'Riordan, F., 1979. Powdery mildew caused by *Oidium begoniae* of Elatior begonias – fungicide control and cultivar reaction. Plant Dis. Reptr. 63: 919–922.

Palmer, L.T. & S.S. Salac, 1977. Reaction of several types of roses to black spot fungus, *Diplocarpon rosae*. Indian Phytopathol. 30: 366–368.

Paternotte, S.J., 1991. *Verticillium albo-atrum* in tomato. A.R. Glasshouse Crops Res. Sta. Naaldwijk, p. 92.

Pelham, J., 1968. TMV resistance. Report of the Glasshouse Crops Research Institute for 1967, pp 45–48.

Pelham, J., J.T. Fletcher & J.H. Hawkins, 1970. The establishment of a new strain of tobacco mosaic virus resulting from the use of resistant varieties of tomato. Ann. appl. Biol. 65: 293–297.

Peng, J.T., 1986. Resistance to disease in *Agaricus bisporus* (Lange) Imbach. PhD Thesis, University of Leeds.

Rademaker, W. & J. de Jong, 1985. Japanese Rust: low susceptibility or resistance in the chrysanthemum. Vakblad voor de Bloemisterij 40: 49.

Rast, A.Th.B., 1972. MII-16, an artificial, symptomless mutant of tobacco mosaic virus for seedling inoculation of tomato crops. Neth. J. Plant Pathol. 78: 110–112.

Rast, A.Th.B., 1988. Pepper tobamoviruses and pathotypes used in resistance breeding. Capsicum Newsletter 7: 20–23.

Rattink, H., 1972. Annual report for flower diseases in the Netherlands at Aalsmeer, p 33–35.

Reinert, R.A. & P.V. Nelson, 1979. Sensitivity and growth of twelve Elatior begonia cultivars to ozone. Hort. Sci. 14: 747–748.

Schepers, H.T.A.M., 1985. Fitness of isolates of *Sphaerotheca fuliginea* resistant or sensitive to fungicides which inhibit ergosterol biosynthesis. Neth. J. Plant Pathol. 91: 65–76.

Semina, S.N. & N.N. Timoshenka, 1979. The resistance of species of wild rose to powdery mildew. Mikologiya i Fitopatologiya 13: 496–500.

Semina, S.N. & V.M. Babkina, 1981. Resistance of chrysanthemums to powdery mildew. Introduktsiya Biologiya i Selektsiya Isvetochuykh Rastenii Yalta, Nikita Botanical Gardens: 115–122.

Semina, S.N. & G.N. Shestachenko, 1981. The resistance of carnations to rust. Mikologiya Fitopatologiya 15: 238–240.

Sezgin, E. & M. Esentepe, 1986. Study on the resistance of some carnation cultivars to *Uromyces caryophyllinus*. J. of Turkish Phytopathol. 15: 43–45.

Spencer, D.M., 1981. Carnation rust caused by *Uromyces dianthi*. Report of the Glasshouse Crops Research Institute, p. 130.

Strider, D.L., 1975. Susceptibility of Rieger elatior begonia cultivars to bacterial blight caused by *Xanthomonas begoniae*. Plant Dis. Reptr. 59: 70–73.

Strider, D.L., 1978a. Alternaria blight of carnation in the greenhouse and its control. Plant Dis. Reptr. 62: 24–28.

Strider, D.L., 1978b. Reactions of recently released Rieger elatior begonia cultivars to powdery mildew and bacterial blight. Plant Dis. Reptr. 62: 22–23.

Strider, D.L., 1978c. Reaction of African violet cultivars to *Phytophthora nicotianae* var *parasitica*. Plant Dis. Reptr. 62: 112–114.

Strider, D.L., 1980. Resistance of African violet to powdery mildew and efficacy of fungicides for the control of the disease. Plant Disease 64: 181–190.

Strider, D.L., 1985. Fusarium wilt of chrysanthemum. Cultivar susceptibility and chemical control. Plant Disease 69: 564–568.

Strider, D.L. & R.K. Jones, 1986. Susceptibility of chrysanthemums to bacterial leaf spot and bud blight caused by *Pseudomonas cichorii*. North Carolina Flower Growers Bulletin 30: 22–24.

Svejola, F.J. & A.J. Bolton, 1980. Resistance of rose hybrids to 3 races of *Diplocarpon rosae*. Can. J. Plant Pathol. 2: 23–35.

van der Mear, Q.P., J. van Bennekom & A.C. van der Giessen, 1978. Gummy stem blight resistance in cucumber (*Cucumis sativus*). Euphytica 27: 861–864.

van Steekelenburg, N.A.M., 1986. *Didymella bryoniae* on glasshouse cucumbers. PhD. Thesis., University of Wageningen.

van Zaayen, A. & B. van der Pol-Linton, 1977. Heat resistance, biology and prevention of *Diehliomyces microsporus* in crops of *Agaricus* species, Neth. J. Plant Pathol. 83: 221–240.

Walker, J.C., 1950. Environment and host resistance in relation to cucumber scab. Phytopathology 40: 1094–1102.

Water, J.K., H.N. Cevat & I.P. Rietskra, 1984. Rust resistant chrysanthemums prove their value in infection trial. Vakblad voor de Bloemisterij 39: 19.

Wilcox, H.J., 1963. Phoma root rot of Chrysanthemums. Proceedings of the 2nd Br. Insectic. Fungic. Conf., Brighton, pp. 291–299.

Wyzogrodska, A.J., P.H. Williams & C.E. Petersen, 1986. Search for resistance to gummy stem blight (*Didymella bryoniae*) in cucumber (*Cucumis sativus*). Euphytica 35: 603–613.

Zhang, B.D. & X.G. Li, 1986. Seasonal incidence, varietal resistance & chemical control of chrysanthemum blight, leaf spot (*Septoria chrysanthemella*). Journal of South China Agricultural University 7: 35–40.

Euphytica **63**: 51–58, 1992.
R. Johnson and G.J. Jellis (eds), Breeding for Disease Resistance

Multiple resistance to diseases and pests in potatoes

G.J. Jellis
Plant Breeding International Cambridge, Maris Lane, Trumpington, Cambridge, CB2 2LQ, UK

Key words: breeding, disease resistance, pest resistance, potato, selection, *Solanum tuberosum*

Summary

The potato has more characters of economic importance that need to be considered by the breeder than any other temperate crop. In Europe these include resistance to at least twelve major diseases and pests. Highest priority has been given to resistance to late blight (*Phytophthora infestans*), virus diseases (particularly those caused by potato leafroll virus and potato virus Y) and potato cyst nematode (*Globodera rostochiensis* and *G. pallida*). Useful sources of resistance are available and early generation screening techniques have been developed to allow positive selection for multiple resistance and the breeding value of clones used as parents to be determined. Progress in restriction fragment length polymorphism technology should result in more efficient selection in the future.

Introduction

The potato (*Solanum tuberosum* ssp. *tuberosum,* hereafter called Tuberosum) has more characters of economic importance that need to be considered by the plant breeder than any other temperate arable crop (Thomson, 1987). Currently, the UK National Institute of Agricultural Botany (NIAB) assesses approximately forty agronomic, quality, pest and disease resistance characteristics (Table 1). Even such an extensive list is not complete; for example, resistance to skin spot (*Polyscytalum pustulans*) and dry rot (*Fusarium* spp.) can be important in some situations. In the breeding programme at the Scottish Crop Research Institute, Dundee (SCRI), more than sixty variates were used as selection criteria (Mackay, 1987a).

Clearly, the plant breeder cannot expect to combine high resistance to all these diseases and pests in one variety using currently available technology, although it is possible that within the not too distant future biotechnologists will develop techniques for introducing general resistance to fungi or bacteria. Furthermore, the breeder has to be mindful of the fact that high disease resistance will rarely compensate for poor agronomic or quality characteristics. It is therefore necessary to set priorities but at the same time to ensure that resistant varieties do not have major weaknesses in other characters.

Priorities in breeding programmes

A number of factors may influence the selection pressure which the breeder decides to apply for resistance to a particular disease or pest. These include: the importance of the disease/pest in the area in which the variety will be grown; the availability of sources of resistance and difficulty in utilizing them in the breeding programme; the development of reliable and simple screening techniques; the availability, effectiveness and public acceptance of other methods of control; statutory requirements (field immunity to wart in new varieties was a requirement in the UK at one time).

The relative importance of the numerous pathogens that attack potatoes is not easy to establish and depends on whether the crop is being grown for seed or ware, and whether the farmer, merchant or retailer is consulted. It is, however, interesting to note that breeders in many European countries have similar priorities in breeding for disease/pest resistance although the emphasis may be slightly different (van Loon, 1987; Mackay, 1987b; Munzert, 1987; Scholtz, 1987; Swiezynski, 1987). Particular emphasis is put on resistance to potato cyst nematodes, late blight and viruses. Effort is also put into breeding for resistance to wart, common scab, the blackleg/soft rot complex and fungal storage rots (*Fusarium* spp. and *Phoma foveata*).

Potato cyst nematode

Resistance to potato cyst nematodes (PCN) is an obvious goal. They are major pests in many of the best potato growing regions and chemical control, although reasonably effective, is causing concern because of its environmental impact. Resistance to both species exists (Table 2), but it is much simpler to breed for resistance to the common pathotype of *G. rostochiensis* (Rol, controlled by a single dominant gene *H1,* derived from *S. tuberosum* ssp. *andigena,* hereafter called Andigena) than pathotype Pa2/3 of *G. pallida* (polygenic inheritance, derived mainly from Andigena and *S. vernei*). The gene *H1* has been available in varieties for over 25 years now and has proved to be extremely durable in the UK where Maris Piper, which possesses the gene, has been widely grown for many years. Simple screening tests for resistance to both species of PCN are available which facilitate early generation screening (Phillips et al., 1980). In some breeding programmes, clones susceptible to *G. rostochiensis* are eliminated at an early stage. Eventually, however, clones need to be trialled in infested fields, as high resistance to multiplication of the pest and tolerance of attack are not necessarily related and only replicated field trials are reliable in assessing the latter (Phillips, Trudgill & Evans, 1988; Evans & Haydock, 1990).

Late blight

Late blight resistance has been a breeding objective since the last century but rose to prominence during the first half of the twentieth century when hypersensitivity genes were introduced, particularly from *S. demissum* (Ross, 1986). Their failure to control the disease led to a renewed interest in 'field resistance' or 'general resistance' which is proving to be more durable and is a complex of many characteristics controlled by several genes. Resistance in the foliage is not necessarily related to resistance in the tubers and must be treated

Table 1. Agronomic, quality and disease/pest resistance characters of economic importance in potato assessed by the NIAB (Anon. 1991)

Agronomic	Quality	Disease/pest resistance
Early yield	Skin colour	Foliage blight } (*Phytophthora infestans*)
Late yield	Skin shape	Tuber blight } (*Phytophthora infestans*)
Outgrades	Flesh colour	Blackleg (*Erwinia carotovora*)
Tuber number	Eye depth	Common scab (*Streptomyces* sp.)
Tuber size	Size uniformity	Powdery Scab (*Spongospora subterranea*)
Dormancy	Attractiveness	Gangrene (*Phoma foveata*)
Emergence	Internal bruising	Wart (*Synchytrium endobioticum*)
Drought tolerance	External damage	Leaf roll (potato leafroll virus)
Foliage cover	Dry matter	Severe mosaic (potato virus Y)
Foliage maturity	Fry colour	Mild mosaic (potato virus X and potato virus A)
Stolon attachment	Texture	Spraing (tobacco rattle virus)
	Discolouration on cooking	Cyst nematode (*Globodera rostochiensis* and *G. pallida*)
	Disintegration on cooking	Slug damage (various spp.)
	Boiling suitability	

separately in screening programmes. Screening techniques were summarised by Wastie (1991a). If at least one parent has high resistance, screening seedlings for foliage resistance in the glasshouse or growthroom is valuable in identifying resistant progenies. Glasshouse or laboratory tests on adult plants or detached leaves can also be very useful, but the final assessment should always be done in the field. Distinguishing between major gene resistance and high levels of field resistance can be a problem.

Tuber tests are usually done by inoculating freshly harvested tubers or tuber slices/discs. There are many sources of field resistance although genes from only a limited number of species of *Solanum* are represented in current varieties (Ross, 1986 and Table 2).

The occurrence of the A2 mating type in Europe (Hohl & Iselin, 1984; Malcolmson 1985; Shaw et al., 1985), which may lead to changes in epidemiology and increase variation in the fungus, and the increasing tolerance of the pathogen to systemic fungicides (Cooke, 1991; Davidse et al., 1991) make late blight resistance an even more important objective now than a decade ago.

Table 2. Sources of resistance to cyst nematode, blight and the major virus diseases, which have been used widely in potato breeding programmes[+]

Disease/Pest	*Solanum* species
Potato cyst nematode	
Globodera rostochiensis	*S. tuberosum* ssp. *andigena*, *S. vernei*, *S. spegazzinii*
G. pallida	as above + *S. gourlayi*, *S. oplocense*, *S. sparsipilum*, *S. multidissectum*
Late blight (*Phytophthora infestans*)	*S. demissum*, *S. stoloniferum*, *S. vernei*, *S. verrucosum*, *S. tuberosum* ssp. *andigena*
Virus diseases	
Leaf roll (PLRV)	*S. demissum*, *S. acaule*
Severe mosaic (PVY)	*S. stoloniferum*, *S. phureja*, *S. tuberosum* ssp. *andigena*, *S. stenotomum*, *S. chacoense*, *S. demissum*, *S. microdontum*
Mild mosaic (PVX)	*S. tuberosum* ssp. *andigena*, *S. acaule*

[+] excluding *S. tuberosum* ssp. *tuberosum*.
Sources of information: Howard et al. (1970), Davidson (1980), Ross (1986), Wastie (1991a).

Virus diseases

Virus diseases are a particularly pressing problem in those countries that do not have designated areas for growing seed but global warming may result in an increase in the incidence of aphid-borne virus in those areas at present best suited for seed production. In western Europe potato leafroll virus (PLRV) and potato virus Y (PVY) cause by far the most important virus diseases, although potato virus X (PVX) can be a particular nuisance in breeding programmes; so much so that at one time resistance to PVX was an early generation selection criterion at the Plant Breeding Institute, Cambridge (PBI) (Howard et al., 1978). For PLRV, the principle type of resistance utilized has been regarded as resistance to infection, as measured by field exposure (Davidson, 1973) or by using viruliferous aphids in the glasshouse, a procedure which must be repeated for several clonal generations to get a reliable result (Swiezynski, 1984). Such resistance is frequently associated with restricted systemic spread of the virus to tubers of those plants which do become infected, and also with low concentration of virus in leaf tissue (Barker, 1987). Such an association of these components of resistance has enabled selection for resistance by graft inoculation of plants, and selecting for low virus concentration, an attractive alternative to field exposure. The major sources of resistance to PLRV in present cultivars are *S. demissum* × *S. tuberosum* hybrids and *S. acaule* (Davidson, 1980; Ross, 1986 and Table 2). The genetics of resistance is unclear but several genes are involved. Resistance to PVY is much easier to achieve than to PLRV. Nowadays, the major effort is towards introducing the *Ry* genes derived from *S. stoloniferum* and Andigena into commercial varieties (Table 2). These are single dominant genes which confer extreme resistance to all strains of PVY and, in the case of some genotypes carrying the gene from *S. stoloniferum,* also potato virus A (Ross, 1986). Selection can be performed at the seedling stage by inoculating with sap from infected tobacco plants using a spray gun. Genes which confer resistance to PVX (*Rx,* from *S. acaule,* Andigena and Tuberosum), potato virus S (*Ns,* from Andigena) and potato virus M ((*Gm,* from *S. gourlayi*) can also be selected in segregating progenies using the spray gun (Ross, 1986; Was et al., 1988). A number of the genes conferring resistance to potato viruses have proved to be durable, even though frequently only a single dominant gene is involved. Probably the best example is the gene *Ry.* At present, no strains of PVY have been reported which can infect varieties carrying this gene. Even genes which are strain-specific can be valuable. The gene *Nx,* found commonly in *S. tuberosum,* only confers resistance to strain groups 1 and 3 of PVX but, in practice, varieties such as King Edward are rarely found to be infected because strain groups 2 (also known as potato virus B) and 4 are rare.

Other diseases and pests

Compared with the major breeding objectives, breeding and selection for resistance to other diseases and pests has been on a small scale and has relied mainly on using adapted parents with moderate to high resistance coupled with selection against undue susceptibility (negative selection). Early generation screening tests have been developed for a number of diseases, including blackleg (Lojkowska & Kelman, 1989), powdery scab (Jellis & Starling, 1990; Wastie, 1991b), gangrene (Wastie et al., 1988, 1990), wart (Frey, 1980) and common scab (Gunn et al., 1983).

When varieties are bred for export, resistance to additional pests and diseases has sometimes to be considered. Many northern European breeding programmes are now aimed at the Mediterranean region, where early blight (*Alternaria solani*) and verticillium wilt (*Verticillium dahliae*) are important diseases (Nachmias et al., 1988).

Screening programmes

Routine testing

Traditionally, screening for disease and pest resistance has generally been done in the second half of a selection programme, i.e. from year 5 onwards, although as all varieties are actually F1s which have been multiplied as clones, reliable testing can be done earlier if seed is available. By year 5, selection for agronomic characters and, to some extent, yield will have been done, and probably less than 5% of the variation in the original population will remain. Breeders generally use the screens to exert negative selection pressure, discarding those clones which are very susceptible. A decision to discard a clone at an advanced stage in the breeding programme will often be taken only when data on a whole range of attributes have been considered and the good features balanced against the bad ones. When resistance to a particular pest or disease is given high priority such a selection procedure is obviously not satisfactory, particularly if the aim is also to select parents for the next generation.

A comparison of screening programmes at SCRI and PBI during the 1980s shows that, overall, there was a broad agreement in the disease tests being done, and the stage in the programme when screening commenced. Early generation screening was being practised for PCN, foliage blight and viruses, at least for some crosses (Table 3).

Jellis et al. (1986) described a scheme for screening for combined resistance to PVY, both species of PCN, and foliage blight within a year of the true seed being grown (year 1 in Table 3). This scheme was for testing the progeny of crosses which combined the genes *Ry* and *H1,* which confer resistance to PVY and *G. rostochiensis* respectively, with genes for resistance to *G. pallida* from Andigena. The Andigena clones used as PCN resistant parents generally have poor resistance to blight, hence the need for an early-generation blight screen. An outline of the scheme is given in Fig. 1.

Parental breeding

Early generation tests on seedlings and their tuber progenies are proving particularly valuable in the development of clones with multiple disease resistance and in the determination of the breeding values of clones used as parents.

Martin (1985a, 1985b), working in eastern Washington and Oregon, USA, developed a novel approach to breeding and selecting parental material with multiple resistance to diseases and pests. His technique ensures that first year clones are exposed to a wide range of pathogens and pests, including PVY, PLRV, common and deep-pitted scab (*Streptomyces* spp.), early blight (*Alternaria solani*), sclerotinia wilt (*Sclerotinia sclerotiorum*), verticillium wilt (*Verticillium dahliae*) powdery mildew (*Erysiphe cichoracearum*), Columbia root knot nematode (*Meloidogyne* sp.) and Colorado beetle (*Leptinotarsa decemlineata*). Firstly, potential multi-resistant germplasm was produced by mass intercrossing of clones identified as having high resistance to one or more specific diseases or pests. Bulked true potato seed (TPS) derived from this intercrossing was then sown directly into soils infested with soil-borne pests and pathogens and cultural conditions and locations were used which provided ideal conditions for foliage diseases. PVY inoculum was applied directly to seedlings by spraying or rubbing and TPS rows were interplanted with PLRV infector rows. Agronomic methods developed for direct-seeded tomatoes were modified slightly and used successfully for growing potato plants from TPS. Plants expressing resistance were identified by spraying their bases with red fluorescent paint, so the tubers could be selected easily at harvest. Survivors were again exposed to a wide range of pests and pathogens. Quality characteristics of tubers were also assessed at this early

Table 3. Screening programmes for resistance to diseases and pests in British breeding stations+

Disease/Pest	Years of testing (Year 1 = seedling year)	
	SCRI*	PBI*
Mild and/or severe mosaic (PVX & PVY)	1#, 3, 7, 8→	1#, 5→
Leaf roll	4, 5	5→
Potato cyst nematode (both species)	1#, 3, 4, 8→	1#→
Late blight (foliage)	1#, 5	1#, 7→
Late blight (tuber)	5→	4→
Common scab	5→	6→
Gangrene	5→	6→
Dry rot	–	7→
Skin spot	6→	–
Wart	5→	6→
Powdery scab	–	8
Blackleg and/or soft rot	8→	7→
Spraing (tobacco rattle virus)	7→	8→
Spraing (potato mop top virus)	8→	–

+ Sources: Jellis et al. (1987) with additional information on cyst nematode; Mackay (1987a).
* Scottish Crop Research Institute, Dundee and Plant Breeding Institute, Cambridge.
# not all families were screened in year 1.
→ and successive years.
– not tested.

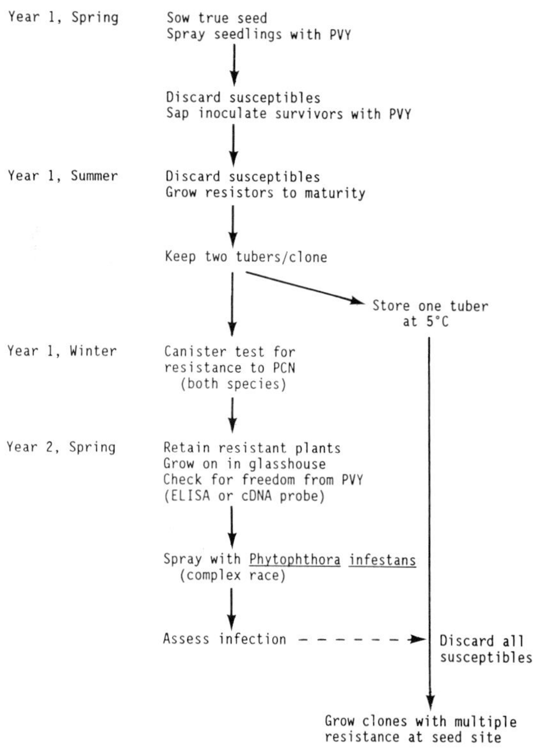

Fig. 1. Scheme for multiple resistance screening. See Jellis et al. (1986) for more detail.

stage. Using these techniques, Martin selected over 200 multi-resistant clones as useful parents.

In Poland, at the Institute for Potato Research, Mlochow, considerable effort has been put into developing parental lines with multiple resistance to viruses. Breeding for high resistance to PVY, PVA, PVX, PVS, PVM and PLRV is being done at both the tetraploid and diploid level (Dziewonska, 1986). In order to achieve this, efficient screening methodology has been developed, as reported by Swiezynski (1984) and Was et al., 1988. Currently, susceptible lines can be eliminated rapidly and genotypes with resistance to all six viruses identified within 4 years of making the cross (Wawrzyczek et al., 1991). This involves spraying seedlings with PVY^o, PVX and PVM and then manually inoculating plants with PVY^n or PVM and graft inoculating with PLRV, PVY^o, PVS, PVM and PVX during the subsequent three clonal generations. In addition, in the third clonal year, resistance to PLRV is tested using aphids. As well as having multiple resistance, clones are selected for high yield, a dry matter content of 16–18% in the tubers, and mid-early to mid-late maturity (Dziewonska, 1986). At the diploid level, an additional objective is to obtain clones homozygous for resistance genes which are then crossed with tetraploids to obtain parental tetraploid clones multiplex for resistance genes. Such clones are very valuable as parents as they will pass on their resistance with a high probability and reduce or eliminate the need to challenge their progeny with the pathogen (Table 4).

Selection of superior parents, i.e. those which give rise to a high proportion of resistant progeny is also an objective of UK breeding stations. This includes breeding clones multiplex for single dominant genes (e.g. *Rx, Ry, H1*) (Mackay, 1987a, 1987c, 1989; Thomson et al., 1987). Tetraploid clones with a single copy of the gene (simplex) are intercrossed and duplex progeny identified by test crossing. Duplex clones are then selfed or intercrossed and triplex or quadruplex progeny identified as before. It is then important to identify within such multiplex breeding material those clones which have other important attributes. This technique is particularly valuable if a decision is made to retain only those clones with a particular resistance gene, such as the gene *H1* which confers resistance to the common pathotype of *G. rostochiensis.*

Genotypic selection is also practised by using early generation screening tests of the type described previously to assess progenies and detect those with a high frequency of resistant clones. Using this information, genetically desirable parents can be identified (Wastie et al., 1988).

The future

The production of variation *in vitro,* by introducing new genes for disease resistance via *Agrobacterium tumefaciens,* for example, could be most effectively exploited if selection can also be done *in vitro.* There has been considerable interest in recent years in the use of fungal exotoxins for this purpose. Results to date have not always been encouraging. This may be due to the pathosystem being studied or the toxins being used. More definitive tests may become available when specific components of the toxin associated with pathogenicity have been identified (Lynch et al., 1991). It is obviously very important that the *in vitro* screen correlates well with field results on adult plants.

Techniques for identifying the presence of resistance genes which do not depend on phenotypic testing with all the ensuing environmental interactions would be highly desirable. Restriction fragment length polymorphism (RFLP) linkage maps provide a direct method for selecting desirable genes (Tanksley et al., 1989). Such linkage maps have been constructed for potato (Bonierbale et

Table 4. Expected ratios when clones with different copy numbers of a dominant resistance gene (R) are crossed with a susceptible (assuming chromosome, not chromatid, segregation[+])

Cross	Expected ratio Resistant : susceptible
Rrrr × rrrr	1 : 1
RRrr × rrrr	5 : 1
RRRr × rrrr	1 : 0
RRRR × rrrr	1 : 0

[+] Howard (1970).

al., 1989; Gebhardt et al., 1989; Jacobs et al., 1990). Many of the resistance genes in modern potato varieties have been introgressed from wild *Solanum* species or cultivated species other than *S. tuberosum* (Table 2), and recently chromosomal fragments of exotic origin have been detected within diploid and monohaploid breeding lines (Debener et al., 1991). Furthermore, it has been shown that an RFLP locus located on such a chromosome fragment is linked to a gene conferring resistance to PVX. Within the near future we can expect to see rapid progress in this field. A major goal is the marking of specific chromosome segments involved in quantitative traits, such as resistance to late blight.

Acknowledgements

I am grateful for helpful comments from R.L. Wastie and R.E. Boulton.

References

Anon, 1991. Potato Variety Handbook, 1992. National Institute of Agricultural Botany, Cambridge.

Barker, H., 1987. Multiple components of the resistance of potatoes to potato leafroll virus. Ann. appl. Biol. 111: 641–648.

Bonierbale, M.W., R.L. Plaisted & S.D. Tanksley, 1989. RFLP maps based on a common set of clones reveal modes of chromosomal evolution in potato and tomato. Genetics 120: 1095–1103.

Cooke, L.R., 1991. Current problems in the chemical control of late blight: the Northern Ireland experience. In: J.A. Lucas, R.C. Shattock, D.S. Shaw & L.R. Cooke (Eds.) *Phytophthora,* pp. 337–348. Cambridge University Press, Cambridge.

Davidse, L.C., G.C.M. van den Berg-Velthuis, B.C. Mantel & A.B.K. Jespers, 1991. Phenylamides and *Phytophthora.* In: J.A. Lucas, R.C. Shattock, D.S. Shaw & L.R. Cooke (Eds.) *Phytophthora,* pp. 349–360. Cambridge University Press, Cambridge.

Davidson, T.M.W., 1973. Assessing resistance to leafroll in potato seedlings. Potato Res. 16: 99–108.

Davidson, T.M.W., 1980. Breeding for resistance to virus disease of the potato (*Solanum tuberosum*) at the Scottish Plant Breeding Station. Rep. Scott. Pl. Breed. Stat. for 1979–80, pp. 100–108.

Debener, T., F. Salamini & C. Gebhardt, 1990. Phytogeny of wild and cultivated *Solanum* species based on nuclear restriction fragment length polymorphisms (RFLPs). Theor. Appl. Genet. 79: 360–368.

Dziewonska, M.A., 1986. Development of parental lines for breeding of potatoes resistant to viruses and associated research in Poland. In: Potato Research of Tomorrow: Drought Tolerance, Virus Resistance and Analytic Methods. Proceedings of an International Seminar, Wageningen, 1985, pp. 96–100. Pudoc, Wageningen.

Evans, K. & P.P.J. Haydcock, 1990. A review of tolerance by potato plants of cyst nematode attack, with consideration of what factors may confer tolerance and methods of assaying and improving it in crops. Ann. appl. Biol. 117: 703–740.

Frey, F., 1980. Screening for resistance against the wart fungus *Synchytrium endobioticum* in potato seedlings with a modified Spieckermann method. Potato Res. 23: 303–310.

Gebhardt, C., E. Ritter, T. Debener, U. Schachtschabel, B. Walkemeier, H. Uhrig & F. Salamini, 1989. RFLP analysis and linkage mapping in *Solanum tuberosum.* Theor. Appl. Genet. 78: 65–75.

Gunn, R.E., G.J. Jellis, P.J. Webb & N.C. Starling, 1983. Comparison of three methods for assessing varietal differences in resistance to common scab disease (*Streptomyces scabies*) of potato. Potato Res. 26: 175–178.

Hohl, H.R. & K. Iselin, 1984. Strains of *Phytophthora infestans* from Switzerland with A2 mating type behaviour. Trans. Brit. mycol. Soc. 83: 529–530.

Howard, H.W., 1970. Genetics of the Potato, *Solanum tuberosum.* Logos Press, London.

Howard, H.W., C.S. Cole, J.M. Fuller, 1970. Further sources of resistance to *Heterodera rostochiensis* Woll. in the Andigena potatoes. Euphytica 19: 210–216.

Howard, H.W., C.S. Cole, J.M. Fuller, G.J. Jellis & A.J. Thomson, 1978. Potato breeding problems with special reference to selecting progeny of the cross of Pentland Crown × Maris Piper. Rep. Pl. Breed. Inst. Camb. for 1977, pp. 22–50.

Jacobs, J.J.M.R., F.A. Krens, W.J. Stiekema, M. van Sponje & M. Wagenvoort, 1990. Restriction fragment length polymorphism in *Solanum* spp. for the construction of a genetic map of *Solanum tuberosum* L.: a preliminary study. Potato Res. 33: 171–180.

Jellis, G.J., C.N.D. Lacey, R.E. Boulton, S.B. Currell, A.M. Squire & N.C. Starling, 1986. Early-generation screening of potato clones for disease and pest resistance. Aspects of App. Biol. 13, Crop Protection of Sugar Beet and Crop Protection and Quality of Potatoes, pp. 301–305.

Jellis, G.J., R.E. Boulton, N.C. Starling & A.M. Squire, 1987. Screening for resistance to diseases in a potato breeding programme. In: G.J. Jellis & D.E. Richardson (Eds.), The Production of New Potato Varieties: Technological Advances, pp. 84–85. Cambridge University Press, Cambridge.

Jellis, G.J. & N.C. Starling, 1990. Resistance to powdery scab disease of potato. In: Abstracts of Conference Papers, 11th Triennial Conference of the EAPR, 1990, pp. 427–428.

Lojkowska, E. & A. Kelman, 1989. Screening of seedlings of wild *Solanum* species for resistance to bacterial stem rot caused by soft rot erwinias. Am. Potato. J. 66: 379–390.

van Loon, J.P., 1987. Potato breeding strategy in the Netherlands. In: G.J. Jellis & D.E. Richardson (Eds.), The Production of New Potato Varieties: Technological advances, pp. 45–54. Cambridge University Press, Cambridge.

Lynch, D.R., R.L. Wastie, H.E. Stewart, G.R. Mackay, G.D. Lyon & A. Nachmias, 1991. Screening for resistance to early blight (*Alternaria solani*) in potato (*Solanum tuberosum* L) using toxic metabolites produced by the fungus. Potato Res. 34: 297–304.

Mackay, G.R., 1987a. Selecting and breeding for better potato cultivars. In: A.J. Abbott & R.K. Atkin (Eds.), Improving Vegetatively Propagated Crops, pp. 181–196. Academic Press, London.

Mackay, G.R., 1987b. Potato breeding strategy in the United Kingdom. In: G.J. Jellis & D.E. Richardson (Eds.), The Production of New Potato Varieties: Technological Advances, pp. 60–67. Cambridge University Press, Cambridge.

Mackay, G.R., 1987c. Screening for resistance to diseases and pests. In: G.J. Jellis & D.E. Richardson (Eds.), The Production of New Potato Varieties: Technological Advances, pp. 88–90. Cambridge University Press, Cambridge.

Mackay, G.R., 1989. Parental line breeding at the tetraploid level. In: K.M. Louwes, H.A.J.M. Toussaint & L.M.W. Dellaert (Eds.), Parental Line Breeding and Selection in Potato Breeding, pp. 131–135. Pudoc, Wageningen.

Malcolmson, J.F., 1985. *Phytophthora infestans* A2 mating type recorded in Great Britain. Trans. Brit. mycol. Soc. 85: 531.

Martin, M.W., 1985a. Breeding multi-resistant potato germplasm. In: G.J. Jellis & D.E. Richardson (Eds.), The Production of New Potato Varieties: Technological Advances, pp. 94–95. Cambridge University Press, Cambridge.

Martin, M.W., 1985b. Field seeding of true potato seed in a breeding programme. In: G.J. Jellis & D.E. Richardson (Eds.), The Production of New Potato Varieties: Technological Advances, pp. 269–270. Cambridge University Press, Cambridge.

Munzert, M., 1987. Potato breeding strategy in the Federal Republic of Germany. In: G.J. Jellis & D.E. Richardson (Eds.), The Production of New Potato Varieties: Technological Advances, pp. 38–44. Cambridge University Press, Cambridge.

Nachmias, A., P.D.S. Caligari, G.R. Mackay & L. Livescu, 1988. The effects of *Alternaria solani* and *Verticillium dahliae* on potatoes growing in Israel. Potato Res. 31: 443–450.

Phillips, M.S., J.M.S. Forrest & L.A. Wilson, 1980. Screening for resistance to potato cyst nematode using closed containers. Ann. appl. Biol. 96: 317–322.

Phillips, M.S., D.L. Trudgill & K. Evans, 1988. The use of single, spaced potato plants to assess their tolerance of damage by potato cyst nematodes. Potato Res. 31: 469–475.

Ross, H., 1986. Potato Breeding – Problems and Prospects. Advances in Plant Breeding 13. Paul Parey, Berlin and Hamburg.

Scholtz, M., 1987. Potato breeding in the German Democratic Republic. In: G.J. Jellis & D.E. Richardson (Eds.), The Production of New Potato Varieties: Technological Advances, pp. 32–37. Cambridge University Press, Cambridge.

Shaw, D.S., A.M. Fyfe, P.G. Hibberd & M.A. Abdel-Sattar, 1985. Occurrence of the rare A2 mating type of *Phytophthora infestans* on imported Egyptian potatoes and the production of sexual progeny with A1 mating type from the UK. Plant Pathol. 34: 552–556.

Swiezynski, K.M., 1984. Early generation selection methods used in Polish potato breeding. Am. Potato. J. 61: 385–394.

Swiezynski, K.M., 1987. Potato breeding strategy in Poland. In: G.J. Jellis & D.E. Richardson (Eds.), The Production of New Potato Varieties: Technological Advances, pp. 55–59. Cambridge University Press, Cambridge.

Tanksley, S.D., N.D. Young, A.M. Paterson & M.W. Bonierbale, 1989. RFLP mapping in plant breeding: new tools for an old science. Bio/technology 7: 257–264.

Thomson, A.J., G.J. Jellis, C.N.D. Lacey, R.E. Boulton, R.M. Negus, E.D. Martlew, A.M. Squire, N.C. Starling & L. Taylor, 1988. Rep. Pl. Breed. Inst. Camb. for 1986, p. 51.

Thomson, A.J., 1987. A practical breeder's view of the current state of potato breeding and evaluation. In: G.J. Jellis & D.E. Richardson (Eds.), The Production of New Potato Varieties: Technological Advances, pp. 336–346. Cambridge University Press, Cambridge.

Was, M., M.A. Dziewonska & H. Butkiewicz, 1988. Postepy W metodach selekcji ziemniakow odpornych na wirusy oraz w diagnostyce wirusów i wiroida wrzecionowatosci bulw ziemniaka. Genetyczne podstawy hodowli ziemniaka. Instytut Ziemniaka, Bonin, Poland, pp. 125–136.

Wastie, R.L., 1991a. Breeding for resistance. In: D.S. Ingram & P.H. Williams (Eds.), *Phytophthora infestans,* The Cause of Late Blight of Potato. Advances in Plant Pathol. 7: 193–224.

Wastie, R.L., 1991b. Resistance to powdery scab of seedling progenies of *Solanum tuberosum.* Potato Res. 34: 249–252.

Wastie, R.L., P.D.S. Caligari, H.E. Stewart & G.R. Mackay, 1988. Assessing the resistance to gangrene of progenies of potato (*Solanum tuberosum* L.) from parents differing in susceptibility. Potato Res. 31: 355–365.

Wastie, R.L., G.R. Mackay, P.D.S. Caligari & H.E. Stewart, 1990. A glasshouse progeny test for resistance to gangrene (*Phoma foveata*). Potato Res. 33: 131–133.

Wawrzyczek, M., M. Dziewonska, E. Mietkiewska & M. Was, 1991. Screening potato genotypes for multiple resistance to viruses. In: Program of the International Symposium on Advances in Potato Crop Protection, September 1991, Wageningen, p. 77.

Euphytica **63**: 59–70, 1992.
R. Johnson and G.J. Jellis (eds), Breeding for Disease Resistance

Resistance to cane and foliar diseases in red raspberry (*Rubus idaeus*) and related species

B. Williamson & D.L. Jennings[1]
Scottish Crop Research Institute, Invergowrie, Dundee DD2 5DA, UK; [1] present address: Clifton, Honey Lane, Otham, Nr Maidstone, Kent ME 15 8RJ, UK

Key words: breeding, cane hairiness, disease resistance, periderms, raspberry, *Rubus idaeus,* interspecific hybrids

Summary

Several types of resistance to fungal pathogens of cane fruits have been identified by controlled infection procedures and incorporated into the UK *Rubus* breeding programme. These are reviewed with a description of the biology of the pathogens involved. Some of them take the form of a mechanical barrier associated with the position and rate of development of suberised periderms. For example, damage to the deep-seated cane polyderm occurs when *Leptosphaeria coniothyrium* (cane blight) infects the phloem and xylem, but forms of resistance which restrict the expansion of these vascular lesions have been found.

Wound periderms, which form rapidly around natural splits in the primary cortex, play an important role in resistance of some genotypes to midge blight (*Resseliella theobaldi* and its associated fungal pathogens). However, they are less effective against *Elsinoe veneta* (cane spot) which attacks canes early in cane differentiation before functional periderms have arisen. Other resistances are associated with distinctive morphological traits, most notably cane pubescence. This character is controlled by gene *H* and is associated with useful resistance to *Botrytis cinerea* (cane botrytis) and *Didymella applanata* (spur blight), but has the disadvantage that it is associated with increased susceptibility to *E. veneta, Sphaerotheca macularis* (powdery mildew) and *Phragmidium rubi-idaei* (raspberry yellow rust). These associations present a dilemma for breeders, but other sources of resistance to these pathogens have been identified that are strong enough to overcome this disadvantage.

From studies of inheritance it is concluded that a major gene determines one form of resistance to *Phragmidium rubi-idaei* and that three major genes determine resistance to *S. macularis.* Minor genes acting additively determine all the other resistances, though major genes may be involved in cane resistance to *Botrytis cinerea* and resistance to *Elsinoe veneta,* which were inherited without diminution over several generations. In two instances a common resistance to two diseases was discovered.

Introduction

The stems (canes) and leaves of red raspberry (*Rubus idaeus*) and its close relatives are prone to infection by several fungal diseases which have the potential to cause serious yield loss worldwide. A comprehensive review of the diseases of raspberries and blackberries has been published (Ellis et al., 1991) and sources of resistance to many of them have been identified and exploited in breeding (Jennings, 1988; Jennings et al., 1990).

This review discusses the biology of several fungal pathogens which predominantly infect leaves and stems and emphasises the physiological and

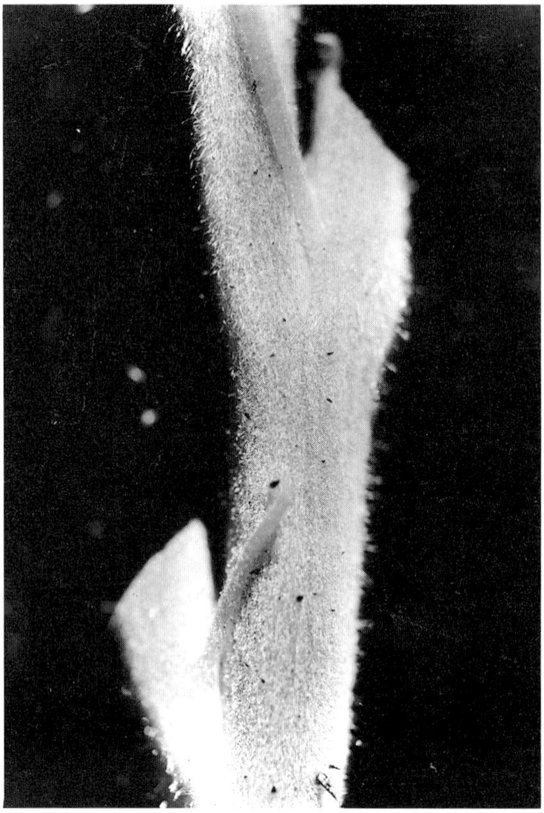

Fig. 1. Primocane of red raspberry (*R. idaeus*) with fine epidermal hairs, a character determined by gene *H*.

histological aspects of pathogenesis which influenced the strategies used in resistance breeding over the past 15 years at the Scottish Crop Research Institute. It gives results from controlled wound inoculation procedures that were used to quantify resistance to cane diseases under field conditions and from either controlled spore applications or precise field scoring of natural infections that were used to assess resistance to the leaf diseases.

Disease resistance is increasing in importance in soft fruit crops because consumers require higher quality fruit at a time when the use of pesticides in food crops is being questioned. However, *Rubus* breeders have limited resources and they can rarely include a primary screen for each fungal disease, because a substantial proportion of their progenies would be discarded and this would reduce their chances of improving other important agronomic characters such as high fruit number, large fruit size, firm texture, thornlessness, cane production, and suitability for machine harvesting, which are among the many requirements of modern cultivars.

For convenience the diseases will be introduced and discussed in three groups; those which cause mainly 1) vascular damage, 2) bud suppression, and 3) defoliation, loss of winter hardiness and vigour. The occurrence and inheritance of the different forms of resistance and the complex effects associated with gene *H,* which confers cane hairiness, on several of the diseases will then be discussed.

Knight & Keep (1958) first reported that raspberry cultivars and selections with fine hairs (pubescent) on canes (Fig. 1) were more resistant to cane botrytis and spur blight than non-hairy (glabrous) canes. Pubescence is determined by gene *H* (genotype *HH* or *Hh*) the recessive allele of which gives glabrous canes (genotype *hh*), but gene *H* is rarely homozygous because it is linked with a lethal recessive gene (Jennings, 1967). It was later found that pubescent canes were more susceptible to cane spot and powdery mildew than glabrous canes (Jennings, 1962; Keep, 1968, 1976; Jennings & McGregor, 1988) and this was seen as a distinct disadvantage of the gene in areas where the diseases caused by these fungi are prevalent. More recently it has also been shown that pubescent canes are also more susceptible to yellow rust populations in Scotland (Anthony et al., 1986) and this relationship was confirmed with Australian populations of the rust (Jennings & McGregor, 1988).

Diseases causing mainly vascular damage

Cane blight

Tissue resistance to cane blight caused by *Leptosphaeria coniothyrium* (anamorph *Coniothyrium fuckelii*) is required because the fungus invades various kinds of wounds (Williamson & Hargreaves, 1978) and can cause yield losses of up to 30% (Ramsay et al., 1985). Infection through wounds inflicted by machine harvesters can be particularly serious. Inoculum placed on the surface of undamaged epidermis of the young canes usually produces only minor infections of the primary cortex with limited sporulation. Inward spread of mycelium is restricted by polyderm, which is a deep-seated multiseriate stem periderm (Fig. 2a) consisting at maturity of alternating layers of suberised phellem and unsuberised phelloid cells (Williamson, 1984). Both cell types are derived from a typical phellogen and take the form of interlocking discs which are polygonal-shaped in tangential longitudinal section and rectangular in transverse or radial section. The suberised phellem layers seem to be impervious to hyphal penetration, except in late spring in the year after inoculation (Seemüller et al., 1988) when limited damage can result.

L. coniothyrium infects wounds of young vegetative raspberry canes (primocanes). The wounds may be caused by abrasion against the stubs of old dead canes at the crown of the plant (Williamson & Hargreaves, 1981b), by hoes or machines, or by the picking and fruit-catching devices of raspberry harvesters (Williamson & Hargreaves, 1976; Hargreaves & Williamson, 1978; Williamson & Ramsay, 1981, 1984). Mycelial inoculation of primocanes at tangential vascular wounds made by scalpel at different times show that tissue resistance increases progressively from July to September (Williamson & Hargreaves, 1978, 1981b; Jennings, 1979) and that little vascular damage results from late infections. Isolation experiments suggest that the pathogen spreads little within the vascular cylinder, but when the primary cortex and polyderm of infected canes are removed by scraping, brown vascular lesions, probably caused by a pathotoxin, can be seen spreading from infected wounds. These spreading 'stripe' lesions block supplies of water and nutrients to lateral shoots causing death of canes above the point of infection if they girdle the cane. The environmental, physiological and genetic factors which influence lesion spread are unknown.

Mycelial inoculation of field-grown primocanes is a reliable method for screening germplasm for resistance to *L. coniothyrium* (Jennings, 1979; Jennings & Brydon, 1989b). *Rubus pileatus* and its F_1 hybrids with raspberry were resistant using this technique, as were two backcross hybrids of *R. coreanus*. However, it appeared that resistance was associated with the hard non-raspberry-like growth of this material, especially when it was grown in pots. There was no evidence that these species contributed resistance independently of this, even though resistant hybrids occurred in the second backcross generation of *R. pileatus*. It seemed that intermediate levels of resistance occur in raspberry germplasm and were identified by large-scale screening; segregants with higher levels of resistance then occurred by recombination of additive genes following crossing among parents with intermediate resistance. The presence of gene *H* caused an increase of resistance of from 10 to 22% in different segregating families and it was therefore essential to evaluate each family separately for the *H* and *h* segregates (Table 1). Selections with gene *H* also had a lower percentage of their cane wounds infected naturally following machine harvesting (Jennings & Brydon, 1989b).

Analysis of lesion sizes in relation to potential yields (estimated from the total number of flowers plus fruits) showed that yield losses occurred only when girdling of the vascular cylinder was almost complete (Williamson & Hargreaves, 1981b). The intermediate levels of resistance identified in the breeding material may therefore provide adequate protection from this disease, especially as observations of a naturally infected plot of selections suggested that the resistance mechanism was more effective against natural spore infection than against the mycelial inoculum used (Jennings & Brydon, 1989b).

Midge blight

Midge blight is a disease complex caused by the raspberry cane midge (*Resseliella theobaldi*) and associated fungi (Pitcher & Webb, 1952; Williamson & Hargreaves, 1979). Raspberry canes are prone to it when the midge emerges at the same time as natural splits occur in their primary cortex thereby providing oviposition sites for successive generations of the pest (Pitcher, 1952; Nijveldt, 1963). Some cultivars show more cane splits than others but the extent and timing of splitting in a cultivar varies each year, and the precise environmental factors which favour splitting and midge attack are still unknown. It is difficult to cause canes to split. The pest is difficult to maintain in a

b

c

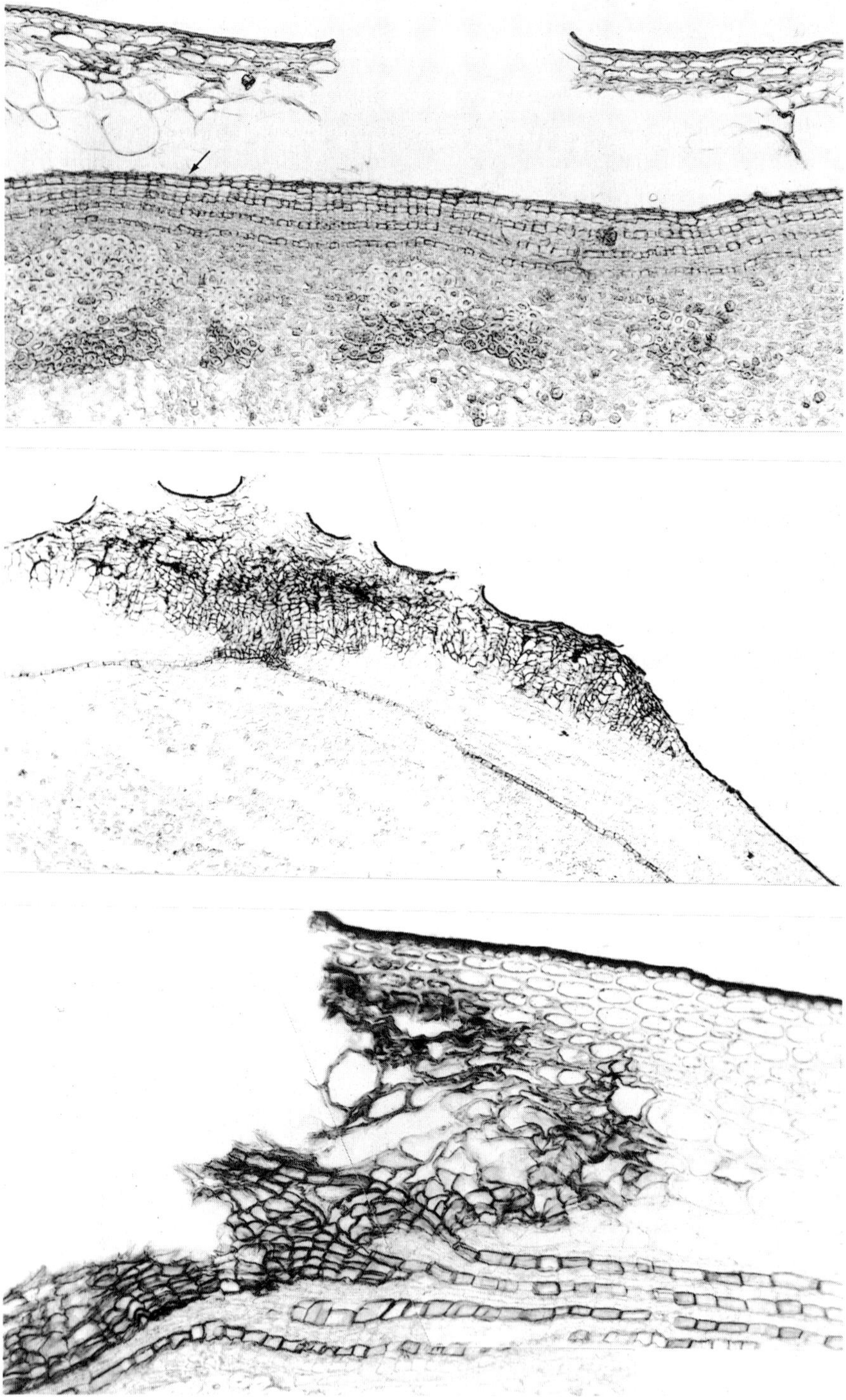

Fig. 2. Transverse sections of primocanes stained to show presence of suberised walls in stem polyderm and wound periderm: a) red raspberry cv. Glen Clova; centre of single natural split in cortex and mature polyderm. Oviposition site of *R. theobaldi* marked (↓). b) *R. crataegifolius,* showing numerous minor splits and extensive wound periderm beneath them. c) *R. crataegifolius* × *R. idaeus* hybrid showing typical polyderm and wound periderm in cortex linking to dedifferentiated polyderm. (McNicol et al., Ann. appl. Biol. 103: 489–495, 1983).

insectaries for screening purposes and natural populations vary in a cyclical manner over *c.* 10 year periods; infestations are consequently variable. Resistance breeding is therefore difficult and progress has been made mainly from observations of naturally infested plots of breeders' selections.

The vascular damage initiated by larvae of the second generation of midges is the most serious. These erode the outer surface of the polyderm which is newly exposed as the epidermis and primary cortex splits and begins to peel (Fig. 2a). The salivary glands of larvae secrete cellulases and esterases which remove the suberin and polysaccharides from the walls of phellem cells in the polyderm (Grünwald & Seemüller, 1979). This predisposes vascular tissues to deep penetration by a range of *Fusarium* and *Phoma* spp., which seem incapable of spreading distally or proximally from the areas first damaged by the pest: hence the lobate 'patch' lesions characteristic of this disease are formed (Williamson & Hargreaves, 1979).

R. crataegifolius has high resistance to the midge and there is evidence that this is associated with a characteristic periderm which is inherited by its hybrids with the raspberry (McNicol et al., 1983). *R. crataegifolius* forms an abundance of shallow natural splits but a substantial wound periderm of lignified and suberised phellem cells forms rapidly beneath the splits (Fig. 2b). Since sap of the species attracts female midges as readily as the raspberry when small wounds are made with a needle, the observed resistance seemed likely to be associated with the characteristic shape, size and texture of the splits and their suitability for oviposition. Histological surveys of F_1 hybrids with red raspberry have supported this and showed that they too were strongly resistant to midge infestation, because wound periderm rapidly formed around splits, fused with the underlying polyderm (Fig. 2c) and often completely enveloped developing midge larvae (McNicol et al., 1983). The distinctive striped bark of *R. crataegifolius* and its hybrids serves as a useful marker for this type of wound healing and facilitates selection for midge resistance in the absence of the pest.

Table 1. Comparisons of mean lesion length of *L. coniothyrium* lesions for pubescent (*H*) and glabrous (*h*) segregates of progenies obtained by crossing two second backcross (2nd BC) hybrids with an F_1 and a first backcross (1st BC) hybrid in 1987 and by crossing an F_1 with first and second backcross hybrids in 1988

F_1 and 2nd BC parent	F_1, 1st and 2nd BC parents					
	Vascular lesions (mm)			Log_e mm		
	7756E4 (F_1)	78210F10 (1st BC)	82224E20 (2nd BC)	7756E4	78210F10	82224E20
Pubescent segregates (*H*)						
82213D18 (2nd BC)	115.3	79.2	–	4.68 (3)*	4.29 (12)	–
82213D22 (2nd BC)	135.4	59.2	–	4.82 (5)	3.78 (15)	–
7756E4 (F_1)	–	50.2	39.8	–	3.83 (8)	3.49 (24)
Glabrous segregates (*h*)						
82213D18	92.4	82.6	–	4.43 (21)	4.28 (16)	–
82213D22	–	–	–	–	–	–
7756E4	–	55.6	36.1	–	3.83 (18)	3.49 (17)
			1987 (837 D.F.)		1988 (301 D.F.)	
S.E.D. between Family/*H* group high numbers			0.282		0.725	
High-low numbers			0.201		0.185	
Low numbers			0.282		0.251	

* Figures in parentheses are the numbers of segregates tested.

Cane spot

Cane spot, or anthracnose, caused by *Elsinoe veneta* (anamorph *Sphaceloma necator*) attacks leaves, fruits and young canes, but it is best known for the deeply-penetrating sunken lesions that it produces on fruiting canes. These damage the vascular tissues and cause yield loss in susceptible cultivars (Harris, 1931). Attacked fruits are misshapen and unsaleable (Harris, 1933; Williamson et al., 1989). Although we include it in this section, it is important to recognise that *E. veneta* also infects foliage and that early and severe infections of emerging fruiting laterals (Travis & Rytter, 1991) can seriously reduce yield in the year of attack.

Like other fungi that cause spot anthracnoses, *E. veneta* is intractable for use in resistance screening because its growth rate is exceptionally slow *in vitro* and it produces conidia unpredictably. Until recently resistance breeding was possible only by observing natural infection of selections when the incidence of the disease was high (Jennings & McGregor, 1988), or by tying pieces of naturally infected canes to test material when it was not (Keep et al., 1977). A method for producing inoculum in greater quantity has recently been devised and screening cultivars for resistance is now possible on a limited scale (Williamson et al., 1989). These tests in glasshouse and field showed that a black raspberry (*R. occidentalis* L.) selection was very susceptible, and confirmed a report (Keep et al., 1977) that a red raspberry derivative of *R. coreanus* (EMRS 2769/9) was resistant.

E. veneta can infect only very young tissues, and has an incubation period under Scottish conditions of 3–4 wk before symptoms develop (Munro et al., 1988; Williamson et al., 1989). It effectively destroys vascular tissues, probably because it penetrates the phloem and xylem near the top of the cane before a functional polyderm has formed (Williamson & McNicol, 1989). Infections caused marked dedifferentiation of parenchyma cells in the primary cortex, phloem and xylem and induce extensive suberisation and lignification in wound periderms. Nevertheless, these host responses did not prevent substantial penetration of vascular tissues in any of the germplasm examined. It seemed that resistance to *E. veneta* was associated with an epidermal characteristic, rather than with the production of periderms, because hyperplasia and suberisation of phloem parenchyma occurred in the most resistant selection which had only a few small lesions.

Gene *H* is associated with such high susceptibility to cane spot in the progenies of European cultivars that it was once thought that cultivars carrying the gene would inevitably be susceptible. However, several North American cultivars are good sources of resistance and Willamette, Nootka, Chilcotin and Meeker are resistant in spite of the disadvantage of carrying the gene. When these cultivars, together with Heritage which is highly resistant, were crossed with a susceptible parent, the progenies obtained contained resistant and susceptible segregates in approximately equal numbers and it seemed that the resistance was controlled by the segregation of a major gene (Table 2). In this work (Jennings & McGregor, 1988) the 0 to 5 scale used to record disease incidence was found to approximate to a logarithmic scale for the range of 0 to 93 disease spots per cane, and although an adjustment was made for the presence or absence of gene *H*, the results provided no evidence for discontinuity in the expression of resistance and so the evidence for the segregation of a major gene was equivocal.

Table 2. Segregation of resistance to *E. veneta* in field progenies exposed to infection in Australia

Type of cross	Number of plants	
	Resistant*	Susceptible
Resistant × susceptible (6 families)	420	403
Resistant × resistant (9 families)	602	180

* Plants were scored 0 to 5 for the frequency of cane spots. To adjust for the effect of gene *H*, which reduces resistance (see Table 5), glabrous plants with scores 0, 1 & 2 and pubescent plants with scores of 0, 1, 2 & 3 were classified as resistant and the remainder as susceptible.

Diseases causing mainly bud suppression

Cane botrytis and spur blight

Both cane botrytis, caused by *Botrytis cinerea* (teleomorph *Botryotinia fuckeliana*), and spur blight, caused by *Didymella applanata* (anamorph is an unnamed *Phoma* sp.), affect mature or senescent leaves on primocanes and they will therefore be discussed together.

Both fungi invade via leaves and spread through petioles to the axillary region, but a primary protective layer of suberised and lignified cells in the adaxial cortex prevents their spread to the subtended adjacent axillary buds (Williamson, 1984). No such layer develops on the abaxial cortex, where hyphae can spread unimpeded through the cane cortex, outside the polyderm. *B. cinerea* grows the more rapidly and may spread to affect several nodes from a single infection. Neither pathogen can penetrate the mature polyderm, and even when inoculated into wounds they rarely spread in vascular tissues.

These unrelated fungi therefore occupy the same ecological niche on raspberry canes, and if one of them dominates in a plantation, it tends to occupy most of the infection sites so that it is difficult to assess the relative resistance of the host to the other (Williamson & Jennings, 1986). Infection of the tissues around the axillary buds by *D. applanata* causes a delay in the development of the buds themselves which are consequently smaller at the end of the season than those at uninfected nodes (Williamson & Dale, 1983; Pepin et al., 1985). The same is true for *B. cinerea,* but bud dwarfing is more severe, even when caused by infections as late as September (Williamson & Jennings, 1986). In the following spring the relatively small buds at infected nodes tend to be suppressed by healthy fast-growing buds and lateral shoots above them (Williamson & Hargreaves, 1981a). This is the most important cause of yield loss, though in the case of *B. cinerea* infections, the sporulating cane lesions later provide a major source of inoculum for fruit infection and indirectly contribute to yield losses caused by fruit grey mould.

Inoculation of leaves with conidial suspensions in the field proved unreliable and a simple wound-inoculation method (Jennings & Williamson, 1982), as described for cane blight (page 2), provided the most dependable method for large-scale screening of germplasm in the field. This method assesses tissue resistance and takes no account of any other form of resistance that may act at the penetration phase.

Symptom expression is markedly affected by the environment and the physiological age of canes. Glasshouse inoculations have resulted in serious bud failure in the complete absence of the characteristic epidermal silvering and fruiting-body development that normally occurs in the field on susceptible cultivars (Williamson & Pepin, 1987). In the field, mycelial inoculations with *B. cinerea* made in mid-August produced lesions which spread more rapidly than those initiated in June or July (Jennings & Williamson, 1982), indicating that resistance of the cortex declines with age, probably due to senescence of this tissue as the polyderm suberises. The brown lesions visible on canes in autumn are longer, and provided that autumn pigments are not prominent, they are easier to measure than the area of epidermis that becomes silvered by the following spring. Consequently, better discrimination between genotypes has been achieved by measuring the lesions in the autumn than in the spring.

Very strong resistance to *B. cinerea* was identified by inoculation in *R. pileatus, R. occidentalis* and in hybrids of these species with the red raspberry and similar hybrids related to *R. crataegifolius* (Jennings & Williamson, 1982; Jennings & Brydon, 1989a). Similarly it was shown that *R. pileatus, R. occidentalis* and *R. coreanus* and hybrids of these three species with red raspberry are highly resistant to *D. applanata* (Jennings, 1982). In subsequent experiments in which three isolates of each pathogen were inoculated to both canes and petioles of seven red raspberry cultivars, it was shown by principal components analysis that common resistance to the two pathogens accounted for most of the variation and there was no evidence for the occurrence of resistance that operated against one pathogen alone (Williamson & Jennings, 1986). This is supported by earlier work which led to similar conclusions from a large-scale study of the segregation of the two resistances in progenies (Jennings, 1983). It was therefore possible to inoculate with *B. cinerea* alone and gain insight to the probable performance of the material if exposed to *D. applanata.*

Inoculation of progenies with *B. cinerea* showed that the high level of resistance identified in *R. pileatus* had been transferred to red raspberry through three generations of backcrossing with little or no diminution in its effectiveness (Table 3). A similar result has been reported for resistance derived from the North American cultivar Chief. It was thought that a major gene for resistance had been transferred from *R. pileatus,* but evidence is lacking because discontinuity in the resistance levels observed was detected in only one group of progenies. However, from a plant breeder's viewpoint the prospects for breeding resistant cultivars remains high regardless of whether the resistance is controlled by a major gene, by a complex of dominant minor genes or by a major gene in combination with minor genes. Gene *H* had a large influence on resistance and assessments of other resistance sources had to be made separately for the hairy and non-hairy segregates in families when the gene was segregating (Jennings & Brydon, 1989a).

Some cultivars, notably Glen Clova, show large silvered lesions with pronounced sporulation when infected by *D. applanata* and yet they rarely suffer bud failure; this can be described as tolerance and was revealed clearly by principal components analysis in another experiment (Pepin et al., 1985).

Diseases that cause mainly leaf infection

Premature and severe defoliation, whether caused by fungal disease, insects or abiotic factors deprives the canes of essential food reserves and causes loss of vigour and marked loss of winter hardiness. Of the diseases which cause serious leaf infection in Rubus, yellow rust and powdery mildew are the only ones to receive attention in resistance breeding, but recent attacks by a downy mildew on raspberries, blackberries and related hybrid fruits in the UK and North America (McKeown, 1988; Wallis et al., 1989; Breese et al., 1991) have raised awareness of the potential of this disease which was hitherto thought to be commercially significant only in New Zealand and South Africa.

Raspberry yellow rust

This macrocyclic autoecious rust (*Phragmidium rubi-idaei*) attacks only red raspberries and came to prominence in the last 20 years on cvs Glen Clova and Malling Delight which are highly susceptible (Anthony et al., 1983, 1985a, b). Several kinds of resistance have been identified from experiments in which use was either made of 'bait' plants exposed in uniformly infected commercial plantations, or of controlled inoculation procedures with bulked mass isolates of urediniospores in glasshouses.

Variation in the pathogenicity of isolates was detected in populations of *P. rubi-idaei* in the UK (Anthony et al., 1985b), but no race structure was found. Complete resistance to *P. rubi-idaei* was

Table 3. Mean lengths of autumn lesions of *B. cinerea* on some resistant genotypes used as parents in successive generations

Parent*	Origin	Mean lesion length	mm $\log_e$
*R. pileatus***	–	19.3	–
77756E4 (*h*)	Glen Prosen × *R. pileatus* = F_1	42.2	3.63
78210F10 (*H*)	Glen Moy × F_1 = 1st backcross (BC)	46.1	3.80
82224D4 (*H*)	1st BC × 789C1 = 2nd BC	32.6	3.47
Susceptible selection			
82224E20 (*H*)	1st BC × 789C1 = 2nd BC	122.2	4.67
	1988 S.E.D. (D.F. 262)		0.199

* *H* and *h* respectively denote that the parent was hairy or glabrous.

** Lesion length for *R. pileatus* from Jennings & Williamson (1982), and for the four selections from Jennings & Brydon (1989b).

shown by inoculation to be inherited undiminished through four generations from cv. Latham to its derivatives, Chief (self of Latham), Boyne (Chief × Indian Summer) and selection K70/23 (Boyne × Fairview), and it was shown that this form of resistance was determined by a major gene, designated *Yr* (Anthony et al., 1986).

A half-diallel cross involving families from the five clonal parent cvs Boyne, Meeker, Glen Prosen, Malling Jewel and Glen Clova was studied at the telial (exposed bait plants) and uredinial (inoculated) stages. Continuous variation was shown by a second form of resistance which was assessed by the telial index, latent period and uredinia production/cm^2 (Table 4). The results indicated a polygenic system of inheritance that was determined predominantly by additive gene action, though with small but significant non-additive effects. This form of resistance is referred to as the 'slow-rusting' type. Cv. Meeker was a particularly potent parent for this form of rust resistance. It had the highest general combining ability value for each component of the resistance, especially for the latent period component, and considerably reduced both the telial index of its progenies and the number of uredinia/cm^2 produced 18 days after inoculation (Table 4) (Anthony et al., 1986). In cv. Malling Jewel a third form of resistance was identified which operated against the fungus only after the aecial stage (Anthony et al., 1985b). This illustrates the importance of assessing disease throughout the growing season.

Powdery mildew

Raspberry genotypes which are highly susceptible to powdery mildew (*Sphaerotheca macularis*) may suffer severe stunting and yield loss (Ellis, 1991). Infected leaves develop light green blotches on the adaxial surface with white mycelial growth on the abaxial surface under them. Infected young shoot tips may become long and spindly with small leaves, often with upturned leaf margins, while infected fruits are unsaleable.

The inheritance of resistance to powdery mildew was reviewed by Keep (1968), who studied the disease in naturally infected plants, and in plants dusted with conidia several weeks before records were taken. She considered that cvs Lloyd George and Burnetholm were heterozygous for three genes that conferred resistance, postulating that Sp_1 and Sp_2 were dominant complementaries and Sp_3 recessive. *R. coreanus* is a source of resistance to powdery mildew and strong resistance was found in F_1 and first backcross (BC_1) derivatives of an accession of this species at East Malling (Keep et al., 1977).

The significance of resistances and susceptibilities associated with gene H

Gene *H* was associated with such large increases in resistance to *B. cinerea, D. applanata* and *L. coniothyrium* and such large decreases in resistance to *E. veneta, P. rubi-idaei* and *S. macularis* that for each disease resistance had to be separately assessed for hairy and non-hairy segregates whenever the gene was segregating (Table 5). We do not know why the gene has such large effects. It is possible that it is closely linked with major genes or with minor-gene complexes that independently contribute to the resistance or susceptibility of the six diseases affected, or possibly with genes that

Table 4. Main effects, which denote general combining ability, of five parents included in a diallel cross for the 'slow-rusting' form of resistance to *P. rubi-idaei**

Parent	Telial index (0–6)	Latent period (days)	Uredinia/ cm^2 at 13 days	Uredinia/ cm^2 at 18 days
Boyne	1.2	12.4	26.1	49.1
Meeker	1.0	13.0	10.7	25.9
Glen Prosen	1.1	12.2	29.4	56.7
Malling Jewel	1.4	12.2	31.1	65.5
Glen Clova	2.2	12.1	41.2	70.0
S.E.D. (D.F. 41)				
Telial index		0.32		
Latent period		0.25		
Uredinia (13 days)		7.53		
Uredinia (18 days)		11.61		

* The parents were included in a diallel cross and the progenies assessed for telial index in the field and for latent period and development of uredinia in the glasshouse.

control one or more 'common resistances' or 'common susceptibilities' to some or all of them.

An alternative explanation is that the gene itself is responsible through pleiotropic effects on each of the resistances – probably yet another example of genetic resistance being common to several pathogens. The gene is known to have other pleiotropic effects besides its main effect on cane pubescence: it is associated with a small increase in spine frequency and a small decrease in spine size (Jennings, 1962; Keep et al., 1977). Hairs and spines are both outgrowths of epidermal cells and their early development is inter-related (Peitersen, 1921). It seems that gene *H* acts early in development and affects several cell characteristics.

Resistance to *B. cinerea* and *D. applanata* is highest in immature tissues and we may therefore postulate that the gene increases resistance by delaying cell maturation; such a delay would be expected to reduce resistance to *E. veneta, P. rubi-idaei* and *S. macularis* because these fungi invade only immature tissues. Resistance to *L. coniothyrium* is also higher in mature tissues and it is surprising that gene *H* increases resistance to it in the same way as it increases resistance to *B. cinerea* and *D. applanata*. However, the situation is poorly understood and it may be significant that *L. coniothyrium* attacks internal vascular tissues whereas the other two fungi attack only the outer stem tissues.

Although the gene's contrasting effects on differ-

Table 5. Average effects of gene *H* on the intensity of diseases caused by five pathogens in segregating families studied from 1979 to 1988

Pathogen	Inoculation year	Plant phenotype*		Reference⁺
		H	*h*	
Botrytis cinerea				
Lesion length (mm) in cortex; autumn	1979	97.5	149.5	a
Lesion length (mm) in cortex; autumn	1987	76.2	115.3	e
Lesion length (mm) in cortex; autumn	1988	49.4	73.8	e
Lesion length (mm) in cortex; spring	1979	50.8	83.5	a
Score for sclerotia; spring (0–5)	1979	1.4	2.5	a
Didymella applanata				
Lesion length (mm) in cortex; autumn	1979	41.5	48.7	a
Lesion length (mm) in cortex; spring	1979	42.7	68.3	a
Score for sclerotia; spring (0–5)	1979	1.7	2.7	a
Leptosphaeria coniothyrium				
Lesion length (mm) in vascular tissue; autumn	1980	60.4	65.2	a
Lesion length (mm) in vascular tissue; autumn	1981	87.7	95.6	a
Lesion length (mm) in vascular tissue; autumn	1985	62.3	65.4	d
Lesion length (mm) in vascular tissue; autumn	1987	60.3	82.0	d
Lesion length (mm) in vascular tissue; autumn	1988	42.4	46.1	d
Phragmidium rubi-idaei				
Latent period (days); Scotland	1983	12.7	13.5	b
Uredinia/cm^2; Scotland	1983	22.9	0.7	b
Telial index (0–6); Scotland	1983	1.3	0.7	b
Telia/cm^2; Australia	1986	17.1	4.1	c
Elsinoe veneta				
Lesion frequency score (0–5); Australia	1985	3.05	2.42	c

* Plant phenotype: *H* and *h* denote hairy and glabrous respectively.

⁺ References: a = Jennings, 1982; b = Anthony et al., 1986; c = Jennings & McGregor, 1988; d = Jennings & Brydon, 1989b; e = Jennings & Brydon, 1989a.

ent resistances may present a dilemma to a plant breeder faced with breeding for many resistances, it is reassuring that for each disease a form of resistance has been identified that is strong enough to overcome the disadvantageous effects associated with *H* or *h* phenotypes.

Acknowledgement

We acknowledge financial support of the Scottish Office Agriculture and Fisheries Department.

References

Anthony, V.M., R.C. Shattock & B. Williamson, 1983. Resistance of raspberry cultivars to yellow rust (*Phragmidium rubi-idaei*). Tests of Agrochemicals and Cultivars No. 4 (Ann. Appl. Biol. 102 Supplement) pp 136–137.

Anthony, V.M., R.C. Shattock & B. Williamson, 1985a. Life-history of *Phragmidium rubi-idaei* on red raspberry in the United Kingdom. Plant Path. 34: 510–520.

Anthony, V.M., R.C. Shattock & B. Williamson, 1985b. Interaction of red raspberry cultivars with isolates of *Phragmidium rubi-idaei.* Plant Path. 34: 521–527.

Anthony, V.M., B. Williamson, D.L. Jennings & R.C. Shattock, 1986. Inheritance of resistance to yellow rust (*Phragmidium rubi-idaei*) in red raspberry. Ann. appl. Biol. 109: 365–374.

Breese, W.A., R.C. Shattock & B. Williamson, 1991. Cross inoculation of downy mildew isolates from *Rubus* and *Rosa.* Phytophthora Newsletter No. 17, pp 44–45.

Ellis, M.A., 1991. Powdery mildew. In: M.A. Ellis, R.H. Converse, R.N. Williams & B. Williamson (Eds.). Compendium of raspberry and blackberry diseases and insects, pp 16–18. APS Press, St Paul, Minnesota.

Ellis, M.A., R.N. Williams, R.H. Converse & B. Williamson, 1991. Compendium of raspberry and blackberry diseases and insects. APS Press, St Paul, Minnesota.

Grünwald, J. & E. Seemüller, 1979. Zerstörung der Resistenzeigenschaften des Himbeerrutenperiderms als Folge des Abbaus von Suberin und Zellwandpolysacchariden durch die Himbeerrutengallmücke *Thomasiniana theobaldi.* Z. Pflanzenkrkh. Pflanzensch. 86: 305–314.

Hargreaves, A.J. & B. Williamson, 1978. Effect of machine-harvester wounds and *Leptosphaeria coniothyrium* on yield of red raspberry. Ann. app. Biol. 89: 37–40.

Harris, R.V., 1931. Raspberry cane spot: its diagnosis and control. J. Pomol. Hortic. Sci. 9: 73–99.

Harris, R.V., 1933. The infection of raspberry fruits by the cane spot fungus. Rep. East Malling Res. Stn. for 1932, pp 86–89.

Jennings, D.L., 1962. Some evidence on the influence of the morphology of raspberry canes upon their liability to be attacked by certain fungi. Hort. Res. 1: 100–111.

Jennings, D.L., 1967. Balanced lethals and polymorphism in *Rubus idaeus.* Heredity 22: 465–479.

Jennings, D.L., 1979. Resistance to *Leptosphaeria coniothyrium* in the red raspberry and some related species. Ann. appl. Biol. 93: 319–326.

Jennings, D.L., 1982. Resistance to *Didymella applanata* in red raspberry and some related species. Ann. appl. Biol. 101: 331–337.

Jennings, D.L., 1983. Inheritance of resistance to *Botrytis cinerea* and *Didymella applanata* in canes of *Rubus idaeus* and relationship between these resistances. Euphytica 32: 895–901.

Jennings, D.L., 1988. Raspberries and blackberries: their breeding, diseases and growth. Academic Press, London.

Jennings, D.L. & E. Brydon, 1989a. Further studies on breeding for resistance to *Botrytis cinerea* in red raspberry canes. Ann. appl. Biol. 115: 507–513.

Jennings, D.L. & E. Brydon, 1989b. Further studies on resistance to *Leptosphaeria coniothyrium* in the red raspberry and related species. Ann. appl. Biol. 115: 499–506.

Jennings, D.L., H.A. Daubeny & J.N. Moore, 1990. Blackberries and raspberries (*Rubus*). In: J.N. Moore & J.R. Ballington (Eds.), Genetic resources of temperate fruit and nut crops, pp 331–389, ISHS, Wageningen.

Jennings, D.L. & G.R. McGregor, 1988. Resistance to cane spot (*Elsinoe veneta*) in the red raspberry and its relationship to resistance to yellow rust (*Phragmidium rubi-idaei*). Euphytica 37: 173–180.

Jennings, D.L. & B. Williamson, 1982. Resistance to *Botrytis cinerea* in canes of *Rubus idaeus* and some related species. Ann. appl. Biol. 100: 375–381.

Keep, E., 1968. Inheritance of resistance to powdery mildew, *Sphaerotheca macularis* (Fr.) Jaczewski in the red raspberry, *Rubus idaeus* L. Euphytica 17: 417–438.

Keep, E., 1976. Progress in *Rubus* breeding at East Malling. Acta Hort. 60: 123–128.

Keep, E., V.H. Knight & J.H. Parker, 1977. *Rubus coreanus* as donor of resistance to cane diseases and mildew in red raspberry breeding. Euphytica 26: 505–510.

Knight, R.L. & E. Keep, 1958. Developments in soft fruit breeding at East Malling. Rep. East Malling Res. Stn. for 1957: 62–67.

McKeown, B., 1988. Downy mildew of boysenberry and tummelberry in the UK. Plant Path. 37: 281–284.

McNicol, R.J., B. Williamson, D.L. Jennings & J.A.T. Woodford, 1983. Resistance to raspberry cane midge (*Resseliella theobaldi*) and its association with wound periderm in *Rubus crataegifolius* and its red raspberry derivatives. Ann. appl. Biol. 103: 489–495.

Munro, J.M., A. Dolan & B. Williamson, 1988. Cane spot (*Elsinoe veneta*) in red raspberry: infection periods and fungicidal control. Plant Path. 37: 390–396.

Nijveldt, W., 1963. I. Biologie, fenologie en bestrijding van de frambozeschorsgalmug (*Thomasiniana theobaldi*) in verband met het optreden van stengelziekte op framboos in Nederland. Neth. J. Plant Path. 69: 222–234.

Peitersen, A.K., 1921. Blackberries of New England: genetic status of the plants. Bull. Vermont Agric. Exp. Stn. No. 218, 34 pp.

Pepin, H.S., B. Williamson & P.B. Topham, 1985. The influence of cultivar and isolate on the susceptibility of red raspberry canes to *Didymella applanata*. Ann. appl. Biol. 106: 335–347.

Pitcher, R.S., 1952. Observations on the raspberry cane midge (*Thomasiniana theobaldi* Barnes). I. Biology. J. Hort. Sci. 27: 71–94.

Pitcher, R.S. & P.C.R. Webb, 1952. Observations on the raspberry cane midge (*Thomasiniana theobaldi* Barnes). II. 'Midge blight', a fungal invasion of the raspberry cane following injury by *T. theobaldi*. J. Hort. Sci. 27: 95–100.

Ramsay, A.M., M.R. Cormack, D.T. Mason & B. Williamson, 1985. Problems of harvesting raspberries by machine in Scotland – a review of progress. The Agricultural Engineer 40: 2–9.

Seemüller, E., S. Kartte & M. Erdel, 1988. Penetration of the periderm of red raspberry canes by *Leptosphaeria coniothyrium*. J. Phytopathol. 123: 362–369.

Travis, J.W. & J. Rytter, 1991. Anthracnose. In: M.A. Ellis, R.H. Converse, R.N. Williams & B. Williamson (Eds.). Compendium of raspberry and blackberry diseases and insects, pp 3–5. APS Press, St Paul, Minnesota.

Wallis, W.A., R.C. Shattock & B. Williamson, 1989. Downy mildew (*Peronospora rubi*) on micropropagated *Rubus*. Acta Hort. 262: 227–230.

Williamson, B., 1984. Polyderm, a barrier to infection of red raspberry buds by *Didymella applanata* and *Botrytis cinerea*. Ann. appl. Biol. 53: 83–89.

Williamson, B. & A. Dale, 1983. Effects of spur blight (*Didymella applanata*) and premature defoliation on axillary buds and lateral shoots of red raspberry. Ann. appl. Biol. 103: 401–409.

Williamson, B. & A.J. Hargreaves, 1976. Control of cane blight (*Leptosphaeria coniothyrium*) in red raspberry following mechanical harvesting. Acta Hort. 60: 35–40.

Williamson, B. & A.J. Hargreaves, 1978. Cane blight (*Leptosphaeria coniothyrium*) in mechanically harvested red raspberry (*Rubus idaeus*). Ann. appl. Biol. 88: 37–43.

Williamson, B. & A.J. Hargreaves, 1979. Fungi on red raspberry from lesions associated with feeding wounds of cane midge (*Resseliella theobaldi*). Ann. appl. biol. 91: 303–307.

Williamson, B. & A.J. Hargreaves, 1981a. Effects of *Didymella applanata* and *Botrytis cinerea* on axillary buds, lateral shoots and yield of red raspberry. Ann. appl. Biol. 97: 55–64.

Williamson, B. & A.J. Hargreaves, 1981b. The effect of sprays of thiophanate-methyl on cane diseases and yield in red raspberries, with particular reference to cane blight (*Leptosphaeria coniothyrium*). Ann. appl. Biol. 97: 165–174.

Williamson, B., L. Hof & R.J. McNicol, 1989. A method for *in vitro* production of conidia of *Elsinoe veneta* and the inoculation of raspberry cultivars. Ann. appl. biol. 114: 23–33.

Williamson, B. & D.L. Jennings, 1986. Common resistance in red raspberry to *Botrytis cinerea* and *Didymella applanata,* two pathogens occupying the same ecological niche. Ann. appl. Biol. 109: 581–593.

Williamson, B. & R.J. McNicol, 1989. The histology of lesion development in raspberry canes infected by *Elsinoe veneta*. Ann. appl. Biol. 114: 35–44.

Williamson, B. & H.S. Pepin, 1987. The effect of temperature on the response of canes of red raspberry cv. Malling Jewel to infection by *Didymella applanata*. Ann. appl. Biol. 110: 295–302.

Williamson, B. & A.M. Ramsay, 1981. Prospects for control of cane blight in machine-harvested raspberries. Proc. Crop Protection in Northern Britain pp 281–285.

Williamson, B. & A.M. Ramsay, 1984. Effects of straddle-harvester design on cane blight (*Leptosphaeria coniothyrium*) of red raspberry. Ann. appl. Biol. 105: 177–184.

Euphytica **63**: 71–83, 1992.
R. Johnson and G.J. Jellis (eds), Breeding for Disease Resistance

Leaf glucosinolate profiles and their relationship to pest and disease resistance in oilseed rape

Richard Mithen
Cambridge Laboratory, Institute of Plant Science Research, John Innes Centre, Colney, Norwich, NR4 7UJ, UK

Key words: Brassica napus ssp. *oleifera,* disease resistance, glucosinolates, isothiocyanates, oilseed rape

Summary

Glucosinolates are sulphur-containing glycosides which occur within vegetative and reproductive tissue of oilseed rape. Following tissue damage, glucosinolates undergo hydrolysis catalysed by the enzyme myrosinase to produce a complex array of products which include volatile isothiocyanates and several compounds with goitrogenic activity. Many of these products have been implicated in the interaction between *Brassica* and their pests and pathogens and some may have a role in defence mechanisms. Low glucosinolate (00) oilseed rape cultivars have been shown to possess similar concentrations of leaf glucosinolates as high glucosinolate (0) oilseed rape cultivars. Likewise, despite considerable speculation to the contrary, 00 cultivars have been shown not to be more susceptible to pests and pathogens than 0 cultivars. The potential to enhance pest and disease resistance of oilseed rape by manipulating the leaf glucosinolate profile without reducing seed quality is discussed.

Glucosinolates: structure, biosynthesis and hydrolysis

Glucosinolates are sulphur-containing glycosides that occur within the order Capparales (which includes the Brassicaceae) and a few other unrelated taxa. They comprise a common glycoside moiety and a variable side chain (Fig. 1). In *Brassica,* the glucosinolates can be divided into three major classes. Firstly, those possessing side chains, which may have aliphatic alkenyl or hydroxyalkenyl groups. Secondly those that have side chains which contain an indolyl group and, thirdly, those that possess aralkyl side chains (Table 1).

The glucosinolate side chains are derived from amino acids as the first steps in the biosynthetic pathway. Glucosinolates possessing aliphatic, indolyl and aralkyl side chains are derived from methionine, tryptophan and phenylalanine respectively. Subsequent to the development of the side chain, the glycone moiety is developed through a complex series of several nitrogenous and sulphur containing intermediates. Details of the biosynthetic pathway, particularly for indolyl glucosinolates, remain to be fully elucidated. A hypothetical pathway for aliphatic glucosinolates biosynthesis is shown in Figure 2. Aldoxime, an intermediate in this pathway between the methionine derivative and the thioglucosinolates is also an intermediate in the biosynthesis of cyanogenic glycosides. The final step in the biosynthesis is the addition of the sulphate to the desulphoglucosinolates. Addition of the hydroxy group to alkenyl glucosinolates such as gluconapin and glucobrassicanapin to produce the hydroxyalkenyl analogues progoitrin and napoleiferin, occurs after the formation of the glucosinolates molecules (Underhill et al., 1973).

The hydrolysis of glucosinolates, which occurs

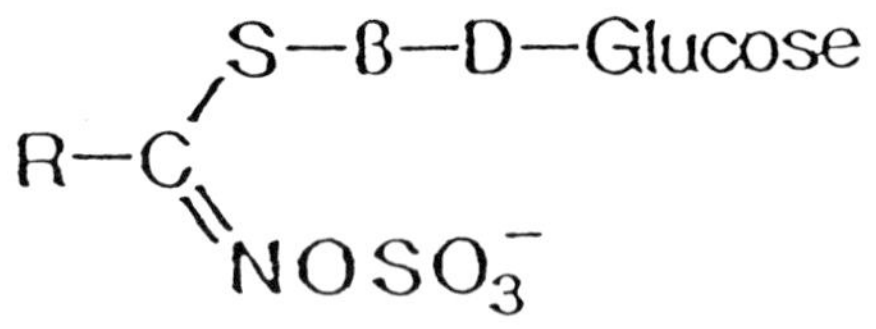

Fig. 1. Molecular structure of glucosinolates. Details of side chain (R) are given in Table 1.

following cellular disruption, is catalysed by the endogenous enzyme myrosinase (see below) and it is the products of hydrolysis which bestow on glucosinolates their biological activity. The enzyme is thought to act as a thioglucosidase to produce an unstable aglucone which then forms several products dependant upon the nature of the glucosinolate side chains and other factors such as pH and the presence of ferrous ions and ascorbic acid (Fenwick et al., 1983).

Alkenyl glucosinolates form stable isothiocyanates (mustard oils) following a Lossen rearrangement of the aglycone (Fig. 3). If the alkenyl side chain contains a hydroxy group the isothiocyanates undergo spontaneous cyclization to form oxazoli-

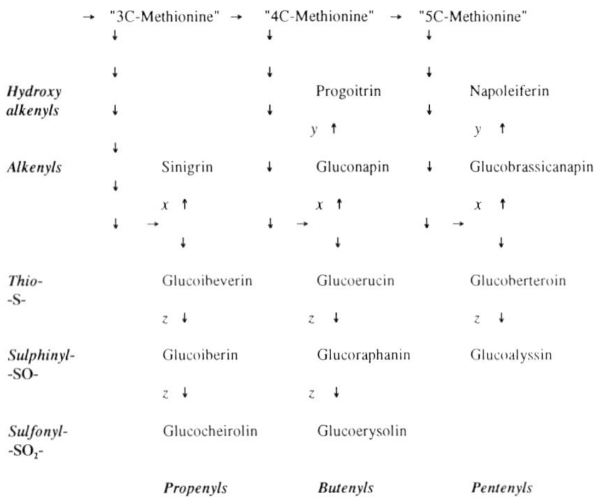

Fig. 2. Hypothetical biosynthetic pathway for alkenyl glucosinolates. x = elimination of terminal H_3CS-group; y = hydroxylation; z = oxidation. (Kraling et al., 1990)

Table 1. Representative glucosinolates from the three major classes. Sinigrin does not occur in oilseed rape

Trivial name	Side chain	Major hydrolysis products
Alkenyls derived from methionine		
Sinigrin	Propenyl	Isothiocyanates
Gluconapin	Butenyl	Isothiocyanates
Glucobrassicanapin	Pentenyl	Isothiocyanates
Progoitrin	Hydroxybutenyl	Oxazolidine-2-thiones
Napoleiferin	Hydroxypentenyl	Oxazolidine-2-thiones
Indolyls derived from tryptophan		
Glucobrassicin	Indolylmethyl	Indolyl-3-carbinol, Thiocyanate, Auxins (?) Phytoalexins (?)
Methoxyglucobrassicin	4-methoxyindolylmethyl	Indolyl-3-carbinol, Thiocyanate, Auxins (?) Phytoalexins (?)
Neoglucobrassicin	1-methoxyindolylmethyl	Indolyl-3-carbinol, Thiocyanate, Auxins (?) Phytoalexins (?)
Hydroxyglucobrassicin	4-hydroxyindolylmethyl	Indolyl-3-carbinol, Thiocyanate, Auxins (?) Phytoalexins (?)
Aralkyls derived from phenylalanine		
Gluconasturtium	Phenylethyl	
Glucotropaeolum	Benzyl	

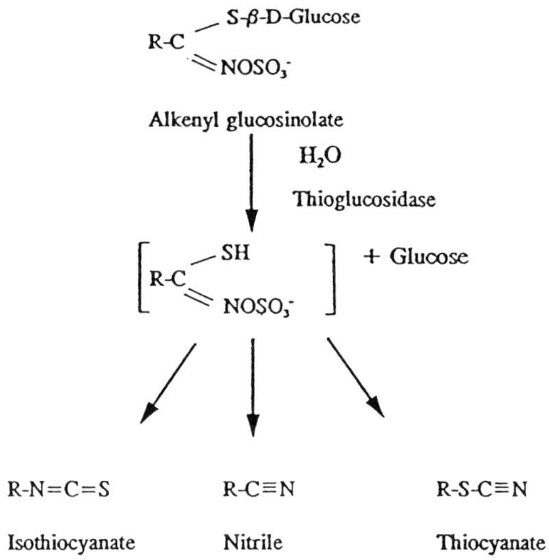

Fig. 3. Hydrolysis of alkenyl glucosinolates.

dine-2-thiones (Fig. 4). Under acidic conditions the predominant hydrolytic products from alkenyl and hydroxyalkenyl glucosinolates are nitriles rather than isothiocyanates. The presence of ferrous ions also promotes the formation of nitriles, either by acting directly by blocking the Lossen rearrangement of the aglycone or by preventing the action of the thioglucosinase by complexing to ascorbic acid which it requires as a cofactor. A further factor influencing hydrolysis is the presence of epithiospecific proteins (ESP). These have been isolated from the seeds of *Crambe abyssinica* and have been shown to promote the transfer of sulphur from the S-Glucose moiety to the alkenyl moiety which results in the formation of epithionitriles (Fenwick et al., 1983).

Glucosinolates possessing indolyl side chains may also form unstable isothiocyanates which degrade to produce the corresponding alcohol and a thiocyanate ion. The alcohol may condense to form diindolylmethane or react with ascorbic acid to form ascorbigen. It is possible that the unstable isothiocyanates may also be precursors to a recently described class of indole phytoalexins that are induced in *Brassica* following biotic or abiotic elicitation (Rouxel et al., 1989). This would involve the addition of methanethiol to the indolyl-3-methyl isothiocyanate (Fig. 5). Currently, this pathway must remain speculative.

Alternatively, indolyl glucosinolates may produce nitrile derivatives which, following nitrilase activity, can produce indole acetic acid. Hence the

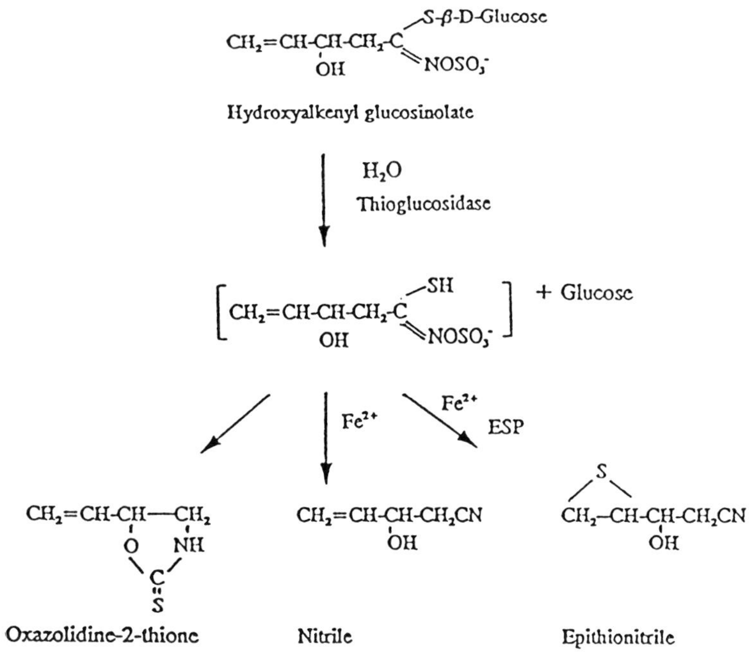

Fig. 4. Hydolysis of hydroxybutenyl glucosinolate (progoitrin).

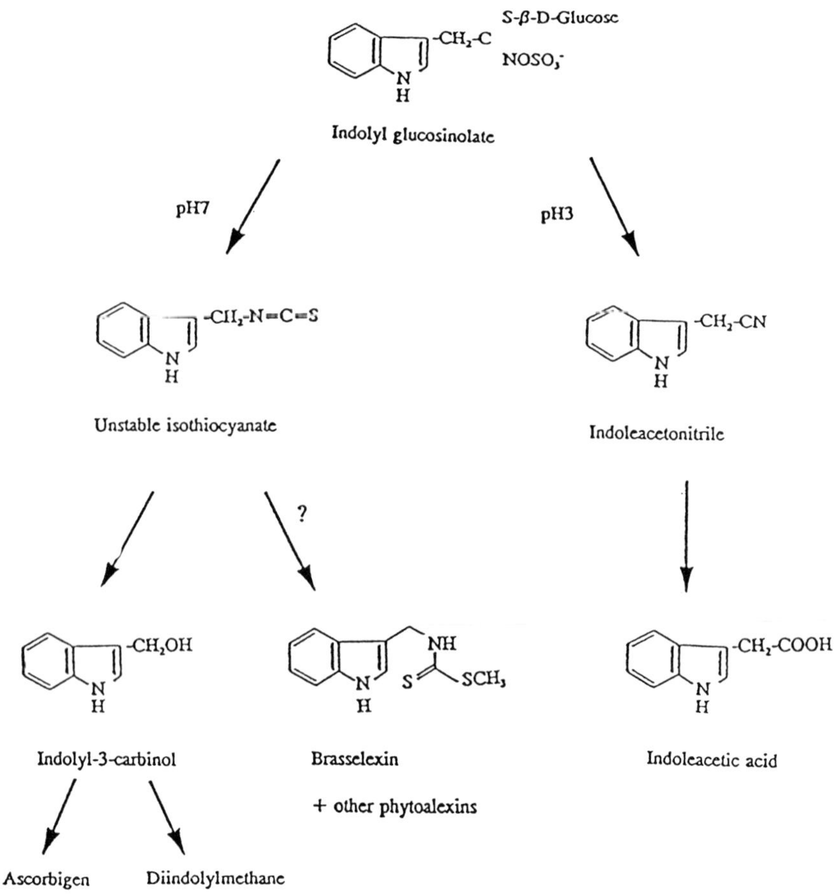

Fig. 5. Hydrolysis of indolyl glucosinolates and the possible production of indoleacetic acid and indole phytoalexins.

indolyl glucosinolates can be an alternative pathway to auxin production in *Brassica.* The induction of auxin via indolyl glucosinolates may be responsible for many of the symptoms produced in *Brassica* following pathogen attack, as described below.

Myrosinase and compartmentalisation of the glucosinolate-myrosinase system

Myrosinase has been shown to occur in multiple forms. All are glycoproteins with two to four subunits with a molecular weight between 125 000 and 153 000. The isoenzymes do not appear to have any specificity towards any particular glucosinolate although they do have different pH optima and show differences in their degree of activation with ascorbic acid.

The compartmentalisation of the glucosinolate-thioglucosinase system has received considerable attention. Upon cellular disruption, the glucosinolates immmediately undergo hydrolysis so it has been thought that these two components are spatially separated, either within different cells or by subcellular compartmentalisation. Historically, myrosinase was thought to be only associated within a group of histochemically distinct cells known as myrosin cells. The presence of myrosinase within these cells has been confirmed with the use of immunochemical techniques (Thangstad et al., 1991) which demonstrated that myrosinase is associated with the tonoplast-like membrane which surrounds the myrosin grains within the myrosin cells. Myrosinase activity has also been demonstrated within tissues that lack myrosin cells (Iverson et al., 1979) and it is likely that myrosinase and myrosin grains occurs within many cells. A subcellular compartmentalisation of glucosinolates and myrosinase has been suggested (Matile, 1980; Luthy & Matile, 1984) in which myrosinase is associated with membrane while glucosinolates and ascorbic acid (which promotes the activity of myrosinase) are located within the vacuole. Further studies with antibodies raised against myrosinase and glucosinolates (Hassan et al., 1988) are required to clarify the distribution of the components of this system.

Glucosinolates in oilseed rape

Glucosinolates have been shown to occur in all tissues of an oilseed rape plant. The major glucosinolates are shown in Table 1. Other glucosinolates such as glucoraphanin and glucoalyssin (Fig. 2) occasionally occur at low concentrations. Within roots and stems indolyl glucosinolates predominate while in leaf tissue alkenyl glucosinolates are the most abundant. There is an increase in the proportion of hydroxyalkenyl glucosinolates, particularly progoitrin, in seeds. The presence of glucosinolates in the meal left after oil extraction from seeds reduces the value of this otherwise high quality animal feed because of several undesirable effects of glucosinolate hydrolysis products. The major problem concerns oxazolidine-2-thiones and thiocyanate ions, derived from progoitrin and indolyl glucosinolates respectively, which have goitrogenic effects when ingested by cattle. Additionally, isothiocyanates from gluconapin and glucobrassicanapin reduce palatability of the meal.

A major component of rape breeding programmes throughout the last two decades has been to reduce the levels of glucosinolates in seeds, in an analogous manner to the reduction in erucic acid. This has largely been successful with older low erucic/high glucosinolate cultivars (0 cultivars) possessing levels of up to 80 μmoles g^{-1} of glucosinolates within seeds while the more recent low erucic/low glucosinolate cultivars (00 cultivars) having 10–20 μmoles g^{-1}. Further decreases in glucosinolate levels are expected. The reduction in glucosinolates has been almost entirely due to a reduction in alkenyl glucosinolates which is under complex polygenic control (Rucker & Rudloff, 1992). The levels of indolyl glucosinolates have remained approximately constant. Recently sources of low indolyl glucosinolates have been found which can be expected to be integrated into breeding programmes (Kraling et al, 1990).

Pest and disease susceptibility of 00 cultivars

The introduction of low glucosinolate cultivars was accompanied by fears that these cultivars would be more susceptible to pest and diseases due to the potential protective effects that these compounds may provide, as described below. While many of the early breeding lines were very susceptible, the present widely grown 00 cultivars appear to be no more susceptible to pests and diseases than previously grown 0 cultivars (Rawlinson et al., 1989; Williams et al., 1992; Inglis et al., 1992). Comparisons between these cultivars have now demonstrated that the reduction in glucosinolates is specific to the seed (Milford et al., 1989; Inglis et al., 1992). Many of the comparisons between 0 and 00 cultivars have thus been undertaken on the false assumption that the glucosinolates levels in the leaves will reflect that in the seed. The only occasion when this does occur is in very young seedlings when there is a good correlation of glucosinolate profiles between vegetative and reproductive tissues (Glen et al., 1990). However, within the 00 breeding lines and cultivars there is significant variation in leaf glucosinolate concentrations (Milford et al., 1989; Schilling & Friedt, 1992) which may partially account for differences in pest and disease susceptibility.

The glucosinolates occurring in the seed are determined by the maternal genotype and not that of the embryo (Kondra & Stefansson, 1970) and it seems likely that glucosinolates are actively transported into the developing embryo from maternal tissue. It can be speculated that in the early low glucosinolate breeding lines the entire glucosinolate pathway may have been disrupted to produce a plant with low levels throughout all tissues and which was susceptible to many pests. In contrast, in later 00 breeding lines and cultivars which retained

Fig. 6. Grazed (arrows) and ungrazed experimental plots of winter oilseed rape. The plots have been grazed by a variety of pests including pigeons, pheasants and rabbits. See Fig. 7 for details of glucosinolate profiles.

their pest resistance, breeders disrupted the transport system into the embryo from maternal tissue and so maintained the glucosinolates in vegetative tissue. Experimental studies to test this hypothesis are required.

The selection of high leaf/low seed glucosinolate lines by plant breeders was done without any evaluation of glucosinolates in vegetative material and arose through selection for other traits such as resistance to pests and disease, winter hardiness and yield parameters. It is thus likely that glucosinolates have an important biological function within vegetative tissues, and the most likely one is their involvement in defence reactions.

Biological activity of glucosinolates and products of hydrolysis

Vertebrates

The detrimental effects of glucosinolates in rape meal (see above) have resulted in many studies on the biological activity of glucosinolate hydrolysis products towards vertebrates. In addition, recent concern about the detrimental effects which oilseed rape may have on wildlife has led to several studies which have sought to understand the relationship between grazing by mammals and birds and leaf glucosinolate content (Askew, 1990). Many of these studies have made the incorrect assumption that the glucosinolates in the leaves of 00 rape cultivars are consistently lower than 0 cultivars.

While it is the goitrogenic activity of the hydrolytic products from progoitrin and indolyl glucosinolates that have the most detrimental effect when large amounts of rape meal is fed to farm animals, it is the effect of the volatile isothiocyanates on palatability of vegetative tissue which is most likely to be of significance in influencing the grazing of oilseed rape crops by pests such as pigeons, rabbits, hares and deer. Subsequent ill effects on wildlife due to the consumption of oilseed rape may result from both the toxic activity of glucosinolates themselves and also from haemolytic anaemia or 'rape sickness' caused by ingestion of S-methyl cysteine sulphoxide (SMCO) which occurs in rape leaves. The

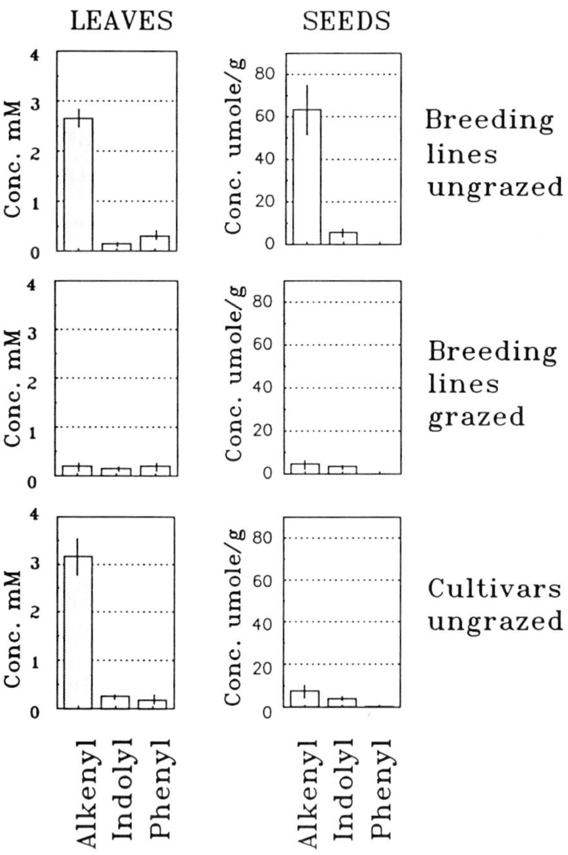

Fig. 7. Leaf glucosinolate profiles (means ± SE) of grazed and ungrazed winter oilseed rape plots.

majority of studies have concluded that there is no difference in the preference of rabbits, hares and pigeons between 0 and 00 cultivars. This is consistent with the lack of variation in leaf glucosinolate profile between these two class of cultivars. Some studies have demonstrated that particular cultivars are preferred by wildlife (Tapper & Cox, 1990). However, in these studies there was no attempt to measure leaf glucosinolate profiles and it may be that significant differences in leaf glucosinolate concentrations would have been revealed (although not correlated with seed profiles).

In contrast to studies with cultivars, plant breeders often observe that some breeding lines are very susceptible to grazing by mammals and birds and are often completely grazed in autumn (Fig. 6). These highly grazed lines, when grown under netting to protect them from grazing have significantly lower levels of alkenyl glucosinolates than ungrazed lines (Fig. 7). This is therefore consistent with the hypothesis that isothiocyanates derived from alkenyl glucosinolates decrease the palatability of tissue to herbivores, in a similar way to decreasing the palatability of meal to farm animals.

Invertebrate

There are many studies on the involvement of glucosinolates and their hydrolysis products in the interaction between insect pests and *Brassica*. Much of this work has recently been reviewed by Chew (1988) and only a summary is presented here.

Insect-plant interactions are complex and involve several distinct phases, the main ones of which are attraction/repulsion of the adults to the host plant, the stimulation of egg laying, and the feeding behaviour of lepidopteran larvae and adult beetles. Many studies have demonstrated that glucosinolates and their hydrolysis products may act as stimuli to attract cruciferous pests and to stimulate egg laying and feeding (Table 2), while generalist insects are often deterred by these compounds. Propenyl isothiocyanate (which is not produced in oilseed rape) has been used for the majority of studies due to its availability. Insects have also been shown to respond to other non-glucosinolate derived volatiles which are also found in the headspace above *Brassica* plants (Evans & Allen-Williams, 1989).

With regard to the potential role of glucosinolates in resistance of oilseed rape to pests, data are required on the threshold levels that will inhibit egg laying and feeding, rather than threshold levels that stimulate behaviour, which has been the subject of the majority of studies. Preliminary studies with plant lines varying in their glucosinolate content suggest that while crucifer specialists may be stimulated to feed by low levels of glucosinolates, at higher concentrations feeding is inhibited (Simmonds & Mithen, unpublished).

Another approach to investigating the interaction between insect pests and glucosinolates has been the application of synthetic isothiocyanate precursors to field and glasshouse grown plants. Application of precursors which released butenyl and phenylethyl isothiocyanate reduced the incidence of *Meligethes* and feeding by *Ceutorhynchus assimilis* (Griffiths et al., 1989; Doughty et al., 1992).

In contrast to studies on insects, there is good evidence to show that high levels of glucosinolates can act as a deterrent to slugs, which can be significant pests on rape during establishment. While 0 and 00 cultivars do not show consistent differences in leaf glucosinolate profiles of adult plants, there is a good correlation between the glucosinolate concentration in seed and that in 1-week-old seedlings. As the palatability of seedlings to slugs is inversely correlated with the concentration of glucosinolates within seedling (Glen et al., 1990), it is likely that the higher levels of isothiocyanates derived from gluconapin and glucobrassicanapin in the seedlings of 0 cultivars deter feeding.

Pathogens

The toxic activity of glucosinolate hydrolysis products towards fungi and bacteria has been frequently reported. These reports include studies of the activities of these compounds towards *Brassica* pathogens. For example, isothiocyanates have been shown to be toxic towards *Peronospora parasitica* (Greenhalgh & Mitchel, 1976), *Mycosphaerella brassicae* (Harthill & Sutton, 1980), *Leptosphaeria maculans* (Mithen et al., 1986) and *Alternaria* spp (Milford et al., 1989). Studies with *L. maculans* suggested that the concentration required for inhibition of fungal growth was similar to that which could be expected to occur in *Brassica* leaves. Toxic activity of the hydrolysis products of indolyl glucosinolates has also been reported (Mithen et al., 1986). In contrast to the toxic activity of isothiocyanates, the hydrolysis products of the hydroxyalkenyl glucosinolate, progoitrin, had no toxic activity towards *L. maculans*.

Although many studies have demonstrated that glucosinolates have toxic effects *in vitro*, few studies have sought to elucidate their role in disease resistance. The high levels of alkenyl glucosinolates in wild forms of *B. oleracea* led both Greenhalgh & Mitchell (1976) and Mithen et al. (1987a) to suggest that these compounds were responsible

Table 2. Examples of involvement of glucosinolates in interactions with insect pests. See Chew (1988) for further details.

Insect	Physiological response	Reference
Response of adults to volatile products		
Meligethes aeneus	Attracted by propenyl isothiocyanates	Free & Williams (1978)
Delia brassicae	Attracted by propenyl isothiocyanates	Finch (1978)
Ceutorhynchus assimilis	Possess olfactory receptors to butenyl and pentenyl isothiocyanate	Blight et al. (1989)
Psylloides chrysocephala	Possess olfactory receptors to butenyl and pentenyl isothiocyanate	
Stimulation of feeding by lepidopteran larvae and adult beetle		
Pieris brassicae	Exogenous applied propenyl isothiocyanates in artificial diets stimulate feeding	Chew (1988)
Phyllotera spp	Feeding stimulated by propenyl, benzyl and indolyl glucosinolates	Hicks (1974)
Psylliodes chryocephala	Host range includes crucifers and the unrelated taxon *Tropaeoleum major,* which contains glucosinolates	Bartlet & Williams (1989)
Ovipositioning responses		
Pieris brassicae	Tarsal chemoreceptors sensitive to glucosinolates	Chew (1988)
Delia brassicae	Tarsal chemoreceptors sensitive to propenyl glucosinolates	Chew (1988)

for the partial resistance of these taxa to *Peronospora parasitica* and *Leptosphaeria maculans* respectively.

Induction of glucosinolates by pests and pathogens

The induction of indolyl and, less frequently, alkenyl glucosinolates, by pests and pathogens has been reported (Table 3). Induction is often associated with abnormal tissue growth which may lead to galls, cankers and twisted leaves which suggests enhanced auxin activity. This may result from the enhanced levels of indolyl glucosinolates which can act as precursors to auxins (Fig. 5). The biological significance of this induction is not understood. It is not associated with high levels of resistance, as infection of resistant lines of *Brassica* with *Plasmodiophora* or *Leptosphaeria* does not result in induction. It may, however, be associated with limiting the fungal growth once infection has occurred, although there is no evidence for this at present. It is possible that the hydrolysis of indolyl glucosinolates does represent an important regulatory event in the interaction between *Brassica* and pests and pathogens. The hydrolysis of these compounds is complex and depends on several factors such as pH (Fig. 5) and may also depend upon the host-pathogen interaction. For example, in susceptible interactions, nitriles and indole acetic acid may be formed leading to the familiar symptoms, while in incompatible interactions, the unstable isothiocyanate may formed and subsequently antifungal compounds such as indole carbinol and its derivatives and, potentially, the indolyl phytoalexins (Fig. 5).

Manipulation of leaf glucosinolate profiles

The evidence provided above suggests that manipulating the leaf glucosinolate profile of oilseed rape may enhance pest and disease resistance. Due to the wide range of biological activity of these compounds the ideal glucosinolate profile may be difficult to predict, and may vary depending upon the most prevalent pests and diseases. Furthermore, within oilseed rape crops there is a considerable interaction between pests and diseases. For example, there is a high correlation between the incidence of stem canker disease and damage by cabbage stem flea beetle and stem weevil (Newman, 1984). Likewise, infestations of *Brassica* pod midge (*Dasineura brassicae*) have been linked to the incidence of seed weevil (*Ceutorhynchus assimilis*), the former utilising the feeding sites of the latter for ovipositioning (Evans & Allen-Williams, 1989). It is also probable that damage to crops by pests such as pigeons and rabbits may provide entry sites for pathogens. The 'optimal' glucosinolate profile with oilseed rape leaves may thus be considerably influenced by these interactions. For example, introducing sinigrin may have detrimental effects by attracting more insects but may have greater beneficial effects by reducing the tissue

Table 3. Brassica pests and pathogens that are known to induce glucosinolates. Other pathogens that cause abnormal growth of host tissue, such as *Peronospora parasitica* and *Albugo candida* are also likely to induce glucosinolates.

Pathogen/pest	Symptom	Reference
Plasmodiophora brassicae	Root galls	Butcher et al. (1974)
Pyrenopeziza brassicae	Leaf curl	Miller, Scholes & Mithen (University of Sheffield, unpublished)
Leptosphaeria maculans	Stem canker	Price, Mithen and Lewis (University of East Anglia, unpublished)
Alternaria brassicae		Doughty et al. (1991)
Delia floralis		Birch et al. (1990)
Brevicoryne brassicae	Leaf curl	Lammerink et al. (1984)
Oryctoloagus cuniculus		Macfarlane et al. (1991)

palatability towards pigeons and hence reducing the incidence of other disease and pest attack. The consequences of these changes are difficult to predict, and ultimately can only be assessed by extensive field trials with oilseed rape lines with contrasting glucosinolate profiles.

However, it is possible to suggest certain changes which may be expected to enhance resistance by reducing the palatability of tissue for pests and by increasing its toxicity towards pathogens. It may be desirable to increase the overall level of leaf glucosinolates without changing the relative proportions of the different glucosinolates. This would have to be achieved without loss of seed quality and thus low seed/high leaf glucosinolate cultivars would need to be developed. There have been few investigations which have sought to study the correlation between leaf and seed glucosinolate concentrations (Schilling & Friedt, 1992; Milford et al., 1989; Fig. 7), and these studies are complicated by environmental factors that may affect both leaf and seed concentrations. Nevertheless, it is apparent that within the range of material currently available to the oilseed rape breeder, leaf and seed glucosinolate concentrations may be under separate genetic control. However, the levels of glucosinolates in currently grown cultivars are clearly insufficient to provide sufficient protection from pests and pathogens and novel breeding material must be sought which has higher concentrations of glucosinolates.

In addition to manipulating total glucosinolate levels, it may be possible to enhance resistance by altering which glucosinolates are present by manipulating the biosynthetic pathway (Fig. 8). For example, it may be desirable to introduce sinigrin which, because it is hydrolysed to produce propenyl isothiocyanate, has the most toxic effects of all glucosinolates (which have been tested) towards pathogens (Mithen et al., 1986) and may be expected to have greatest effect on palatability due to its strong flavour and pungent nature. In contrast to sinigrin, glucobrassicanapin may have the least effect on pest and pathogen interactions. Therefore it may be desirable to eliminate the part of the pathway which leads to pentenyl glucosinolates (Fig. 2) which would result in an equivalent increase in the

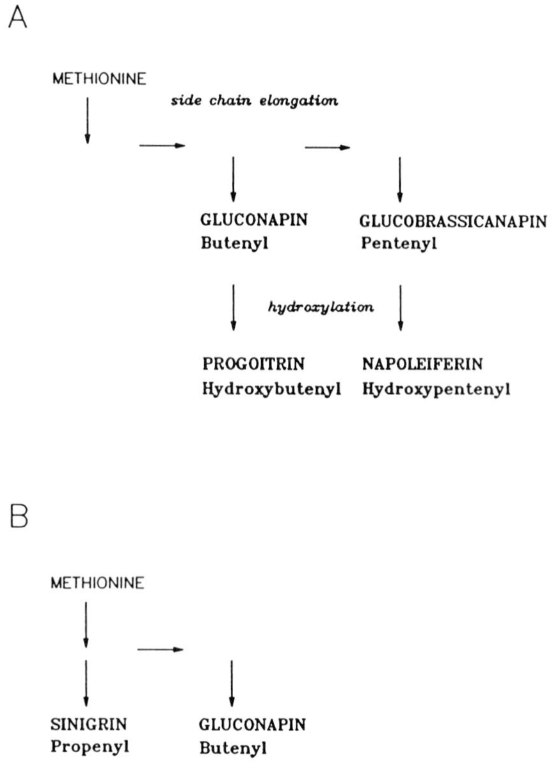

Fig. 8. Existing (A) and desirable (B) alkenyl glucosinolate biosynthetic pathways in oilseed rape leaves.

concentration of butenyl glucosinolates (and propenyl glucosinolates, if present) without changing the overall level of alkenyl glucosinolates. Likewise, it may also be desirable to prevent the hydroxylation of gluconapin and glucobrassicanapin to produce progoitrin and napoleiferin which have less biological activity. This would result in an increase in gluconapin and glucobrassicanapin and hence an increase in the amount of isothiocynates produced following tissue damage without changing total levels of alkenyl glucosinolates. This change would have the added benefit of enhancing seed quality. Finally, it may be possible to introduce other glucosinolates such as methylthioalkyls and methylsulphinylalkyls homologues of the three classes of side chain lengths (Fig. 2). Unfortunately, at present, there is not the genetic variation within *B. napus* to introduce these characters into oilseed rape breeding lines.

In addition to altering glucosinolate profiles, it may be possible to manipulate the glucosinolate-myrosinase system through other approaches. For example, myrosinase is known to exisit in several different forms which may have different levels of activity, either towards specific glucosinolates or to glucosinolate in general. By manipulating the myrosinase isozymes which occur in leaves, it may be possible to increase the levels of toxic products which would be released following tissue damage. It may also be productive to investigate the phenomenon of glucosinolate induction in more detail so that instead of an increase in total levels of alkenyl glucosinolates throughout the plant, there would be localised induction in the part of the plant which is under attack.

Wild *Brassica* species: a source of genetic diversity for oilseed rape breeding programmes

B. napus (2n = 38, genome AACC) is an amphidiploid species derived from the spontaeneous hybridisation of *B. oleracea* (2n = 18, CC) and *B. rapa* (2n = 20, AA) followed by chromosome doubling to restore fertility. It is likely that this hybridisation event occurred very rarely and the entire genetic variation within *B. napus* is derived from a few diploid genotypes. This may partly explain the restricted glucosinolate profiles observed in *B. napus* when compared to the diverse profiles observed in *B. oleracea* and *B. rapa* (Mithen et al., 1987b). Wild accessions of the diploid species have very high concentrations of leaf glucosinolates, presumably as a defence against pests. Synthetic forms of *B. napus* can now be made routinely by artificial cross pollination of *B. rapa* and *B. oleracea* followed by *in vitro* ovary culture and embryo rescue techniques (Chen & Heneen, 1989; Mithen & Herron, 1992). It has been shown that glucosinolate profiles of the synthetic *B. napus* results from an interaction of the A and C genomes (Gland, 1982; Mithen & Magrath, 1992). By selecting specific accessions, it is possible to develop synthetic *B. napus* lines with all the changes described above (e.g. an increase in alkenyl glucosinolates, an increase in the proportion of sinigrin, and a reduction in pentenyl and hydroxalkenyl glucosinolates). These synthetic lines are interfertile with oilseed rape cultivars and thus can be used within backcrossing programmes to introduce these desired changes into oilseed rape breeding lines. The efficiency of these backcrossing programmes can be expected to increase significantly with the development of molecular markers within *Brassica* (King, 1990).

Acknowledgement

The author thanks Dr J Bowman for Fig. 6 and provision of oilseed rape breed lines.

References

Askew, M.F., 1990. Rapeseed 00 and Intoxication of Wild Animals. Commission of the European Community.

Bartlet, E. & I.H. Williams, 1989. Host plant selection by the cabbage stem flea beetle (*Psylliodes chrysocephala*). Aspects of Applied Biology 23. Production and Protection of Oilseed rape and Other Brassica Crops, pp. 335–338.

Birch, A.N.E., D.W. Griffiths & W.H. Macfarlane Smith, 1990. Changes in forage and oilseed rape (*Brassica napus*) root glucosinolates in response to attack by turnip root fly (*Delia floralis*). J. Sci. Food Agric. 51: 309–320.

Blight, M.M., J.A. Pickett, L.J. Wadhams & C.M. Woodcock, 1989. Antennal responses of *Ceutorhynchus assimilis* and *Psylloides chrysocephala*. Aspects of Applied Biology 23. Production and Protection of Oilseed rape and Other Brassica Crops, pp. 329–334.

Butcher, D., S. El-Tigani & D. Ingram, 1974. The role of indole glucosinolates in the clubroot disease of the Cruciferae. Physiol. Plant. Path. 4: 127–140.

Chen, B.-Y. & W.K. Heneen, 1989. Resynthesised *Brassica napus:* A review of its potential in breeding and genetic analysis. Hereditas 111: 255–263.

Chew, F.S., 1988. Biological effects of glucosinolates. In: H.G. Cutler (Ed.), Biologically active natural products – potential use in Agriculture. ACS symposium, pp. 155–181.

Doughty, K.L., A.J.L. Porter, A.M. Morton, G. Kiddle, C.H. Bock & R. Wallsgrove, 1991. Variation in glucosinolate content of oilseed rape (*Brassica napus*) leaves. II. Response to infection by *Alternaria brassicae*. Ann. Appl. Biol. 118: 469–477.

Doughty, K.J., A.J. Hick, B.J. Pye & L.E. Smart, 1992. Effect of field application of isothiocyanate precursors on pests and diseases of oilseed rape. Proceedings of the Eighth International Rapeseed Congress. Saskatoon, Canada, pp. 489–494.

Evans, K.A. & L.J. Allen-Williams, 1989. The response of the cabbage stem weevil (*Ceutorhynchus assimilis*) and the brassica pod midge (*Dasineura brassicae*) to flower colour and volatile of oilseed rape. Aspects of Applied Biology 23. Production and Protection of Oilseed Rape and Other Brassica Crops, pp. 347–353.

Fenwick, G.R., R.K. Heaney & W.J. Mullin, 1983. Glucosinolates and their breakdown products in food and food plants. Crit. Rev. in Food Sci. Nutrit. 18: 123–301.

Finch, S. 1978. Volatile plant chemicals and their effect on host finding by the cabbage root fly, (*Delia brassicae*). Entomol. Exp. Appl. 24: 150–167.

Free, J.B. & I.H. Williams, 1978. The response of the pollen beetle, *Meligethes aeneus,* and the seed weevil, *Ceuthorhynchus assimilis,* to oilseed rape, *Brassica napus,* and other plants. J. Appl. Ecol. 15: 761–744.

Gland, A., 1982. Gehalt und muster der glucosinolate in Samen von resynthetisierten Rapsformen. Z. Pflanzenzüchtg 88: 242–254.

Glen, D.M., H. Jones & J.K. Fieldsend, 1990. Damage to oilseed rape seedlings by the field slug *Deroceras reticulatum* in relation to glucosinolate concentration. Ann. Appl. Biol. 117: 197–207.

Greenhalgh, J.G. & N.D. Mitchell, 1976. The involvement of flavour volatiles in the resistance to downy mildew of wild and cultivated forms of *Brassica oleracea.* New Phytol. 77: 391–398.

Griffiths, D.C., A.J. Hick, B.J. Pye & L.E. Smart, 1989. The effects on insect pests of applying isothiocyanate precursors to oilseed rape. Aspects of Applied Biology 23. Production and Protection of Oilseed Rape and Other Brassica crops, pp. 359–364.

Harthill, W.F.T. & P.G. Sutton, 1980. Inhibition of germination of *Mycosphaerella brassicola* ascospores on young cabbage and cauliflower leaves. Ann. Appl. Biol. 96: 153–161.

Hassan, F., N.E. Rothnie, S.P. Yeung & M.V. Palmer, 1988. Enzyme-linked immunosorbent assays for alkenyl glucosinolates. J. Agric. Food. Chem. 36: 398–403.

Hicks, K.L., 1974. Mustard oil glucosides: Feeding stimulants for adult cabbage stem flea beetles, *Phyllotreta cruciferae.* Ann. of the Entomol. Soc. of America 67: 261–264.

Inglis, I.R., J.T. Wadsworth, A.N. Meyer & C.J. Feare, 1992. Vertebrate damage to 00 and 0 varieties of oilseed rape in relation to SMCO and glucosinolate concentration in the leaves. Crop Protection 11: 64–68.

Iverson, T-H., C. Baggerud & T. Beisvag, 1979. Myrosin cells in Brassicaceae roots. Z. Pflanzenphysiol. 94: 143–154.

King, G.J., 1990. Molecular genetics and breeding of vegetable brassicaes. Euphytica 50: 97–112.

Kondra, Z.P. & B.R. Stefansson, 1970. Inheritance of the major glucosinolates in rapeseed (*Brassica napus*) meal. Can. J. Plant. Sci. 50: 643–647.

Kraling, K., G. Robbelen, M. Thies, M. Herrmann & M.R. Ahmadi, 1990. Variation in seed glucosinolates in lines of *Brassica napus.* Plant Breeding 105: 33–39.

Lammerink, J., D.B. Macgibbon & A.R. Wallace, 1984. Effect of the cabbage aphid (*Brevicoryne brassicae*) on total glucosinolate in the seed of oilseed rape. New Zealand J. Agric. Res. 27: 89–92.

Luthy, B. & P. Matile, 1984. The mustard oil bomb: rectified analysis of the subcellular organisation of the myrosinase system. Biochem. Physiol. Pflanzen 179: 5–12.

Macfarlane Smith, W.H., D.W. Griffiths & B. Boag, 1991. Overwintering variation in glucosinolate content of green tissue of rape (*Brassica napus*) in response to grazing by wild rabbit (*Oryctolagus cuniculus*). J. Sci. Food Agric. 56: 511–521.

Matile, P., 1980. Die Senfolbombe: zur Kompartimentierung des Myrosinasesystems. Biochem. Physiol. Pflanzen 175: 722–731.

Milford, G.F.J., J.K. Fieldsend, A.J.R. Porter, C.J. Rawlinson, E.J. Evans & P. Bilsborrow, 1989. Changes in glucosinolate concentrations during the vegetative growth of single- and double-low cultivars of winter oilseed rape. Aspects of Applied Biology 23. Production and Protection of Oilseed Rape and other Brassica Crops, pp. 83–90.

Mithen, R.F., B.G. Lewis & G.R. Fenwick, 1986. *In vitro* activity of glucosinolates and their products against *Leptosphaeria maculans.* Trans Brit. Mycol. Soc. 87: 433–440.

Mithen, R.F., B.G. Lewis, G.R. Fenwick & R.K. Heaney, 1987a. Resistance of *Brassica* species to *Leptosphaeria maculans.* Trans Brit. Mycol. Soc. 88: 525–531.

Mithen, R.F., B.G. Lewis, G.R. Fenwick & R.K. Heaney, 1987b. Glucosinolates of wild and cultivated *Brassica species.* Phytochem. 26: 1969–1973.

Mithen, R.F. & C. Herron, 1992. Transfer of disease resistance to oilseed rape from wild *Brassica* species. Proceedings of the 8th International Rapeseed Congress, Saskatoon, Canada, pp. 244–250.

Mithen, R.F. & R. Magrath, 1992. Glucosinolates and resistance to *Leptosphaeria maculans* in wild and cultivated *Brassica* species. Plant Breeding 108: 60–68.

Newman, P.L., 1984. Screening for resistance in winter oilseed rape. Aspects of Applied Biology 6. Agronomy, Physiology, Plant Breeding and Crop Protection of Oilseed Rape, pp. 371–380.

Rawlinson, C.J., K.J. Doughty, C.H. Bock, V.J. Church, G.F.J. Milford & J.K. Fieldsend, 1989. Diseases and response to disease and pest control on single- and double-low cultivars of oilseed rape. Aspects of Applied Biology 23. Production and Protection of Oilseed Rape and other Brassica Crops, pp. 393–400.

Rouxel, T., A. Sarniguet, A. Kollman & J-F Bosquet, 1989. Accumulation of a phytoalexin in *Brassica* spp. in relation to a hypersensitive reaction of *Leptosphaeria maculans.* Physiol. Mol. Plant Path. 34: 507–517.

Rucker, B. & E. Rudloff, 1992. Investigations on the inheritance of the glucosinolate content in seeds of winter oilseed rape (*Brassica napus*). Proceedings of the Eighth International Rapeseed Congress, Saskatoon, Canada, pp. 191–196.

Schilling, W. & W. Friedt, 1992. Breeding 00-rapeseed (*Brassica napus*) with differential glucosinolate content in the leaves. Proceedings of the Eighth International Rapeseed Congress, Saskatoon, Canada, pp. 250–255.

Tapper, S. & R. Cox, 1990. Selection of 00 oilseed rape crops by rabbits, hares, and deer grazing at night. In: M. Askew (Ed.) Rapeseed 00 and Intoxication of Wild Animals. pp. 152–160. Commission of the European Community, Brussels.

Thagstad, O.P., K. Evjen & A. Bones, 1991. Immunogold-EM localisation of myrosinase in Brassicacae. Protoplasma 161: 85–93.

Underhill, E.W., L.R. Wetter & M.D. Chisholm, 1973. Biosynthesis of glucosinolates. Biochem. Soc. Symp. 38: 303–326.

Williams, I.J., K.J. Doughty, C.H. Bock & C.J. Rawlinson, 1992. Incidence of pests and diseases and effects of crop protection on double- and single-low winter rape cultivars. Proceedings of the Eighth International Rapeseed Congress, Saskatoon, Canada, pp. 518–523.

Euphytica **63**: 85–93, 1992.
R. Johnson and G.J. Jellis (eds), Breeding for Disease Resistance

Resistance of cowpea and cereals to the parasitic angiosperm *Striga*

J.A. Lane & J.A. Bailey
Department of Agricultural Sciences, University of Bristol, AFRC Institute of Arable Crops Research, Long Ashton Research Station, Long Ashton, Bristol BS18 9AF, UK

Key words: breeding for resistance, cowpea, parasitic angiosperm, resistance mechanisms, sorghum, *Striga, Vigna unguiculata, Sorghum bicolor*

Summary

Striga species are parasitic angiosperms that attack many crops grown by subsistence farmers in sub-Saharan Africa and India. Control of the parasite is difficult and genetically resistant crops are the most feasible and appropriate solution. In cowpea, complete resistance to *Striga gesnerioides* has been identified. Breeding for resistance in sorghum has identified varieties with good resistance to *S. asiatica* in Africa and India. One variety was also resistant to *S. hermonthica* in W. Africa. No such resistance to *Striga* has been found in maize or millets.

Resistant varieties have usually been sought by screening germplasm in fields naturally infested with *Striga*. However, laboratory techniques have also been developed, including an *in vitro* growth system used to screen cowpeas for resistance to *S. gesnerioides*. Two new sources of resistance in cowpea have been identified using the system. The technique has also been used to investigate the mechanisms of resistance in this crop. Two mechanisms have been characterised, both were expressed after penetration of cowpea roots by the parasite.

The resistance of some sorghum varieties to *Striga* is controlled by recessive genes. In cowpea, resistance to *Striga* is controlled by single dominant genes. The genes for resistance are currently being transferred to cowpea varieties which are high yielding or adapted to local agronomic conditions. One *Striga* resistant cowpea variety, Suvita-2, is already being grown widely by farmers in Mali. Reports of 'breakdown' of resistance in cowpea to *Striga* have not yet been confirmed, but a wider genetic base to the resistance is essential to ensure durability of *Striga* resistance.

Abbreviations: ICRISAT – International Crops Research Institute for the Semi-Arid Tropics, IITA – International Institute of Tropical Agriculture, LARS – Long Ashton Research Station, SAFGRAD – Semi-Arid Food Grain Research and Development.

Introduction

Striga species ('witchweeds') are flowering plants that parasitise the roots of their hosts. Many crops grown by smallholders in the semi-arid regions of Africa and India are devastated following parasitism by *Striga*. The crops affected and the distribution of the main species of *Striga* of economic importance are shown in Table 1. The principal crop hosts are maize (*Zea mays.*), millets, (*Pennisetum typhoides* and *Eleusine coracana*), sorghum (*Sorghum bicolor*) and cowpea (*Vigna unguiculata*). These crops provide at least 70 per cent of the diet of people in the semi-arid regions of sub-Saharan Africa. Cowpea alone provides about 50 per cent of the total protein in such diets (Anon., 1991a). Crop losses arising from *Striga* infestation are commonly in excess of 30 per cent (Aggarwal & Ouedraogo, 1989; Sauerborn, 1991), though total yield losses have also been recorded (Parker, 1991). An important consequence of such infestations is that land can sometimes be abandoned following severe *Striga* damage.

Control of *Striga* is difficult because the parasite's life-cycle is closely linked to that of the host. Like mildews and rust fungi, all species of *Striga* are obligate pathogens and successful infection depends on a series of co-ordinated processes. After a two-week period of imhibition, naturally provided by the onset of a rainy season, *Striga* seeds need a stimulus from host roots to initiate germination. This ensures that only those seeds near to the roots germinate, and hence are near to sites of infection. Other parasite seeds remain dormant. *Striga* radicles penetrate host roots and parasite tubercles ('haustoria') develop on the host root surface. These organs facilitate transfer of essential nutrients and water from the host to the parasite. During the early stages of its life-cycle, while it remains below soil level, *Striga* is completely dependent on the host. It is during this time that maximum damage occurs to crop growth and subsequent yield. Once *Striga* stems emerge above ground, the parasite is able to photosynthesise, but even at this stage the majority of assimilates continue to be derived from the host plant (Press et al., 1991). Flowering and seed production is completed within six to eight weeks after the emergence of parasite stems. Depending upon the species, each *Striga* plant can produce between 20,000 and 90,000 seeds, which can remain viable in the soil for up to 20 years. As a result, high levels of inoculum can be quickly established and readily sustained in soils where susceptible crops are grown repeatedly.

Table 1. Distribution and host range of *Striga* species of major economic importance

Striga species	Distribution	Susceptible crops
S. asiatica	West, East and southern Africa; Indian sub-continent; Near East; Far East; USA.	Maize, finger and pearl millets, sorghum, sugar cane, upland rice.
S. hermonthica	West and East Africa.	Maize, finger and pearl millets, sorghum, sugar cane.
S. gesnerioides	West and southern Africa; Indian sub-continent; Near East; USA.	Cowpea and tobacco in southern Africa.

Several control strategies have been proposed, including application of herbicides or germination stimulants and cultural practices based on reducing the *Striga* seed population (reviewed by Lane, 1989; Parker, 1991). None of these methods is completely effective nor economically feasible for the low-input farming systems of the semi-arid regions. Potentially, the safest, most appropriate and most effective method for control of *Striga* is the use of genetically resistant crop plants.

Genetic resistance to Striga

Resistance to *Striga* has been defined by Ramaiah (1987), in the context of field trials, as the absence of emerged parasite stems or a very low number of infections compared with susceptible varieties. The definition is clearer if those varieties which support few emerged parasite stems are termed partially resistant. A third term, tolerant, is used to describe varieties which support similar numbers of emerged parasite stems to those observed on susceptible varieties, but which give a significantly greater yield than the susceptible varieties in parasite infested fields.

At the present time all types of resistance are being considered as there is an urgent demand for any type of resistance. Crops with complete resistance are most desirable as they control the disease, as well as reducing the amount of inoculum in the soil. This contrasts with the use of partially resistant and especially tolerant varieties which, though providing significantly better yields and thus alleviating short-term needs, support reproduction of *Striga* and hence continue to add to the seed population.

The need for *Striga*-resistant crops was first identified in the 1930s and over the last 50 years there have been extensive searches for resistant germplasm, especially in sorghum (reviewed by Ramaiah, 1987). Early advances were the discovery of partial resistance to *S. asiatica* in some sorghum varieties. This was considered to be due to the failure of host roots to stimulate good parasite germination and hence is referred to as 'low-stimulant' resistance. At ICRISAT in India, a laboratory technique was used to screen 15,057 sorghum varieties for this type of resistance (Vaidya et al., 1991). Of the 672 varieties that were identified as being 'low-stimulant', 80 varieties were resistant to *S. asiatica* under field conditions. One sorghum variety (SAR-1) which has the 'low-stimulant' type of resistance to *S. asiatica* has been released to farmers in India (Vasudeva Rao et al., 1989). Some of these resistant sorghum varieties were assessed in field trials in Africa for resistance to both *S. asiatica* and *S. hermonthica*. Results from southern Africa are encouraging as resistance to *S. asiatica* was observed in several of those varieties (Obilana et al., 1991). One variety (IS-7777) resistant to *S. asiatica* was also completely resistant to *S. hermonthica* in W. Africa and several others have revealed partial resistance to this species (Olivier et al., 1991b). Recently, two other varieties with partial resistance to *S. hermonthica* have been released for use in Sudan. They are SRN-39, which is suitable for mechanised farming and IS-9830, which is for traditional farming systems (Anon, 1991b).

Progress in breeding *Striga*-resistant maize and millets has been slow and no complete resistance has been identified. Semi-wild 'shibra' millets may have partial resistance to *S. hermonthica* (Parker & Wilson, 1983). Maize varieties grown in the USA showed a range of responses to *S. asiatica*, from susceptibility to emergence of only a few parasite stems (Ransom et al., 1990). These maize varieties were subsequently assessed in field trials in Kenya for resistance to *S. hermonthica*. No resistance to this species was observed. However, the *Striga* inoculum was extremely high in the Kenyan trials and thus any partial resistance to the parasite may have been obscured (J.K. Ransom, personal communication). Maize varieties with some tolerance to *S. hermonthica* were found by screening over 15,000 varieties in *Striga*-infested fields in Nigeria (Kim, 1991).

Unlike other crops parasitised by *Striga*, there are several cowpeas that are completely resistant to *S. gesnerioides*. The first resistant varieties to be identified were Suvita-2 and 58-57. They were discovered to be resistant in Burkina Faso, but when grown in Nigeria they were susceptible (Aggarwal et al., 1986). Parker & Polniaszek (1990) established that these differences were due to variation in the pathogenicity of *S. gesnerioides* populations across W. Africa. Furthermore, they identified the variety B301, a cowpea collected in Botswana by C. Riches, as resistant to races of *S. gesnerioides* from throughout Africa. Subsequently, a second variety, IT82D-849, was found to be resistant to all races of the parasite (Singh & Emechebe, 1991). The variation in pathogenicity of *S. gesnerioides* and interaction with resistant cowpeas is summarised in Table 2.

Identification of *Striga*-resistant germplasm

Screening plants in fields with natural infestations of *Striga* has been the principal method for identifying new sources of resistance. However, as indicated earlier, the amount of *Striga* inoculum within a field can vary greatly. *Striga* seeds can be added to fields in an attempt to solve this problem, but variability of infestation is still one of the major prob-

Table 2. Interaction of cowpea varieties with *S. gesnerioides* populations from West Africa

Cowpea variety	Origin of *S. gesnerioides*		
	Burkina Faso	Mali	Niger or Nigeria
Blackeye	S[a]	S	S
58-57	R[b]	S	S
Suvita-2	R	R	S
B301	R	R	R
IT82D-849	R	R	R

[a] S = susceptible.
[b] R = resistant, no successful development of parasite stems.

Fig. 1. In-vitro growth system for cowpea and *S. gesnerioides*, 2 weeks after placing germinated parasite seed on cowpea roots. Full details of the system are given in Lane et al. (1991a). Bar = 3 cm.

lems in field screening of new germplasm (reviewed by Vasudeva Rao, 1987). Alternatively, varieties can be assessed by growing them in pots containing soil which has been artificially infested with known amounts of *Striga* seed. This is more reliable than field assessments. As indicated earlier, a large amount of sorghum germplasm has been screened for resistance in laboratories to identify reduced ability to stimulate parasite germination (Vasudeva Rao, 1987). These methods were based solely on the use of root exudates and did not involve inoculating host plants. As a result it was not possible to determine whether any of the resistance was associated with events that took place after the penetration of host roots by *Striga*.

Other laboratory techniques which allow more comprehensive studies of host resistance mechanisms have been developed (Visser et al., 1977; Parker & Dixon, 1983; Lane et al., 1991a). The *in vitro* growth system described by Lane et al. (1991a) has also been used to screen cowpeas for resistance to *S. gesnerioides* (Fig. 1). Cowpeas were grown in shallow plastic trays on moist filter and tissue paper. *Striga* seeds that had imbibed for two weeks were placed close to host roots and after three days, germinated *Striga* seedlings were placed on cowpea roots. Two to three weeks later, parasite tubercles, 0.5 to 5 mm in diameter, from which stems were developing, had formed on the roots of susceptible cowpea plants. Such tubercles do not form on resistant plants. The technique has a number of advantages for screening germplasm. All stages of the parasite infection process, from germination to tubercle formation, can be easily observed and thus information can be gained about the expression of host resistance. This is a significant advantage over the use of screens based on field experiments or those done in pots, where the reason for failed infections cannot be assessed. The screen is rapid: susceptibility can be determined within three weeks compared to the 10 to 12 weeks required for pot or field trials. The material is easily contained during testing and readily destroyed after assessment. This means that extensive testing could be done in Africa, rather than having to rely on experiments being done in temperate regions, e.g. Europe, where *Striga* poses no risks. Currently, cowpea varieties have to be tested in several African countries because *Striga* seed cannot be moved between countries due to the risk of cross-contamination. Finally, the tested cowpea plants can be transferred to pots containing soil and grown to produce seed, or propagated clonally. This is essential for plant breeders using single plant selection methods.

Two new cowpea varieties with resistance to *S. gesnerioides* were identified using the *in vitro* growth system (Lane, Child & Moore, unpublished data). Both varieties are landraces from W. Africa, one (87-2) was from Niger and the other (APL-1) from Nigeria. The landraces were both completely resistant to *S. gesnerioides* from Burkina Faso and Cameroon, while the landrace from Nigeria was also resistant to parasite samples from Mali. In addition, the landrace from Niger was partially resistant to *Striga* from Mali, Niger and Nigeria. Despite reservations expressed earlier about the use of partial resistance, this material could be of great immediate value in Niger and Nigeria as there are currently only two cowpea varieties that are completely resistant to *Striga* in these countries. Most other cowpeas are very susceptible (Emechebe et al., 1991).

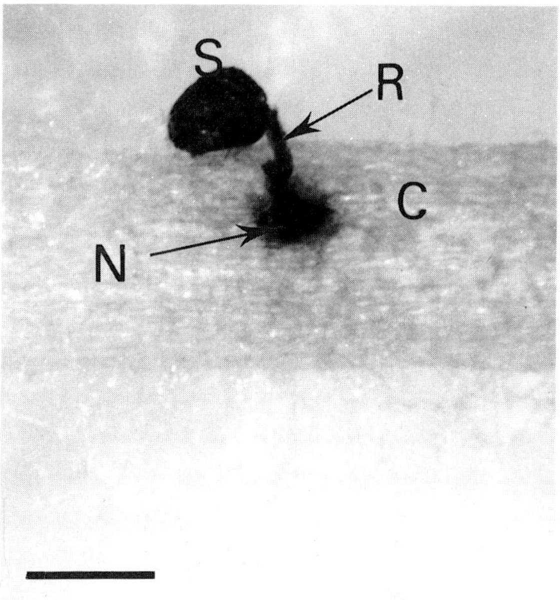

Fig. 2. Resistance of cowpea variety 58-57. Necrosis of cowpea root tissue around site of penetration of *S. gesnerioides* radicle, accompanied by the death of the *Striga* seedling. C = cowpea root; N = necrosis of cowpea root; R = *S. gesnerioides* radicle; S = *S. gesnerioides* seed. Bar = 0.5 mm.

Mechanisms of resistance

Two mechanisms of resistance have been characterised in cowpea using the *in vitro* growth system (Lane & Bailey, 1991). In neither type is resistance due to reduced parasite germination. On variety 58-57, penetration of host roots by *Striga* was associated with necrosis of host cells surrounding the invading parasite radicles, a process analogous to the hypersensitive response shown by plants to fungi (Wood, 1982). Five to six days after inoculation the parasite radicle was black, indicating death (Fig. 2). On variety B301, a small percentage of those *Striga* radicles that penetrated the roots exhibited a similar hypersensitive reaction. However, a second mechanism of resistance was more common in this variety. The majority of *Striga* seedlings penetrated the cortex and successfully merged with the cells of the host stele. Tubercles began to develop on the host root surface, but these did not enlarge on the roots of B301 and remained < 0.5 mm in diameter. No stem developed, even after three to four weeks. Similarly small tubercles occurred on roots of B301 grown in soil for 12 weeks (Fig. 3).

As discussed earlier, research on the mechanisms of resistance of sorghum to *S. asiatica* has focused on 'low-stimulant' varieties (Ramaiah, 1987). However, some varieties e.g. N-13, stimulate germination of *S. asiatica* but support fewer parasite stems than a susceptible variety. Maiti et al. (1984) suggested that in six partially resistant varieties, various anatomical features of the roots prevent or delay penetration of the parasite. For example, variety N-13 had thicker cell walls in the endodermis and pericycle than the susceptible variety. Silica was also present in the endodermis of variety N-13. These structural features were evident in variety N-13 prior to parasite invasion. One of the resistant varieties, IS-5106, had similar root anatomy to the susceptible varieties, indicating that other mechanisms must be responsible for host resistance. Another resistant variety, IS-7777, accumulated cellulose-rich wall layers in those root cells in contact with invading parasite tissue (Oli-

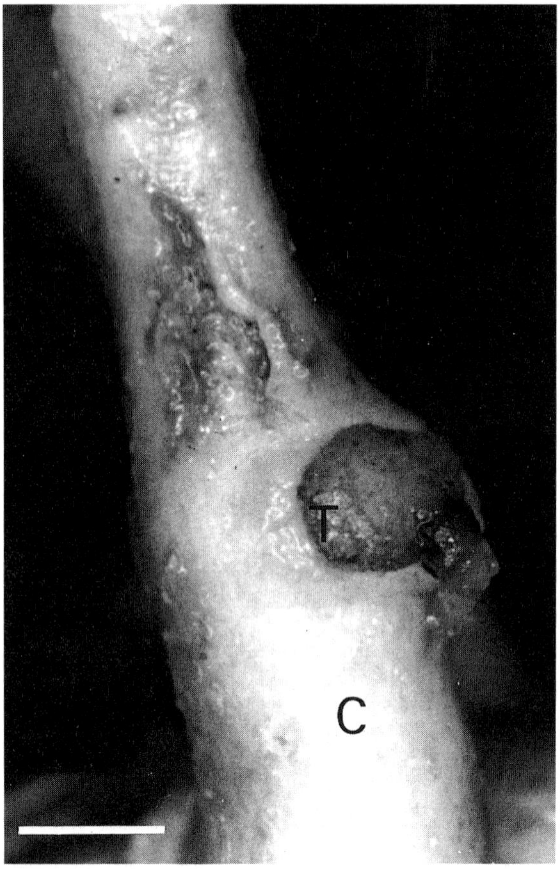

Fig. 3. Resistance of cowpea variety B301 to *S. gesnerioides*. Restricted development of *S. gesnerioides* tubercle, 12 weeks after inoculating soil with *Striga* seed and sowing the cowpea seeds in pots. Bar = 1 mm.

vier et al., 1991a). Host cell walls in contact with parasite tissue also showed alterations in structure as indicated by the absence of labelling with an exoglucanase-gold complex which bound to uninfected tissue. It was suggested that both these responses restricted the spread of *Striga* tissue to the outer cortical layers of the host roots. The susceptible variety Bimbiri showed similar alterations in host cell wall structure following parasite invasion and tyloses formed in the lumen of the xylem. Olivier et al. (1991a) suggested that the changes in the susceptible variety occurred too slowly to halt successful penetration of host roots.

Genetic basis of resistance

Studies on the inheritance of resistance of cowpea to *S. gesnerioides* showed that the resistance of varieties Suvita-2 and 58-57 was controlled by single dominant genes (Aggarwal, 1991). Similarly, Aggarwal (1991) indicated that resistance of variety B301 to the Burkina Faso race of the parasite is also controlled by a single dominant gene, a conclusion that was confirmed by more extensive studies in Nigeria (Singh & Emechebe, 1990). Studies on the inheritance of resistance in sorghum revealed that resistance is controlled by a complex of recessive genes. Single recessive genes appear to control 'low-stimulant' resistance (Ejta et al., 1991; Ramaiah et al., 1990) whilst resistance of sorghum variety L187 to *S. hermonthica* is controlled by between two and five recessive genes (Obilana, 1984).

Breeding for resistance

One cowpea variety, Suvita-2, is agronomically acceptable and has good resistance to drought. It is already being grown by farmers in Mali (Anon., 1991a). However, several cowpeas which are resistant to *S. gesnerioides*, e.g. B301 and 58-57, are not acceptable to farmers as they are low yielding and have unsuitable seeds. For example, variety B301 produces low yields of small brown seeds. Breeding programmes have thus been established at international institutes to produce elite varieties by transferring genes for resistance into a number of different genetic backgrounds. At IITA in Nigeria, varieties B301 and IT82D-849 were crossed with variety IT84S-2246-4, which has resistance to several diseases and three insect pests (aphids, bruchids and thrips). Recurrent backcross selection produced advanced breeding lines (F_6) which were resistant to *Striga* in field trials in 1991 (Singh & Emechebe, 1991). Crosses are also being made with the F_6 material to transfer the *Striga* resistance of B301 into local varieties of cowpea which differ in plant type, seed characteristics and date of maturity. In the SAFGRAD/IITA programme in Burkina Faso, several varieties, including Suvita-2, some tolerant varieties and B301, were crossed together to produce *Striga*-resistant breeding lines with a wide range of plant types. Advanced breeding lines (F_6) were evaluated in field trials in 1990 (N. Muleba, personal communication). In future, the most promising lines will be tested by national programmes across W. Africa.

Within the national programmes of at least seven countries in W. Africa, breeding of *Striga*-resistant cowpeas is concentrated on producing cowpeas which are suitable for local environmental conditions and agronomic practices. For example, in Niger, farmers grow cowpeas as an intercrop with millet to provide seeds for human consumption and fodder for animals. The main objective of these plant breeders is to transfer the resistance of varieties B301 and IT82D-849 into local varieties which have good vegetative growth and grain yield when grown in intercrop situations (M. Toure, personal communication).

Parasite variation

Breeding for resistance to *Striga* in cowpea has to encompass the constraint of the variation in parasite pathogenicity across W. Africa. Three different races, each specifically associated with one country, were described by Parker & Polniaszek (1990) and by Aggarwal (1991). Effective use of resistance genes by plant breeders will require accurate knowledge of the geographic distribution of the parasite races. In 1990, samples of *S. gesnerioides* were collected in the main cowpea growing areas of W. Africa (Nigeria across to Mali) through a joint IITA/LARS survey in collaboration with national scientists. The pathogenicity of the parasite samples was assessed using the *in vitro* system of Lane et al. (1991a) against a differential series of cowpea varieties. Initial results identified two new races and confirmed the existence of the three others. In addition, the country specificity was shown to be invalid as some parasite samples from the borders of countries were of a different race to those in the centre. A map of the geographic variation in pathogenicity has been produced and this will be made available to plant breeders in Africa (Lane & Cardwell, unpublished data). Since selection for *Striga* resistance is usually only performed in a few locations within a country, this map will help plant breeders to predict the geographic areas where the resistance will be effective.

Breeding programmes for resistance to *Striga* in cowpea are currently based on only a few dominant genes. The resistant cowpeas have been used widely by plant breeders for several years and the resistance to *Striga* has continued to be effective. However, there have been a few reports that B301 can be susceptible, e.g. at Bakura in Nigeria in 1989 (reported by Parker, 1991). Cowpea is an autogamous species, but in humid conditions cross-pollination by insect vectors can occur (Purseglove, 1968). It was thus proposed that there had been some outcrossing of B301 with susceptible varieties at Bakura in 1987, followed by selfing in 1988 and that this had resulted in a proportion of the F_2 progeny being susceptible to the parasite in 1989 (B.B. Singh, personal communication). However, when new seed of B301 was grown at this site in 1990, there was no significant parasite development (Lagoke et al., 1991). Furthermore, when B301 plants, grown using the *in vitro* system of Lane et al. (1991a), were inoculated with parasite seed collected from Bakura no significant parasite development occurred (Lane & Child, unpublished data). B301 also appeared susceptible to *S. gesnerioides* in one location in southern Benin in 1990 (K. Cardwell, personal communication). *Striga* seeds from this site were inoculated on to the roots of B301. All plants were completely susceptible to the parasite (Lane, Child & Moore, unpublished data). More extensive field trials are being conducted in southern Benin in the 1992 season to establish the distribution of this new race of the pathogen.

Post infection resistance in cowpea is extremely effective. Coupled with the facts that *Striga* only reproduces once or twice a year, and that its spread is limited because it is a soil pathogen, this indicates that resistance should be durable. Even if varieties are used which have resistance based on only a single gene, the likelihood of new virulent forms of this parasite arising and then becoming widely established is low. Nevertheless, the durability of resistance would be prolonged if growth of resistant varieties were associated with sensible agronomic practises, e.g. hand weeding and destruction of all emerging *Striga* plants prior to flowering. Ideally, the genetic basis of resistance in cowpea should also be broadened. The primary centre for diversification of cowpea was probably sub-Saharan Africa (Smart, 1990). These regions may contain a wide variety of cowpea genotypes, including more examples of resistance and additional genes for resistance may come from landraces and wild relatives of cowpea.

Over the last ten years much progress has been made in studying the mode of inheritance of resistance in cowpeas and in transferring the resistance to agronomically acceptable varieties. There are no similar examples of complete resistance in cereals. One reason for the relatively slow progress in finding resistant cereal germplasm may be the concentration on 'low-stimulant' partial resistance and the lack of appropriate methods for detecting resistance which is expressed after penetration of host roots. The *in vitro* growth system of Lane et al. (1991a) has recently been adapted to study the parasitism of maize and sorghum by *S. asiatica* and *S. hermonthica* (Lane et al., 1991b). This system provides a means of searching for new types of resistance in cereals. In addition, the mechanisms of resistance of varieties identified during screening or in field trials in Africa could be characterised using this method. As in the research on cowpeas, these approaches are more likely to succeed if close co-operation is maintained between agronomists, plant breeders and laboratory-based scientists interested in the fundamental basis of interactions between plants and their parasites.

Acknowledgements

This research was largely financed by the UK Overseas Development Administration, Natural Resources Institute (contract no. X0075). The authors acknowledge discussions with P.J. Terry and the research assistance of D.V. Child and T.H.M. Moore.

References

Aggarwal, V.D., 1991. Research on cowpea-*Striga* resistance at IITA. In: S.K. Kim (Ed.), Combating *Striga* in Africa, pp. 90–95. IITA, Ibadan.

Aggarwal, V.D., S.D. Haley & F.E. Brockman, 1986. Present status of breeding cowpeas for resistance to *Striga* at IITA. In: S.J. ter Borg (Ed.), Proceedings of a workshop on biology and control of *Orobanche*, pp. 176–180. LH/VPO, Wageningen.

Aggarwal, V.D. & J.T. Ouedraogo, 1989. Estimation of cowpea yield loss from *Striga* infestation. Trop. Agric. 66: 91–92.

Anon., 1991a. The SAFGRAD Networks. Serving National Agricultural Research Systems and Food Grain Farmers in sub-Saharan Africa. OAU/STRC – SAFGRAD, Ouagadougou.

Anon., 1991b. Semi-Arid Tropics News, ICRISAT 8: 3.

Ejta, G., L.G. Butler, D.E. Hess & R.K. Vogler, 1991. Genetic and breeding strategies for *Striga* resistance in sorghum. In: J.K. Ransom, L.J. Musselman, A.D. Worsham & C. Parker (Eds.), Proceedings of the 5th International Symposium of Parasitic Weeds, pp. 539–544. CIMMYT, Nairobi.

Emechebe, A.M., B.B. Singh, O.I. Leleji, I.D.K. Atokple & J.K. Adu, 1991. Cowpea-*Striga* problems and research in Nigeria. In: S.K. Kim (Ed.), Combating *Striga* in Africa, pp. 18–28. IITA, Ibadan.

Kim, S.K., 1991. Breeding maize for *Striga* tolerance and the development of a field infestation technique. Combating *Striga* in Africa, pp. 96–108. IITA, Ibadan.

Lagoke, S.T.O., J.Y. Shebayan, G. Weber, O. Olufajo, K. Elemo, J.K. Adu, A.M. Emechebe, B.B. Singh, A. Zaria, A. Awad, L. Ngawa, G.O. Olaniyan, S.O. Olfare & A.A. Adeoti, 1991. Survey of *Striga* problem and evaluation of *Striga* control methods and packages in the Nigerian savannah. Report of the FAO-IAR *Striga* project, IAR, Ahmadu-Bello University, Zaria.

Lane, J.A., 1989. Prospects for the control of *Striga*, a noxious parasitic weed of tropical crops. Rural Dev. Pract. 1: 9–10.

Lane, J.A. & J.A. Bailey, 1991. Resistance of cowpea to *Striga gesnerioides*. In: K. Wegmann & W. Forstreuter (Eds.), Progress in *Orobanche* research. Proceedings of the International Workshop on *Orobanche* Research, Obermarchtal 1989, pp. 344–350. Erberhard-Karls-Universität, Tubingen.

Lane, J.A., J.A. Bailey & P.J. Terry, 1991a. An *in vitro* growth system for studying the parasitism of cowpea (*Vigna unguiculata*) by *Striga gesnerioides*. Weed Res. 31: 211–217.

Lane, J.A., M.J. Kershaw, T.H.M. Moore, D.V. Child, P.J. Terry & J.A. Bailey, 1991b. An *in vitro* system to investigate the parasitism of cereals by *Striga* species and resistance of cowpea to *Alectra vogelii*. In: J.K. Ransom, L.J. Musselman, A.D. Worsham & C. Parker (Eds.), Proceedings of the 5th International Symposium of Parasitic Weeds, pp. 237–240. CIMMYT, Nairobi.

Maiti, R.K., K.V. Ramaiah, S.S. Bisen & V.L. Chidley, 1984. A comparative study of the haustorial development of *Striga asiatica* (L.) Kuntze on sorghum cultivars. Ann. Bot. 54: 447–457.

Obilana, A.T., 1984. Inheritance of resistance to *Striga* (*Striga hermonthica* Benth.) in sorghum. Protec. Ecol. 7: 305–311.

Obilana, A.T., W.A.J. de Milliano & A.M. Mbwaga, 1991. *Striga* research in sorghum and millets in Southern Africa: status and host plant resistance. In: J.K. Ransom, L.J. Musselman, A.D. Worsham & C. Parker (Eds.), Proceedings of the 5th International Symposium of Parasitic Weeds, pp. 435–441. CIMMYT, Nairobi.

Olivier, A., N. Benhamou & G.D. Leroux, 1991a. Cell surface interactions between sorghum roots and the parasitic weed *Striga hermonthica:* cytochemical aspects of cellulose distribution in resistant and susceptible host tissues. Can. J. Bot. 69: 1679–1690.

Olivier, A., K.V. Ramaiah & G.D. Leroux, 1991b. Selection of sorghum (*Sorghum bicolor* L. Moench) varieties resistant to the parasitic weed *Striga hermonthica.* Weed Res. 31: 219–225.

Parker, C., 1991. Protection of crops against parasitic weeds. Crop Protec. 10: 6–22.

Parker, C. & N. Dixon, 1983. The use of polyethylene bags in the culture and study of *Striga* spp. and other organisms on crop roots. Ann. Appl. Biol. 103: 485–488.

Parker, C. & T.I. Polniaszek, 1990. Parasitism of cowpea by *Striga gesnerioides:* variation in virulence and discovery of a new source of host resistance. Ann. Appl. Biol. 116: 305–311.

Parker, C. & A.K. Wilson, 1983. *Striga*-resistance identified in semi-wild 'shibra' millet (*Pennisetum* sp.). Med. Fac. Land. Rijks. Univ. Gent 48: 1111–1117.

Press, M.C., J.D. Graves & G.R. Stewart, 1991. Physiology of the interaction of angiosperm parasites and their higher plant hosts. Plant Cell Environ. 13: 91–104.

Purseglove, J.W., 1968. Tropical Crops, Dicotyledons. Longmans, London.

Ramaiah, K.V., 1987. Breeding cereal grains for resistance to witchweed. In: L.J. Musselman (Ed.), Parasitic Weeds in Agriculture. Vol. 1. *Striga*, pp 227–242. CRC Press, Boca Raton.

Ramaiah, K.V., V.L. Chidley & L.R. House, 1990. Inheritance of *Striga* seed-germination stimulant in sorghum. Euphytica 45: 33–38.

Ransom, J.K., R.E. Eplee & M.A. Langston, 1990. Genetic variability for resistance to *Striga asiatica* in maize. Cereal Res. Commun. 18: 329–333.

Sauerborn, J., 1991. The economic importance of the phytoparasites *Orobanche* and *Striga*. In: J.K. Ransom, L.J. Musselman, A.D. Worsham & C. Parker (Eds.), Proceedings of the 5th International Symposium of Parasitic Weeds, pp. 137–143. CIMMYT, Nairobi.

Singh, B.B. & A.M. Emechebe, 1990. Inheritance of *Striga* resistance in cowpea genotype B301. Crop Sci. 30: 879–881.

Singh, B.B. & A.M. Emechebe, 1991. Breeding for resistance in *Striga* and *Alectra* in cowpea. In: J.K. Ransom, L.J. Musselman, A.D. Worsham & C. Parker (Eds.), Proceedings of the 5th International Symposium of Parasitic Weeds, pp. 303–305. CIMMYT, Nairobi.

Smart, J., 1990. Grain Legumes: Evolution and Genetic Resources. Cambridge University Press, Cambridge.

Vaidya, P.K., B. Raghavender & S.Z. Mukuru, 1991. Progress in breeding resistance to *Striga asiatica* in sorghum at the ICRISAT Centre. In: S.K. Kim (Ed.), Combating *Striga* in Africa, pp. 81–89. IITA, Ibadan.

Vasudeva Rao, M.J., 1987. Techniques for screening sorghums for resistance to *Striga*. In: L.J. Musselman (Ed.), Parasitic Weeds in Agriculture. Vol. 1. *Striga*, pp. 281–304. CRC Press, Boca Raton.

Vasudeva Rao, M.J., P.K. Vaidya, V.L. Chidley, S.Z. Mukuru & L.R. House, 1989. Registration of 'ICSV 145' *Striga asiatica* resistant sorghum cultivar. Crop Sci. 29: 488–489.

Visser, J.H., I. Dörr & R. Kollmann, 1977. On the parasitism of *Alectra vogelii* Benth. (Scrophulariaceae). 1. Early development of the haustorium and initiation of the stem. J. Plant Physiol. 84: 213–222.

Wood, R.K.S., 1982. Active Defense Mechanisms in Plants. Plenum Press, New York.

Euphytica **63**: 95–102, 1992.
R. Johnson and G.J. Jellis (eds), Breeding for Disease Resistance

The role of resistance breeding in the integrated control of downy mildew (*Bremia lactucae*) in protected lettuce

Ian R. Crute
Crop Protection Department, Horticulture Research International (East Malling), West Malling, Kent, ME19 6BJ, UK

Key words: Bremia lactucae, downy mildew, integrated control, *Lactuca sativa,* lettuce, resistance

Summary

Over the last 30 years, six resistance alleles (*Dm2, Dm3, Dm6, Dm7, Dm11* and *Dm16*) located in two linkage groups, have contributed to the control of downy mildew in lettuce crops grown under protection (glass or polythene) in northern Europe. More recently, an as yet genetically uncharacterised resistance factor, R18, has also begun to assume importance. The occurrence of the various combinations of these resistance alleles that exist in commercial cultivars has been dictated by the pathotypes of *Bremia lactucae* used in their selection but also restricted by linkage in repulsion. In the UK, a pathotype of *B. lactucae* insensitive to phenylamide fungicides, such as metalaxyl, emerged in 1978 and became prevalent throughout lettuce production areas in subsequent years. The specific virulence of this pathotype was identical to the previously described phenylamide sensitive pathotype NL10 and cultivars carrying *Dm11, Dm16* or R18 were resistant. Consequently, an integrated control strategy based on the utilisation of metalaxyl on cultivars carrying *Dm11* provided effective control in UK until 1987 when a new phenylamide insensitive pathotype began to cause problems. The specific virulence of this second pathotype, which was first reported in the Netherlands and France, was identical to the previously described phenylamide sensitive pathotype NL15. Cultivars carrying *Dm6, Dm16* or R18, but not *Dm11,* were resistant to NL15; consequently an appropriate change in the cultivar recommendations for use in the integrated control strategy was successfully promulgated. It is predicted that variations of this integrated control strategy involving the use of appropriately selected *Dm* gene combinations may prove effective for some time. This prediction is based on studies of the status of the avirulence loci in the two phenylamide insensitive pathotypes and of the specific virulence characteristics of phenylamide sensitive components of the pathogen population.

Introduction

Lettuce downy mildew, caused by the obligately biotrophic oomycete fungus *Bremia lactucae,* is a troublesome disease of crops grown in glasshouses and under other forms of protection in northern Europe particularly in autumn and spring. Outbreaks of disease on crops close to maturity can often render them unmarketable or cause them to be downgraded as a consequence of the need to remove diseased leaves at harvest. Reviews of the biology and options for the control of *B. lactucae* have recently appeared elsewhere (Crute, 1989; 1991).

This paper describes the role that breeding for resistance has played in the control of lettuce downy mildew in protected crops and the way in which it has become possible to integrate the deployment of resistant cultivars with the use of phenylamide fungicides, such as metalaxyl, to combat control problems created by the emergence of fungicide insensitive pathotypes.

The history of breeding for downy mildew resistance in protected lettuce and combinations of resistance genes in available cultivars

Six downy mildew resistance alleles that commonly occur among European protected lettuce cultivars have been important in providing disease control and have been consciously selected in breeding programmes. These six alleles are not the only ones that exist within the protected lettuce genepool but are the ones that have most frequently provided resistance to prevalent pathotypes of *B. lactucae* in northern Europe. The six alleles occur in two linkage groups (I and III; Table 1) (Norwood & Crute, 1980; Hulbert & Michelmore, 1985; Farrara et al., 1987; Landry et al., 1987; Kesseli et al., 1990) and since no recombinants between *Dm7* and *Dm11* have ever been recovered, it is possible they may be alleles at the same locus. A further uncharacterised resistance factor, *R18*, has recently assumed some significance but has yet to be assigned to a linkage group. It is instructive to consider briefly the history of downy mildew resistance breeding in protected lettuce and the way in which the gene combinations available have been determined by the pathotypes used in the selection process and linkage relationships.

The history of breeding for resistance to lettuce downy mildew in protected lettuce started when MacPherson (1932) reported on some trials conducted near Blackpool, UK over the winter of 1931–1932. Two French cultivars were observed to be completely resistant to downy mildew: Gotte à Graine Blanche de Loos and Rosée Printanière. These were the days of uncontrolled synonymy and further information on the fate of these two resistant cultivars has not been possible to trace. In addition however, only a few plants of the cultivar May Queen (obtained from the Netherlands) became infected. May Queen is synonymous with several other cultivars of the May King Forcing type that were subsequently reported as resistant in trials in Germany conducted from 1935–1937 (Schultz & Röder, 1938) and had also been found to be resistant by Jagger at two out of three locations in the USA (Jagger & Whittaker, 1940; Whittaker et al., 1958). Of equal significance, Macpherson also reported that the Blackpool grower, J.W. Cardwell had produced a downy mildew resistant selection of the then standard cultivar French Frame Forcing (otherwise called Early French Frame). This selection or further derivatives of it came to be known variously as Forcing Mildew Resistant (or Resisting), the 'Blackpool Strain' of Early French Frame, or Resistant Early French Frame. Proeftuin's Blackpool was a further selection from the 'Blackpool Strain' made in the Netherlands.

Table 1. The linkage relationships of *Dm* genes of importance in European protected lettuce cultivars (1950–1990)

Linkage group I	2-3-6-16
Linkage group III	7-11 (may be allelic)
Unknown	*R18*

Proeftuin's Blackpool (contributing *Dm3*) and selections from the German type, Maikönig Treib (May King; May Queen; Interrex) (contributing *Dm2*) were the starting material for a breeding programme conducted in the Netherlands in the early 1950s, at the Institute for Horticultural Plant Breeding (IVT), Wageningen (now Centre for Plant Breeding and Reproduction Research; CPRO-DLO) (Rodenburg & Huyskes, 1962). Efforts were made to develop improved lettuce genotypes for protected cropping from autumn and winter sowings and downy mildew resistance was an important attribute for such cultivars. IVT released to Dutch private breeding companies some important germplasm developed from crosses between these two traditional European lettuce types. It is now known that *Dm2* and *Dm3* were linked in repulsion (Table 1) yet one of the important IVT germplasm releases (Type 57) carried this gene combination and still represents the only example of recombination between *Dm* genes in linkage groups I and III present in commercial cultivars (Table 5).

Through the 1960s, four pathotypes (NL1, NL2, NL3 and NL4) were characterised (Tjallingii &

Rodenburg 1967; 1969) (Table 2) and in order to provide resistance to all of these, genotypes from US breeding programmes carrying *Dm7* (originating from cv. Romaine Blonde Lente à Monter) were used as parents in subsequent breeding (Table 2). The result was the emergence of numerous cultivars carrying different combinations of these three genes.

In the late 1960s, further breeding material carrying resistance derived from a *Lactuca serriola* accession was released from IVT; this is described variously in literature as: F8 Hilde × Botanische, H × B, and H × *L. serriola*. This resistance is now assigned to *Dm11* and was used extensively. All gene combinations involving *Dm2, Dm3* and *Dm11* but notably not *Dm7* with *Dm11* (because of linkage in repulsion) exist among commercial cultivars which were selected on the basis of resistance to further pathotypes of *B. lactucae* (NL5, NL6, NL7 and NL9) identified during this period (Table 3) (Blok & Eenink, 1974; Rodenburg, 1975; Blok & Van der Schaf-van Waadenooyen Kernkamp, 1977; Blok & Van Bakel, 1976; Groenwold, 1978).

Resistance gene *Dm6* (probably derived from the *L. serriola* accession PI 91532) (Crute, 1992) is now also represented in the genepool from which new glasshouse cultivars are being developed as are two further, but less well documented sources of resistance (*Dm16* and *R18*), which probably have their origins in *L. serriola* accessions identified in private breeding companies. During the 1980s, further pathotypes of *B. lactucae* were characterised (Eenink & Blok, 1980; Eenink et al., 1983) and new *Dm* gene combinations, restricted by linkage in repulsion, were produced (Table 4). A summary of the available combinations of important *Dm* genes in protected lettuce cultivars is shown in Table 5 together with the pairs of *Dm* genes that have not been combined because of linkage or allelism.

In passing, it is worth noting the value of the clear and complete documentation of the early history of lettuce breeding when innovation was primarily the responsibility of publicly funded organisations. To the detriment of informed use of resistance, this contrasts with the paucity of similar information to become available over the last 15 years or so now that private breeding companies

Table 2. Important *Dm* genes utilised in protected lettuce breeding during the 1960s, gene combinations produced and the specific virulence characteristics of Dutch pathotypes used to select for resistance

Dm2[a] × *Dm3*[a]
Dm2
Dm3 × *Dm7*[a]
Dm2 + *Dm3*
Dm7
Dm2 + *Dm7*
Dm3 + *Dm7*
Dm2 + *Dm3* + *Dm7*

Year of discovery	Pathotype	*Dm* gene 2	3	7
1964	NL1	+	–	–
1964	NL2	+	+	–
1965	NL3	–	–	+
1968	NL4	+	–	+

+ = virulent/susceptible.
– = avirulent/resistant.
[a] See text for details of origin of *Dm* gene.

Table 3. Important *Dm* genes utilised in protected lettuce breeding during the 1970s, gene combinations produced and the specific virulence characteristics of Dutch pathotypes used to select for resistance

Dm2
Dm3
Dm7
Dm2 + *Dm3* × *Dm11*[a]
Dm2 + *Dm7*
Dm3 + *Dm7*
Dm2 + *Dm3* + *Dm7*
Dm2 + *Dm11*
Dm3 + *Dm11*
Dm2 + *Dm3* + *Dm11*

Year of discovery	Pathotype	*Dm* gene 2	3	7	11
1972	NL5	–	+	+	–
1973	NL6	+	–	–	+
1976	NL7	+	+	+	–
?	NL9	–	+	+	–

+ = virulent/susceptible.
– = avirulent/resistant.
[a] See text for details of origin of *Dm* gene.

are as much involved in early germplasm development as they are in final cultivar production.

Insensitivity to phenylamide fungicides in *Bremia lactucae*

The phenylamide fungicide, metalaxyl, was first commercially cleared for the control of downy mildew in UK lettuce crops during 1978. Fungicide formulations containing metalaxyl provided exceptional disease control for five full seasons (Crute, 1987), but in autumn 1983 failure of metalaxyl to exert control of downy mildew in a major production region in north west England was shown to be due to the occurrence of a highly insensitive pathotype of *B. lactucae*. During the next three seasons, this pathotype rapidly dominated the pathogen population and by the end of 1986 had been recorded in all major UK lettuce production areas. This pathotype was shown to have an identical virulence phenotype to the previously described sensitive pathotype NL10 (Tables 4 and 6) and in particular was avirulent on cultivars carrying *Dm11*. At the time of the emergence of the phenylamide insensitive 'NL10 type' a sensitive pathotype, with identical virulence phenotype, was the most frequently encountered pathotype in the UK population (Table 8). The circumstances surrounding the first occurrence of phenylamide resistance in *B. lactucae*, details of how this insensitive pathotype spread and the characteristics of insensitive isolates have been fully documented elsewhere (Crute et al., 1987).

Although pathotypes of *B. lactucae* virulent on cultivars carrying *Dm11* were not infrequent over the period 1983–1986, they were all phenylamide sensitive (Table 8). Consequently, good downy mildew control was re-established by the combined

Table 4. Important *Dm* genes utilised in protected lettuce breeding during the 1980s, gene combinations produced and the specific virulence characters of Dutch pathotypes used to select resistance

Dm2		
Dm3		
Dm7		
Dm2 + *Dm3*		
Dm2 + *Dm7*	× *Dm6*[a]	
Dm3 + *Dm7*		
Dm2 + *Dm11*		
Dm3 + *Dm11*	× *Dm16*[a]	
Dm2 + *Dm3* + *Dm7*		
Dm2 + *Dm3* + *Dm11*		
		Dm16
		Dm6 + *Dm11*
		Dm7 + *Dm16*
		Dm11 + *Dm16*

Year of discovery	Pathotype	*Dm* genes 2	3	6	7	11	16	18
1980	NL10[b]	+	+	+	+	−	−	−
1983	NL11	−	−	+	+	−	+	−
1983	NL12	−	−	+	+	+	+	−
1983	NL13	−	+	−	+	+	−	−
?	NL14	+	+	+	−	+	−	−
?	NL15[b]	+	+	−	+	+	−	−
?	NL16	+	+	+	+	+	+	−

+ = virulent/susceptible.
− = avirulent/resistant.
[a] See text for details of origin of *Dm* gene.
[b] Phenylamide sensitive and insensitive forms of the pathotype occur.

Table 5. Combinations of important *Dm* genes available or unavailable in protected lettuce cultivars (*Dm* genes in linkage group I are shown in bold type)

Available *Dm* gene combinations	*Dm* gene pairs not available due to linkage in repulsion or allelism
2 **3**	
	2 **6**
2 7	
2 11	
	2 **16**
	3 **6**
3 7	
3 11	
	3 **16**
6 7	
6 11	
	6 **16**
	7 11
7 **16**	
11 **16**	
2 **3** 7	
2 **3** 11	

use of fungicides containing metalaxyl with cultivars carrying *Dm11*. One of the most important attributes of a successful lettuce cultivar in the UK at this time was that it should carry *Dm11*. Under the influence of this control regime, the phenylamide insensitive pathotype declined markedly in frequency and has not been recorded in the course of pathogen population monitoring since 1988 when it was recorded on a single occasion from Northern Ireland. However, the frequency of virulence to match *Dm11* in the phenylamide sensitive component of the pathogen population increased markedly from 1983 onwards such that this resistance gene now no longer exerts any significant influence on disease control.

During 1986, in the Netherlands and France, a new phenylamide resistant pathotype emerged to create control problems (Leroux et al., 1988). The virulence phenotype of this second insensitive pathotype was identical to the previously described sensitive pathotype NL15 (Table 6) which was at about this time among the most frequently encountered pathotypes in the UK (Table 8). This insensitive pathotype was first recorded in the UK in

Table 6. Virulence phenotypes of European pathotypes of *Bremia lactucae* highly insensitive to phenylamide fungicides

Year of discovery	*Dm* gene							
	2	3	6	7	11	16	18	SCT[a]
1983 (NL10 type)	+	+	+	+	−	−	−	B2
1986 (NL15 type)	+	+	−	+	+	−	−	B1
1989	+	+	+	+	+	−	−	B2

[a] Sexual compatibility type.
+ = virulent/susceptible.
− = avirulent/resistant.

1987 and, as with the 'NL10 type', spread to all production regions over a three year period. Because cultivars carrying *Dm11* provided no control of the new insensitive 'NL15 type' this demanded the promulgation of a modified integrated control strategy. Genes *Dm6, Dm16* and *R18* provided effective control of the 'NL15 type' (Table 6) and the latter two genes were also effective against the 'NL10 type'. Hence, control of both phenylamide insensitive pathotypes could be effected by utilising cultivars carrying *Dm6* + *Dm11, Dm16* or *R18.* With the exception of *R18,* sensitive pathotypes capable of rendering all other gene combinations ineffective had been recorded in the UK so the use of metalaxyl in combination with cultivars carrying appropriate resistance genes was still required to ensure control (Table 8). As might have been anticipated however, the increased utilisation of culti-

Table 7. Virulence phenotypes of pathotypes of *Bremia lactucae* from the UK with an intermediate response to phenylamide fungicides

Year of discovery	*Dm* gene							
	2	3	6	7	11	16	18	SCT[a]
1986	+	−	−	+	+	−	−	B1
1987	+	−	+	+	−	−	−	nt
1987	+	+	+	+	−	+	−	B2
1989	+	+	+	+	+	−	−	B1

[a] Sexual compatibility type.
nt = not tested.
+ = virulent/susceptible.
− = avirulent/resistant.

Table 8. Virulence phenotypes of phenylamide sensitive pathotypes recorded in the UK five times or more over the period 1983–1990 (excluding 1986)

Dm gene							Year (Number of records of pathotype)						
2	3	6	7	11	16	18	83	84	85	87	88	89	90
+	+	+	+	−	−	−	7	10	3	0	0	0	0 (NL10)
+	+	+	+	+	−	−	2	3	3	1	1	1	0
+	+	+	+	−	+	−	2	2	3	0	0	0	0
−	−	+	+	+	+	−	1	1	5	0	1	6	4 (NL12)
−	+	+	+	+	+	−	0	3	1	0	0	1	0
+	+	−	+	+	−	−	0	0	5	4	12	2	0 (NL15)
−	−	+	+	−	+	−	0	0	6	2	0	1	0 (NL11)

(17 further pathotypes were recorded 1–4 times only).
+ = virulent/susceptible.
− = avirulent/resistant.

vars carrying *Dm6* or *Dm16* resulted in an increase in the frequency of sensitive pathotypes with matching virulence (Table 8).

In addition to the two fully phenylamide insensitive pathotypes that have created serious control problems in UK lettuce production, a few pathotypes with intermediate fungicide sensitivity have been isolated in the course of population monitoring over the past 9 years (Table 7). Isolates with a similar intermediate response to metalaxyl were recovered in the laboratory among F_1 progeny from crosses between fully insensitive and sensitive parental isolates (Crute & Harrison, 1988). The isolates of intermediate sensitivity are considered to be heterozygous at the locus controlling response to phenylamide fungicide. The occurrence of intermediate isolates among field isolates is suggestive of the movement of the allele or alleles for phenylamide insensitivity through the fungus population as a consequence of natural sexual recombination. In 1989, a third fully insensitive pathotype was found at a single location in north-west England (Table 6). This new pathotype differed from the 'NL15 type' by its virulence on *Dm6* but the use of cultivars carrying *Dm16* and *R18* would still have provided effective control had it become a well established pathotype. In the event this pathotype has only been recorded on one further occasion early in 1992.

Phenylamide resistance has also been reported in California, USA and it is apparent that a similar integrated control strategy to that effected in the UK is also likely to be appropriate (Schettini et al., 1991).

Variation for specific virulence in the phenylamide sensitive population of *Bremia lactucae*

Mention has already been made of the increase in the frequency of phenylamide sensitive *B. lactucae* pathotypes carrying virulence to match *Dm* genes being promoted through the integrated control strategy. However, while pathotypes carrying matching virulence for genes *Dm6, Dm11* and *Dm16* markedly increased in frequency between 1984 and 1990, the frequency of pathotypes carrying virulence to match genes *Dm2* and *Dm3* declined. *Dm2* and *Dm3* were present among the earliest of protected lettuce cultivars to emerge from modern breeding programmes but linkage with *Dm6* and *Dm16* has resulted in their absence from more recently produced cultivars (Table 5). In contrast to virulence to match *Dm7,* which has remained present at a consistently high frequency in the *B. lactucae* population for more than 15 years, there is some evidence that the recent decline in the use of cultivars carrying *Dm2* and *Dm3* has been associated with a comparable decline in the frequency of matching virulence. Virulence to match *R18* (as determined on cv. Mariska) has not yet been located among isolates of the pathogen from UK lettuce crops tested since 1985.

Integrated control of downy mildew in protected lettuce production – the future

Some tentative predictions are possible about the future application of the integrated control strategy for downy mildew in protected lettuce that has hitherto proved so effective in the UK. These predictions are based on a consideration of data from pathotype monitoring and genetic studies conducted in the laboratory.

There is every reason to believe that new phenylamide insensitive pathotypes of *B. lactucae* will emerge through selection of *de novo* mutations or through the recombination of existing alleles for insensitivity with new virulence gene combinations. Experience has shown that the selection of a *de novo* mutation is most likely to occur from the sensitive pathotype that is most prevalent at the time. Consequently, it might be predicted on the basis of this logic that the next insensitive pathotype to emerge is most likely to lack virulence for *Dm2* and *Dm3* (Table 8) and therefore be controlled by the use of cultivars carrying these genes.

The analysis of F_1 progeny from a cross between insensitive 'NL10 type' and 'NL15 type' isolates allowed the status of pathogen loci expressing avirulence for particular *Dm* genes to be determined (avirulence is dominant in this diploid fungus). The 'NL10 type' was heterozygous for *Avr11* and *Avr16* while the 'NL15 type' was homozygous for *Avr6* and *Avr16*. One or both isolates were homozygous for avirulence to match *R18*. It is not yet possible to determine from the data currently available whether phenylamide insensitivity is controlled by alleles at the same locus in both insensitive pathotypes. However, these data indicate that, should a natural hybrid between these pathotypes become established, *Dm6*, *Dm16* and *R18* would still be expected to provide effective control. Similarly, it is also likely that at least one of *Dm2, Dm3, Dm6, Dm16* or *R18* would prove effective if a natural hybrid between either of these insensitive pathotypes and any frequently occurring sensitive pathotype should become established. This conclusion is borne out by the observed virulence phenotypes of putative naturally occurring hybrid isolates heterozygous at the locus or loci controlling response to phenylamide fungicides (Table 7). Since combinations of these resistance genes are well represented in the genepool from which new lettuce cultivars for protected cropping are continually being selected (Table 5) there is every reason to be confident that the integrated control strategy with appropriate modification can be sustained for some years.

Conclusions

It has often been assumed that pathotype specific resistance genes are of no further value once matching virulence has been recorded. In reality, the frequency with which different pathotypes occur within the pathogen population change over time for reasons that are sometimes readily explicable. Such was the case with the intense selection of the phenylamide insensitive 'NL10 type' created by extensive fungicide use followed by intense selection against this pathotype created by the deployment of *Dm11*. In other circumstances, changes in frequencies may be less dramatic or the reasons less apparent. Nevertheless, it is clear that sustained monitoring of the pathogen population and informed deployment of resistance genes in combination with fungicides provides a dynamic approach to disease control that it should be possible to maintain effectively. This approach to control will be aided in future by the utilisation of newly identified downy mildew resistance genes being incorporated into well adapted genotypes to become part of the genepool from which breeders will select new cultivars in the future.

Acknowledgements

I am grateful to Miss Pamela Gordon, Horticulture Research International, Wellesbourne for recovering and testing sexual progeny isolates of *B. lactucae* and for determining the specific virulence and response to phenylamide fungicides of field isolates. The financial support of the Ministry of Agriculture, Fisheries and Food of England and Wales is gratefully acknowledged.

References

Blok, I. & A.H. Eenink, 1974. *Bremia* in latucae sorten, in het bijzonder in *L. sativa*. Zaadbelangen 28: 138–40.

Blok, I. & J.J.M. Van Bakel, 1976. Wit in sla. Groenten en Fruit, No. 31638, 172.

Blok, I. & Van Der Schaf-Van Waadenooyen Kernekamp, 1977. Problemen rond *Bremia* in sla. Zaadbelangen 31: 57–58, 60.

Crute, I.R., 1987. The occurrence, characteristics, distribution, genetics and control of a metalaxyl-resistant pathotype of *Bremia lactucae* in the United Kingdom. Plant Disease 71: 763–767.

Crute, I.R., 1989. Lettuce downy mildew: a case study in integrated control. In: K.J. Leonard & W.E. Fry (Eds.) Plant Disease Epidemiology Vol. 2, pp. 30–53. McGraw-Hill, New York.

Crute, I.R., 1991. Downy mildew of lettuce. In: A.N. Mukhopadhyay, H.S. Chaube, J. Kumar & U.S. Singh (Eds.) Plant Diseases of International Importance Vol. II, pp. 165–185. Prentice Hall, New Jersey.

Crute, I.R., 1992. From breeding to cloning (and back again?): a case study with lettuce downy mildew. Ann. Rev. Phytopathol. 30: 485–506.

Crute, I.R. & J.M. Harrison, 1988. Studies on the inheritance of resistance to metalaxyl in *Bremia lactucae* and on the stability and fitness of field isolates. Plant Pathol. 37: 231–250.

Crute, I.R., J.M. Norwood & P.L. Gordon, 1987. The occurrence, characteristics and distribution in the United Kingdom of resistance to phenylamide fungicides in *Bremia lactucae* (lettuce downy mildew). Plant Pathol. 36: 297–315.

Eenink, A.H. & I. Blok, 1980. Valse meeldauw in sla. Zaadbelangen, 34: 46–48.

Eenink, A.H., I. Blok & R. Groenwold, 1983. Onderzoek van de relatie tussen *Lactuca* en *Bremia lactucae.* Verslag van een voorlichtingsbijeenkomst voor slaveredelaars. Zaadbelangen 37: 95–96.

Farrara, B.F., T.W. Ilot & R.W. Michelmore, 1987. Genetic analysis of factors for resistance to downy mildew (*Bremia lactucae*) in species of lettuce (*Lactuca sativa* and *L. serriola*). Plant Pathol. 36: 499–514.

Groenwold, G. 1978. Het *Bremia* probleem in sla. Zaadbelangen 32: 132–134.

Hulbert, S.H. & R.W. Michelmore, 1985. Linkage analysis of genes for resistance to downy mildew (*Bremia lactucae*) in lettuce (*Lactuca sativa*). Theor. Appl. Genet. 70: 520–528.

Jagger, I.C. & T.W. Whitaker, 1940. The inheritance of immunity from mildew (*Bremia lactucae*) in lettuce. Phytopathology 30: 427–433.

Kesseli, R.V., I. Paran & R.W. Michelmore, 1990. Genetic map of lettuce (*Lactuca sativa,* 2N = 18). In: S.J. O'Brien (Ed.) Genetic Maps, pp. 6.100–102. Cold Spring Harbor Press, New York.

Landry, B.S., R.V. Kesseli, B. Farrara & R.W. Michelmore, 1987. A genetic map of lettuce (*Lactuca sativa* L.) with restriction fragment length polymorphism, isozyme, disease resistance and morphological markers. Genetics 116: 331–337.

Leroux, P., B. Maisonneuve & Y. Bellec, 1988. Detection en France de souches de *Bremia lactucae,* agent du mildiou de la laitue, resistantes au metalaxyl et a l'oxadixyl. P.H.M. Revue-Horticole 292: 37–40.

MacPherson, N.J., 1932. The cultivation of lettuce under glass with special reference to varieties resistant to downy mildew. J. Min. Agric. (Lond.) 38: 998–1003.

Norwood, J.M. & I.R. Crute, 1980. Linkage between genes for resistance to downy mildew (*Bremia lactucae*) in lettuce. Ann. appl. Biol. 94: 127–135.

Rodenburg, C.M., 1975. Identificatie onderzoek bij sla. Zaadbelangen 29: 319–321.

Rodenburg, C.M. & J.A. Huyskes, 1962. Ten years of Dutch lettuce breeding. Grower 57: 300–301.

Schettini, T.M., E.J. Legg & R.W. Michelmore, 1991. Insensitivity to metalaxyl in California populations of *Bremia lactucae* and resistance of Californian lettuce cultivars to downy mildew. Phytopathology 81: 64–70.

Schultz, H. & K. Röder, 1938. Die Anfälligkeit verschiedener Varietäten und Sorten von Salat (*Lactuca sativa* L. und *Lactuca serriola* L.) gegen den Falschen Meltau (*Bremia lactucae* Regel). Züchter 10: 185–194.

Tjallingii, F. & C.M. Rodenburg, 1967. Onderzoek van slarassen op vatbaarheid voor drie fysio's van valse meeldauw (*Bremia lactucae*). Zaadbelangen 21: 104–105.

Tjallingii, F. & C.M. Rodenburg, 1969. Onderzoek van slarassen op vatbaarheid voor vier fysio's van valse meeldauw. Zaadbelangen 23: 436–438.

Whitaker, T.W., G.W. Bohn, J.E. Welch & R.G. Grogan, 1958. History and development of head lettuce resistant to downy mildew. Amer. Hort. Soc. Sci. Proc. 72: 410–416.

Euphytica **63**: 103–113, 1992.
R. Johnson and G.J. Jellis (eds), Breeding for Disease Resistance

Pre-emptive breeding to control wheat rusts

R.A. McIntosh
University of Sydney Plant Breeding Institute, Cobbitty, Cobbitty Road, Cobbitty, N.S.W., Australia 2570

Key words: durable resistance, genetic diversity, pathogenic variation, pre-emptive breeding, *Puccinia* spp., resistance breeding, *Triticum aestivum,* wheat rust

Summary

Pre-emptive or anticipatory breeding for resistance is breeding for resistance to future pathotypes. It is assumed that these will be derivatives of currently frequent pathotypes that need to mutate with respect to single host resistance genes in order to attack widely-grown cultivars. Success in this approach depends on relevant knowledge of the pathogenicity phenotypes and host resistance genes that occur throughout the wheat-growing areas. Because durability of resistance cannot be assumed, resistance breeding strategies are usually supported with the maintenance of genetic diversity to provide buffering against extreme crop losses in the event of significant pathogenic changes.

Introduction

The aim of pre-emptive or anticipatory breeding to control wheat rusts is to have sources of resistance available to combat future pathotypes. Its effectiveness (success) depends on the ability of the breeder to predict the likely pathogen phenotypes (pathotypes) that will be important at some future time.

Important future pathotypes will be those with ability to overcome one or more of the genes currently deployed to confer resistance. In practice it will be advantageous to know the arrays of genes for avirulence and virulence (described by pathotype, strain, race, or pathogenicity formula) that will be associated with the critical virulence attribute(s) conferring the immediate selective advantage. Under Australian conditions mutation in the pathogen is the most important cause of variability that can be anticipated. The likelihood of mutant pathotypes occurring and becoming established is related to the frequency of potential parental genotypes and the numbers of avirulence genes which must change to virulence in order to enable a mutant pathotype to overcome the resistance of a particular host cultivar. In addition, the area sown to a particular host genotype (one or more cultivars) will contribute in providing the substrate for potential mutants to be selected and to become established.

In order to understand pathogen populations and to manage a diversified host population, it is necessary to know the resistance genes that are deployed and to conduct relevant pathogenicity surveys that monitor variability with respect to those genes.

Relevance of pathogenicity surveys and of pathotypes in selection nurseries

In a resistance breeding context, a relevant pathogenicity survey is one that monitors variability in pathogen populations in terms of the resistance genes that are currently deployed or likely to be deployed in the immediate future. If resistances are based on genes effective only at the adult plant stages, then adult plant tests should be conducted on a routine basis. Many pathogenicity surveys monitor an arbitrary array of seedling resistance factors and the data obtained are of only limited use to breeders who then adopt the strategy of using all available pathotypes in selection nurseries. In Australia, pathogenicity surveys have had two major objectives. Firstly, pathogenicity genes are used as markers in monitoring the origins, movement and epidemiology of rust pathogen populations and secondly, they monitor the disease responses conferred by the resistance genes actually deployed. The majority of current resistances to stem rust and leaf rust can be explained by the presence of genes effective at the seedling stage although there is increasing evidence, especially for leaf rust, that additional factors effective only at the adult stage, may be involved. Effective stripe (yellow) rust (caused by *Puccinia striiformis* f. sp. *tritici*) resistances are mainly of the post-seedling type and surveys based on seedling tests are used for epidemiological studies and to monitor the responses of cultivars which carry seedling resistance factors, including *YrA, Yr6, Yr7, Yr9* and *Yr17*. A cultivar with *Yr10* was recently registered in South Australia.

With relevant information on pathogenicity from surveys and genetic information from host studies, it should be possible to choose limited numbers of pathotypes for release in selection nurseries. Whereas breeding objectives may be satisfied using pathotype mixtures, genetic studies are most appropriately conducted with single pathotypes in order to avoid the complexities that occur when resistance to a pathotype mixture is due to gene combinations. For example, a mixture of *Puccinia recondita* f. sp. *tritici* (*P. r. tritici*) pathotypes (pt.) 104–2,3,6,(7) and 53–1,6,(7),10,11 is used at the Plant Breeding Institute (PBI), Cobbitty, to avoid selection of lines with only *Lr13* (which is overcome by pt. 53–) or lines with genes such as *Lr1, Lr3, Lr14a* and *Lr23* to which pt. 53– is avirulent, but 104– is virulent. The Thatcher derivative, Manitou, with gene *Lr13* (Dyck et al., 1966) is resistant to this mixture, whereas Thatcher and certain lines with *Lr13* such as Sunstar and a Red Bobs + *Lr13* line, are susceptible. While a previously undocumented gene could be postulated to explain the resistance of Manitou, adult plant studies of the above wheats in the greenhouse using separate inoculations showed that Thatcher was actually resistant to pt. 53–, but susceptible to pt. 104–. Thatcher is known to carry the adult plant resistance gene *Lr22b* which is not widely effective in North America (Dyck, 1979). It seems likely that in the Australian situation the combination of *Lr13* and *Lr22b* confers resistance to the pathotype mixture. It is of interest that had a genetic study of Manitou been conducted using the same inoculum mixture, the F2 segregation ratio would have been 9 resistant : 7 susceptible and the hypothesis would have been that resistance was conferred by complementary genes (Table 1). An important principle from this study is that care must be exercised when genetic data from one geographical area are applied in another geographical area. A resistance gene considered unimportant in one area can make a significant contribution in another area.

Pathogenic variation

Watson (1970) reviewed the contributions of the major evolutionary forces to pathogenic variability in *Puccinia graminis* f. sp. *tritici* (*P. g. tritici*) in Australasia. In reviewing these factors greatest attention will be given to changes that have occurred since 1970.

Migration. Australia and New Zealand comprise a single geographical area with respect to pathogenic variability in the cereal rusts. The introduction of inoculum (and successful establishment of introduced clones) from outside the area is a rare event. In the case of *P. g. tritici* there were three occasions when distinctive pathotypes were introduced, viz. 1926, 1954 and 1968. The sources of the first two introductions are unknown, but the 1968 introductions most probably came from southern Africa and wind-borne entry is suspected (Watson & de Sousa, 1982). Once introduced, further introductions of identical or similar pathotypes cannot be recognized. Upon establishment within Australasia further spread is largely dependent on weather patterns which are predominated by air movement from west-to-east. Pathogens that become established in Western Australia rapidly appear in eastern Australia and ultimately, New Zealand, whereas urediniospore movements in the opposite direction are less likely events, although they are recorded from time to time (Luig, 1977).

The introduction of *P. striiformis* f. sp. *tritici* (*P. s. tritici*) to Australia in 1979 (O'Brien et al., 1980), and its subsequent establishment and spread, serves as an example of global movement of plant pathogens; the introduction of *P. s. hordei* to South America is a further example (Dubin & Stubbs, 1986). Within a year of its discovery in Victoria, stripe rust was found over most of the wheat-growing areas of eastern Australia extending from central Queensland to the western wheat-growing areas of South Australia, and appeared in New Zealand. However, it has not been reported in Western Australia despite the widespread cultivation of susceptible cultivars. Pathotype 104 E137 A- was almost certainly man-borne from Europe (Wellings et al., 1987), probably France (Stubbs & Wellings, personal communication).

Information on exotic sources of *P. recondita* f. sp. *tritici* (*P. r. tritici*) in Australia is not so clear. There is little doubt that the isozymically-distinctive *Lr13*-virulent pathotype 53–1,6,(7),10,11 was an exotic introduction to New Zealand where it was first identified in early 1981 (Luig et al., 1985). This pathotype appeared in eastern Australia in 1984 where it became established on cultivars such as Sunstar with *Lr13*. Several avirulence and virulence attributes as well as their specific combination appear to be unique. A North American origin for this pathotype appears unlikely because it is also isozymically distinctive from North American pathotypes. In addition, this pathotype was avirulent for *Lr15* and *Lr22b* which are extremely rare features in North American isolates of the pathogen.

Mutation. Much of the pathogenic variation found in cereal rust pathogens in Australia can be explained on the basis of mutation. New variants usually differ from pre-existing pathotypes on the basis of single genes. Evolutionary pathways can be constructed to show the occurrence of mutational changes over time and space (for example, Wellings & McIntosh, 1990). Evolutionary pathways were also developed for *P. g. tritici.* In the early 1970s similar pathways were demonstrated in the laboratory using sequential treatments with mutagenic chemicals (Watson, Luig & Gow, unpublished data).

During the early 1980s stem rust appeared on triticale in south eastern Australia. The pathotype involved, 34-2,12, was similar to less evolved variants mainly located in Western Australia where susceptible wheats continued to be grown. However, it was uniquely virulent for several recommended

Table 1. Genotypes and phenotypic responses of F2 individuals from a cross of Manitou (*Lr13Lr13 Lr22bLr22b*) and a susceptible line (*lr13lr13 lr22blr22b*) to single cultures and a mixture of *P. r. tritici* pathotypes

Genotype		Ratio	Response to *P. r. tritici* pathotypes		
			104–2,3,6,(7)	53–1,6,(7),10,11	Mixture
Lr13Lr13	*Lr22bLr22b*	1	Res[1]	Res	Res
	Lr22blr22b	2	Res	Res	Res
	lr22blr22b	1	Res	Sus	Sus
Lr13lr13	*Lr22bLr22b*	2	Res	Res	Res
	Lr22blr22b	4	Res	Res	Res
	lr22blr22b	2	Res	Sus	Sus
lr13lr13	*Lr22bLr22b*	1	Sus	Res	Sus
	Lr22blr22b	2	Sus	Res	Sus
	lr22blr22b	1	Sus	Sus	Sus
Ratio Res:Sus			3:1	3:1	9:7

[1] Res = resistant; Sus = susceptible.

triticale cultivars including Coorong (McIntosh et al., 1983). Coincidentally, this pathotype was virulent for *Sr27* which had been earlier transferred to wheat from Imperial rye. In 1982 severe losses in triticale crops were experienced by farmers in northern New South Wales and Queensland. As a consequence, cultivar Satu and others with a different gene were recommended, but these succumbed to a variant of the 'Coorong' pathotype 34-2,12,13 within 2 years. These changes within *P. g. tritici* were independent of those relating to wheat with the consequence that in north eastern Australia quite different pathotypes were occurring on wheat and triticale. R.P. Singh (personal communication) has demonstrated pathogenic polymorphism in Mexican *P. r. tritici* populations with respect to triticale genotypes. This variability is additional to, and independent of, the variability that occurs with respect to wheat.

The role of mutation in the evolution of *P. r. tritici* in Australia is less clear, mainly because this pathogen has not had the detailed attention that has been given to *P. g. tritici.* Although mutation events have undoubtedly occurred, some variants seem to differ from their putative parental types by more than a single characteristic. During the mid-1980s, pathotype 104–2,3,6,(7),11 with virulence for *Lr16* appeared as a putative mutant of 104–2,3,6,(7), but this 'mutant' appeared somewhat more avirulent than its assumed parent with respect to host lines with *Lr2a* and *Lr26.* The emergence of pt. 104–2,3,6,(7),11 occurred at a time of staff changes within the Institute and, because the avirulent seedling response conferred by *Lr16* requires relatively high temperatures, some early isolates were probably not correctly identified. Consequently, there is no reliable record of where this pathotype first became established, but it has spread throughout eastern Australia, and more recently appeared in Western Australia (Park & Wellings, 1992). It and a mutant derivative, 104–1, –2,3,6,(7),11, now predominate throughout the country. *Lr16* is not present in any Australian wheat cultivar.

Recombination. Sexual recombination does not contribute to pathogenic variability in wheat rust fungi in Australia. Watson (1981) described instances where asexual recombination (somatic hybridization) gave rise to an important related group of pathotypes which are now extinct. In addition, somatic hybridization between *P. g. tritici* and *P. g. secalis* apparently generated variants that appeared to survive on certain grasses and barley rather than on wheat or rye (Luig, 1977). Although asexual recombination has been reported in *P. s. tritici* (Little & Manners, 1969; Goddard, 1976; Wright & Lennard, 1980) there is no evidence that it has had a significant impact on the evolution of this species or of *P. r. tritici* within Australasia.

Selection. Selection of pathotypes by host genotypes with relevant resistance factors is a major force by which newly introduced or mutant pathotypes increase in frequency. The rapid increases in *P. g. tritici* pathotypes with virulences for triticales in Australia was due to selection on triticales with unique resistance genes; firstly pathotype 34–2,12 with virulence for *Sr27* in Coorong, and subsequently 34–2,12,13 with virulence for Sr_{Satu} in Satu and certain additional cultivars. Neither of these pathotypes could attack the majority of wheats in the areas where they occurred, nor could the concurrent pathotypes found on wheat attack triticale.

Chance. Chance may be a neglected factor in the long term survival of pathogens that undergo massive population explosions and contractions during annual cropping and non-cropping cycles. In order to survive during the non-cropping periods the obligate wheat rust pathogens must find congenial wheat, barley or grass genotypes. This is now extremely difficult for *P. g. tritici* in the traditional oversummering summer rainfall areas of northern New South Wales and Queensland where, except for very small areas, only resistant cultivars are recommended and grown. A major consequence is that virtually all self-sown out-of-season wheat and stubble regrowth is also resistant.

P. g. tritici pathotype 343–1,2,3,5,6 capable of attacking Oxley, a predominantly Queensland cultivar, first appeared in Victoria in 1973. This pathotype subsequently spread throughout mainland

Australia, including Western Australia, either spreading from the initial source, or alternatively, arising as a consequence of repetitive mutation events (Zwer et al., 1992). Oxley finally succumbed to damaging stem rust infections in 1982 and 1983. Although eventual crop losses in Oxley were predicted, the genetic 'cost' of rust on Oxley was the occurrence in 1984 of a mutant pathotype, 343–1,2,3,4,5,6, capable of attacking cv. Cook. Despite the presence of the virulent pathotype, no crop of Cook recorded significant crop loss. An active publicity campaign alerting farmers to the vulnerability of Cook at a time when both Oxley wheat and triticales were being damaged, and the availability of Cook derivatives with added genes for resistance, resulted in the complete withdrawal of Cook in a single season. As a consequence, the Cook-attacking pathotype also declined. Apparently it had not become established over an adequate area to ensure its survival in the absence of a selective host. The Cook derivatives that are currently cultivated have not only the added genes, but also the additional benefit of gene *Sr36* which has been used in Australian wheat cultivars since the 1960s.

Figure 1 shows the numbers of *P. g. tritici* isolates processed in national pathotype surveys for the period 1957–1990 and the number of different pathotypes identified in each year. In addition to the correlation between numbers of isolates and pathotypes, a notable aspect is the low numbers of isolates obtained over recent years. This seems to be the outcome of a long-term policy of recommending only stem rust resistant cultivars in the more rust-prone areas of New South Wales and Queensland. With a very low disease incidence, the probability of pathogen mutants becoming established also appears to be low, and genes formerly believed to lack adequate durability, such as *Sr9e,* appear to contribute to genetic diversity and the provision of long-term resistance. Because self-sown wheat and stubble regrowth are also resistant, opportunities for oversummer survival of *P. g. tritici* appear to be minimized and stem rust isolates from barley, *Agropyron scabrum* and from non-inoculated research stations often yield one of the triticale-attacking pathotypes, *P. g. secalis* or putative hybrids of *P. g. tritici* and *P. g. secalis.*

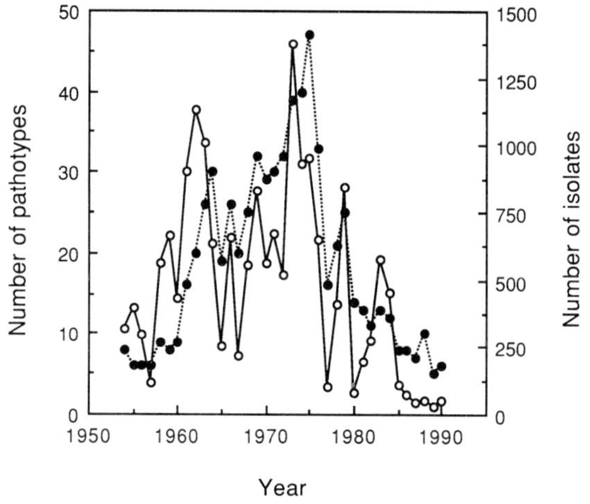

Fig. 1. Frequencies of *P. graminis tritici* (solid points) and isolates (open points) for the period 1950–1990.

In southern New South Wales, Victoria and South Australia, several isolates of pathotype 343–1,2,3,5,6,8,9 were identified in 1988. This pathotype is virulent for *Sr30* which seems to be the main effective gene in the widely grown cultivars Osprey, Rosella, Vulcan and Sunfield. Farmers were alerted by the allocation of these cultivars to the 'recommended but outclassed category', but due to the lack of suitable alternative cultivars, they continue to be grown. Because of the extremely low incidence of stem rust, it is unlikely that pathogen populations could increase sufficiently to cause damage in a single season. Meanwhile, shifts to alternative cultivars such as Janz and Sunbri are being encouraged and a very close monitoring of rust incidence and pathotypes is being maintained so that farmers can be alerted in the event of significant population changes.

Features of rust resistance

In discussing resistance, three factors need to be considered, viz. genetic diversity, durability and effectiveness.

Genetic diversity. Because mutational changes in pathogens usually involve single genes, host genetic diversity is encouraged to provide a buffering influence in the event of a rust outbreak. Information on genetic diversity, or a listing of recommended cultivars in diversity groups, should be available to extension personnel and farmers to assist in advisory and decision-making processes. This necessitates the availability of genetic information on recommended cultivars. Despite concern that such policies will exhaust the available sources of resistance, breeders continue to find and exploit further genes or gene combinations, particularly those which gain a reputation for durability as well as new sources emerging from cytogenetical investigations. The use of genetic diversity in this way is an inherent admission that resistance may not be stable over time; hence diversity is insurance against genetic vulnerability.

Rust epidemics are analogous to bush or scrub fires – excessive amounts of dry fuel over large areas (susceptible hosts) combined with hot weather and dry winds (favourable environment) are the ingredients of devastating fires periodically experienced in Australia, western U.S.A. and elsewhere. Indeed, fires (and epidemics) can become so fierce that they are self-driven and even relatively green grass and timber lands (resistant cultivars) will be affected. The extent of damage caused by fires can be curbed by the removal or reduction of dry fuel, or by the provision of fire breaks (resistant cultivars or alternative crops). Obviously, a fire hazard will also be reduced if green materials (resistant genotypes) are mixed with dry fuel (i.e., multilines).

Durability. A durable resistance source is one that has remained effective with prolonged widespread use (Johnson, 1984 for review). Past effectiveness alone is sufficient incentive to use it in the future. Obviously, if a source of resistance could be assumed to be permanently durable (non-specific resistance) and provided adequate resistance was conferred, then only one source of resistance would be required. Characteristics often associated with durable resistance sources are that they are more likely to be of the adult plant, rather than seedling type, that they are conferred by more than a single gene acting additively, and that they are not associated with genes conferring hypersensitive responses.

The successes of wheat rust resistance breeding programmes have been aided by the use of resistance sources that have proved durable. In Australia and elsewhere the use of adult plant resistance to stem rust derived from Hope and H–44, and determined largely by the gene *Sr2* on chromosome 3BS, has had a major impact on the control of stem rust. Many of the wheats selected from the multilocational testing strategy adopted by the International Maize and Wheat Improvement Centre (CIMMYT) in Mexico carry this gene in combination with an array of seedling resistance factors. The gene *Sr26* transferred to wheat from *Agropyron elongatum* (Knott, 1961) has been widely used in Australian wheat cultivars since the release of Eagle in 1969. No *P. g. tritici* variant with virulence for this seedling-effective factor has been sampled from the field. A second seedling-effective gene, *Sr24,* from *A. elongatum* has been deployed in Australian cultivars since 1983 (Table 2); however, pathotypes virulent for *Sr24* have been recorded in South Africa (Le Roux & Rijkenberg, 1987) and India (S.K. Nayar, personal communication). The gene *Sr36* can be described as having provided durable resistance since the 1960s, despite the occurrence of virulent pathotypes on at least three occasions, the most recent of which was described above. The use of these genes in combination with others has undoubtedly reduced the vulnerability of the Australian wheat crop to stem rust.

Whereas leaf rust is considered internationally to be the most important wheat rust disease, this is not a commonly held view in Australia. Although yield reductions as high as 40% have been associated with leaf rust infection (Samborski, 1985), Australian farmers generally do not perceive significant losses. A number of Australian wheats have at least some resistance to leaf rust. The gene *Lr24* occurs in several cultivars, *Lr13* and/or gene combinations involving *Lr13* provide resistance in several other wheats. Apart from the exotic pt. 53–1,6,(7),10,11 which is avirulent for a number of well known previously 'defeated' genes including *Lr1,* all *Lr2* alleles, *Lr3, Lr14a* and *Lr22b,* all Australian *P. r. tritici* pathotypes are avirulent for *Lr13* and no local mutant with *Lr13* virulence has been identified since the release of Egret in 1973. Pt. 53– is virulent on wheats such as Sunstar with only *Lr13*. Combinations of *Lr13* and any of the previously defeated genes listed above confer resistance to the entire spectrum of Australian pathotypes. No mutant variant of pt. 53– has been identified. The effectiveness of gene combinations involving *Lr13* in Australia has been verified for several wheats considered susceptible in other countries, such as the Indian cultivars WL711 (*Lr13* + *Lr14a*) and Sonalika (*Lr13* + *Lr14a*), the Canadian cultivar Manitou (*Lr13* + *Lr22b*) and the CIMMYT-derived line, Inia 66 (*Lr13* + *Lr17* + *Lr14a*). The resistance in Australia of many wheats from a range of sources can be explained on the basis of gene combinations including *Lr13*. Moreover, recent evidence from local work (McIntosh, unpublished) and from Dr. R.P. Singh in Mexico indicate that additional adult plant resistance genes are involved in many instances. For example, the Australian cultivar Hartog (= Pavon 'S') appears to carry two genes for adult plant resistance in addition to *Lr1* and *Lr13* which combined, confer resistance to all pathotypes.

Work in Canada (Dyck, 1987; Dyck & Samborski, 1982; Shang et al., 1986), U.S.A. (Roelfs, 1988) and Mexico (Sing & Rajaram, 1991) has in-

Table 2. Australian cultivars approved for sowing in the summer rainfall areas

Cultivar	Stem rust		Leaf rust		Stripe rust	
	Response	Effective *Sr*	Response	Effective *Lr*	Response	Effective *Yr*
Banks	R[1]	*12, 30*[2]	R	*13*[3]	MR	*A**[4], APR[5]
Batavia	R	?	R	?	MR	APR
Cunningham	R	*24*	R	*24*	R	APR
Dollarbird	R	*2, 9g, 30*	R	*1, 30*	R	*7*, APR
Hartog	R	*2, 9g, 30*	R	1, 13 APR	R	*6, 7*, APR
Hyb. Comet	R	*26*	S	–	MR	*A*, APR
Hyb. Meteor	R	*26*	R	*13*	R	*A*, APR
Janz	R	*24*	R	*24*	MR	APR
Kamilaroi[6]	R	?[7]	R	?	R	APR
Kite	R	*26*	S	–	MR	APR
Miskle	R	*13*, 30*	R	*2a, 13*	MR	APR
Osprey	MR	*12, 30*	S	–	MR	APR
Owlet	R	*7b, 26*	R	?	R	APR
Perouse	R	*24*	R	*24*	R	*6*, APR
Rosella	R	*7b, 12, 30*	S	–	MR	*A*, APR
Sunbri	R	*36, 38*	R	*3, 37*	R	*11*, APR
Sunco	R	*24, 36*	R	*24*	R	APR
Suneca	R	*2*	R	*1, 13*	MS	APR
Sunelg	R	*24, 26*	R	*24*	R	APR
Sunfield	R	*12, 30*	R	*1, 13*	MR	*A*, APR
Sunkota	R	*2, 9g*	R	*17*, +	MR	*7*, APR
Takari	R	*26*	R	*3, 13*	MR	*6*, APR
Vasco	R	*24*	R	*24*	R	*6*, APR
Vulcan	MR	*30*	R	*13*	MR	*A*, APR
Yallaroi[6]	R	?	R	?	R	APR

[1] R = resistant, MR = moderately resistant, MS = moderately susceptible, S = susceptible.
[2] Pathotype virulent for *Sr30* is extremely rare.
[3] Although *Lr13* is ineffective against pt. 53–1,6,(7),10,11, there have been no reports of severe rusting from farmers.
[4] Asterisk indicates that the cultivar is heterogeneous for this gene.
[5] Unidentified gene(s) for adult plant resistance.
[6] *T. turgidum* group durum.
[7] ? unidentified gene(s).

dicated the role of *Lr34* as a source of durable resistance to leaf rust. Recent studies in Australia (McIntosh, 1992) and Mexico (Singh, 1992a) have shown that lines carrying *Lr34* also possess resistance to stripe rust conferred by a factor recently designated *Yr18*. Moreover, this gene combination appears to be present in Chinese Spring (Dyck, 1991) and Bezostaja (Dyck, personal communication), as well as a wide range of spring and winter wheats possibly including Cappelle Desprez and Hybride de Bersee (McIntosh, 1992). Canadian workers (German & Kolmer, 1992) have shown that *Lr34* is highly interactive with a number of genes which alone, confer only moderate levels of resistance.

Examples of durable resistance to stripe rust are listed in Johnson (1988). Genetic studies of at least some of these sources and others by McIntosh and colleagues indicate that several sources of adult plant resistance to stripe rust are conferred by combinations of two to several genes. Hence the apparent durability of certain sources of adult plant resistance to stripe rust may be due to this multiple gene basis.

Effectiveness. Under field conditions, responses to rust diseases cover a continuous spectrum ranging from no visible symptoms to extremely susceptible. A concept of susceptibility varies with the interest and objectives of the scientist. From the standpoint of plant pathology, a susceptible genotype is one that sustains an unacceptable degree of loss when infected with a particular pathogen. Rather than being loss based, assessment of susceptibility is generally empirical and based on experience. A genetic definition of a susceptible plant might be one that has no measurable degree of resistance relative to an arbitrarily designated reference base or might simply be a contrasting phenotype to one considered resistant. Genotypes considered 'susceptible' or moderately susceptible in an agricultural context often have considerable degrees of resistance compared to some extreme suscepts.

Plants possessing genes for resistance may be subject to yield loss even when infected with avirulent pathotypes. For example, wheat lines possessing *Sr13* gave mean reductions in grain weight of 20% compared with rust-free controls. By comparison, susceptible control cultivars had mean grain weight reductions of 55% (McIntosh, Latter, Rees & Platz, unpublished). Although these differences might not be as great under commercial growing conditions, the gene *Sr13* is considered inadequately effective when used alone. A further example involves gene *Lr34* which in certain genetic backgrounds may not confer adequate resistance. Some wheats with this gene, such as Oxley and Opata 85, may reach terminal leaf rust levels of 50–60S under Australian field conditions. Such levels of disease may be considered unacceptably high by breeders whose objective is to minimise fungal population levels as well as to protect the crop against loss.

Again, wheat cultivars show a very wide range of responses to stripe rust varying from no visible symptoms to complete defoliation typical of cv. Morocco and certain selections of Avocet. Many wheats considered too susceptible in agriculture possess levels of resistance relative to Avocet. Pope (1968) described transgressive segregation in intercrosses of susceptible and moderately susceptible cultivars. Similar results were reported by others (Henriksen & Pope, 1971; Reinhold et al., 1983; Wallwork & Johnson, 1984). Following the finding that adult plant stripe rust resistance in Australian wheats is conferred by gene combinations with additive effects, attempts are currently being made to identify lines that differ from Avocet by single genes for adult plant resistance. None of these lines is expected to be as resistant as the most resistant Australian cultivars such as Cook and Oxley.

Until recently, CIMMYT breeders had a policy of selecting wheats with moderate levels of adult plant resistance (slow rusters) as a means of avoiding less durable hypersensitive resistances controlled by single genes. Several genes of this slow rusting type were identified in a range of lines developed in Mexico (R.P. Singh, personal communication). Deliberate attempts to combine these genes resulted in rust-free lines that cannot be distinguished from lines with hypersensitive responses. Such lines have the advantage of producing no inoculum as well as a reduced likelihood of being rapidly overcome by mutational change in the pathogen. Because these resistances are based on gene combinations they are more likely to confer low coefficients of infection over wide areas. Moreover, lines with *Lr34/Yr18* could be deliberately used as a component of such resistance, and so provide an assurance that a considerable degree of stripe rust resistance could be achieved even in the absence of selection for that disease.

Use of genetic linkages in breeding for rust resistance

A major objective in mapping molecular markers in agriculturally important crops is the expectation of discovering close genetic linkages involving these easily classified, usually co-dominant, markers and genes controlling quantitative traits or resistances to diseases whose responses are difficult to score or expensive to measure. For example, resistance to eyespot (caused by *Pseudocercosporella herpotrichoides*) transferred to the wheat line VPM1 from *Aegilops ventricosa* is completely associated with the isozyme marker *Ep-D1b* in chromosome 7D (McMillin et al., 1986; Worland et al., 1988). Similarly, the presence of a unique glucose phosphate isomerase (*Gpi-R1*) allele (Gale & Sharp, 1988) or of the presence of secalin gene markers can be used to 'tag' a 1BL.1RS translocation chromosome which carries a number of genes for disease resistance.

Similarly, there are several instances where genes for rust resistance are linked with other disease resistance factors, or with distinctive morphological markers. Wheats possessing the above 1RS chromosome from Petkus rye always carry *Sr31* (resistance to stem rust), *Lr26* (leaf rust), *Yr9* (stripe rust) and possibly *Pm8* (powdery mildew) and selection for any gene results in the presence of the others, or can be assisted by selection for one of the others. For purposes of gene postulation, the positive identification of any one gene permits postulation of the others. The gene *Lr24* derived from *Agropyron elongatum* is always associated in wheat with the presence of *Sr24*. In addition to eyespot resistance in chromosome 7D, VPM1 wheat and its U.K. derivative, Rendezvous, and North American derivatives, Hyak and Madson, carry the linked genes *Sr38, Yr17* and *Lr37* in chromosome 2A (Bariana, 1991). The presence of gene *Sr38,* and probably *Lr37,* in these lines was presumably a consequence of selection for stripe rust resistance conferred by *Yr17.*

For many years, it has been known that wheats with *Sr2* develop a melanin pigmentation on spikes and lower leaves (Hare & McIntosh, 1979). More recently it has been shown in our laboratory (Brown & Atkinson, unpublished) that seedlings of genotypes with *Sr2* develop a distinctive mild necrosis when grown at temperatures above about 22° C. This feature enables selection and confirmation of the presence of this usually recessive adult plant resistance gene. *Lr34/Yr18* is associated with a distinctive leaf tip necrosis (Dyck, 1991; Singh, 1992b). As wheat breeders become more dependent on alien sources of disease resistance usually involving large chromosome segments, an increasing number of genetic linkages involving genes for resistance to other diseases, as well as to morphological and molecular markers, can be anticipated.

Conclusions

Cereal rust research in Australia has been ongoing since the late 1800s. The first phase of this work involved a search for resistance and a basic understanding of the epidemiology of the diseases involved. The first successful rust resistant cultivars such as Eureka (*Sr6*) and Gabo (*Sr11*), released in the 1940s, conferred resistance for less than five years. An understanding of the relationship between host response and pathogenicity, made possible by Flor's work with flax rust (Flor, 1956), showed the need for genetic diversity as a basis for continued resistance. The use of oligogenic resistances and the incorporation of more durable resistances conferred by genes such as *Sr2, Sr36* and *Sr26* extended the post-release periods of effective resistance for individual cultivars. The construction of gene combinations was aided by increasing knowledge of host genetics and by the knowledge that probable future pathotypes would be mutants of currently important pathotypes. To overcome difficulties in assembling certain gene combinations, chemical mutagens were used to construct pathotypes suitable for selection purposes. One such culture, 74–L–1 (pt. 34–1,2,3,4,5,6,7) is still a major test culture in our laboratory after 18 years. With increasing numbers of effective resistance genes for deployment, a recognition of those with potential for more durable resistance and an increasing area under cultivation with stem rust resistant cultivars, it became possible finally to adopt a policy of recommending only resistant wheats throughout the more rust-prone summer rainfall areas. This has resulted in a decline in frequencies of both *P. g. tritici* populations and pathotypes. A sound knowledge of the resistance genes that are deployed, assists in selecting future resistance sources, the actual pathogen cultures to be used for breeding, and ability to determine which cultivars will be affected if a significant pathotype change were to occur. The decline in the *P. g. tritici* population in the summer rainfall areas has had a carryover effect in the southern wheat areas where the probability of rust epidemics is lower. In addition, there has been a gradual increase in the use of resistant cultivars in these latter areas combined with attempts to recommend resistant genotypes in local areas considered more rust-prone.

Another strategy attempted in the late 1960s was to derive a global inventory of pathogenicity patterns that would enable the construction of effective gene combinations at appropriate international locations. This procedure had potential to achieve the same objectives as the multilocational field testing operations practiced by CIMMYT breeders. Unfortunately, human, economic and plant quarantine constraints precluded the likelihood of success.

With the decline of stem rust in Australia there have been more serious attempts to make similar achievements with leaf rust and stripe rust as well as other diseases. Most cultivars recommended in the summer rainfall areas have adequate resistance to both leaf rust and stripe rust. Despite attempts to introduce minimum disease standards aimed at avoiding cultivars with excessive levels of disease susceptibility, further increases in disease resistance levels are required in the south. In Western Australia there is a need for cultivars with improved levels of stripe rust resistance prior to the almost inevitable introduction of the pathogen from eastern Australia and its establishment on currently very susceptible cultivars.

Acknowledgements

The generosity of several colleagues in sharing results and providing manuscripts prior to publication is gratefully acknowledged. Wheat rust research in the PBI has been generously supported by the Grains Research and Development Corporation, formerly the Australian Wheat Research Council, since 1957.

References

Bariana, H.S., 1991. Genetic studies on stripe rust resistance in wheat. Ph.D. Thesis, The University of Sydney, 194 pp.
Dubin, H.J. & R.W. Stubbs, 1986. Epidemic of barley stripe rust in South America. Plant Disease 70: 141–144.
Dyck, P.L., 1979. Identification of the gene for adult plant leaf rust resistance in Thatcher. Can. J. Plant Sci. 59: 499–501.
Dyck, P.L., 1987. The association of a gene for leaf rust resistance with the chromosome 7D suppressor of stem rust resistance in common wheat. Genome 29: 467–469.
Dyck, P.L., 1991. Genetics of adult-plant leaf rust resistance in 'Chinese Spring' and 'Sturdy' wheats. Crop Sci. 31: 309–311.
Dyck, P.L. & D.J. Samborski, 1982. The inheritance of resistance to *Puccinia recondita* in a group of common wheat cultivars. Can. J. Genet. Cytol. 24: 273–283.
Dyck, P.L., D.J. Samborski & R.G. Anderson, 1966. Inheritance of adult plant leaf rust resistance derived from the common wheat varieties Exchange and Frontana. Can. J. Genet. Cytol. 8: 665–671.
Flor, H.H. 1956. The complementary genic systems in flax and flax rust. Adv. Genet. 8: 29–54.
Gale, M.D. & P.J. Sharp, 1988. Genetic markers in wheat – developments and prospects. In: T.E. Miller & R.M.D. Koebner (Eds). Proc. 7th Int. Wheat Symp. pp. 469–475, IPSR, Cambridge, U.K.
German, S.E. & J.A. Kolmer, 1992. Effect of resistance gene *Lr34* in the enhancement of resistance to leaf rust of wheat. Theor. Appl. Genet. 84: 97–105.
Goddard, M.V., 1976. The production of a new race, 105E137 of *Puccinia striiformis* in glasshouse experiments. Trans. Br. Mycol. Soc. 67: 395–398.

Hare, R.A. & R.A. McIntosh, 1979. Genetic and cytogenetic studies of durable adult-plant resistances in 'Hope' and related cultivars to wheat rusts. Z. Pflanzenzüchtg. 83: 350–367.

Henriksen, G.B. & W.K. Pope, 1971. Additive resistance to stripe rust in wheat. Crop Sci. 11: 825–827.

Johnson, R., 1984. A critical analysis of durable resistance. Ann. Rev. Phytopathol. 22: 309–330.

Johnson, R., 1988. Durable resistance to yellow (stripe) rust in wheat and its implications in plant breeding. In: N.W. Simmonds & S. Rajaram (Eds). Breeding strategies for resistance to the rusts of wheat. pp. 63–75. CIMMYT, Mexico.

Knott, D.R., 1961. The inheritance of rust resistance VI. The transfer of stem rust resistance from *Agropyron elongatum* to common wheat. Can. J. Plant Sci. 41: 109–123.

Le Roux, J. & F.H.J. Rijkenberg, 1987. Pathotype of *Puccinia graminis* f. sp. *tritici* with increased virulence for *Sr24*. Plant Disease 71: 1115–1119.

Little, R. & J.G. Manners, 1969. Somatic recombination in yellow rust of wheat (*Puccinia striiformis.*) I. The production and possible origin of two new physiologic races. Trans. Br. Mycol. Soc. 53: 251–258.

Luig, N.H., 1977. The establishment and success of exotic strains of *Puccinia graminis tritici* in Australia. Proc. Ecol. Soc. Aust. 10: 89–96.

Luig, N.H., J.J. Burdon & W.M. Hawthorn, 1985. An exotic strain of *Puccinia recondita tritici* in New Zealand. Can. J. Plant Pathol. 7: 173–176.

McIntosh, R.A., 1992. Close genetic linkage of genes conferring adult plant resistance to leaf rust and stripe rust in wheat. Plant Pathology 41: 523–527.

McIntosh, R.A., N.H. Luig, D.L. Milne & J. Cusick, 1983. Vulnerability of triticales to wheat stem rust. Can. J. Genet. Cytol. 5: 61–69.

McMillin, D.E., R.E. Allan & D.E. Roberts, 1986. Association of isozyme locus and strawbreaker foot rot resistance derived from *Aegilops ventricosa* in wheat. Theor. Appl. Genet. 72: 743–747.

O'Brien, L., J.S. Brown, R.M. Young & I. Pascoe, 1980. Occurrence and distribution of wheat stripe rust in Victoria and susceptibility of commercial wheat cultivars. Australasian Plant Pathology 9: 14.

Park, R.F. & C.R. Wellings, 1992. Pathogenic specialisation of wheat rusts in Australia and New Zealand in 1988 and 1989. Australasian Plant Pathology 21: 61–69.

Pope, W.K., 1968. Interaction of minor genes for resistance to stripe rust in wheat. In: K.W. Findlay & K.W. Shepherd (Eds). Proc. 3rd Int. Wheat Genetics Symp. pp. 251–257. Aust. Acad. Sci., Canberra.

Reinhold, M., E.L. Sharp & Z.K. Gerechter-Amitai, 1983. Transfer of additive 'minor effect' genes for resistance to *Puccinia striiformis* from *Triticum dicoccoides* into *Triticum durum* and *Triticum aestivum*. Can. J. Bot. 61: 2702–2708.

Roelfs, A.P., 1988. Resistance to leaf and stem rusts in wheat. In: N.W. Simmonds & S. Rajaram, (Eds). Breeding strategies for resistance to the rusts of wheat. pp. 10–22. CIMMYT, Mexico.

Samborski, D.J., 1985. Wheat leaf rust. In: A.P. Roelfs & W.R. Bushnell (Eds). The Cereal Rusts Vol II. pp. 39–59. Academic Press, Orlando.

Shang, H.S., P.L. Dyck & D.J. Samborski, 1986. Inheritance of resistance to *Puccinia recondita* in a group of resistant accessions of common wheat. Can. J. Plant Pathol. 8: 123–131.

Singh, R.P., 1992a. Genetic association of leaf rust resistance gene *Lr34* with adult plant resistance to stripe rust in bread wheat. Phytopathology 82: 835–838.

Singh, R.P., 1992b. Genetic association between gene *Lr34* for leaf rust resistance and leaf tip necrosis in bread wheats. Crop Sci. 32: 874–878.

Singh, R.P. & S. Rajaram, 1992. Genetics of adult plant resistance to leaf rust in 'Frontana' and three CIMMYT wheats. Genome 35: 24–31.

Singh, R.P. & S. Rajaram, 1991. Genes for low reaction to *Puccinia recondita* f. sp. *tritici* in 50 Mexican *Triticum aestivum* cultivars. Crop Sci. 31: 1472–1479.

Wallwork, H. & R. Johnson, 1984. Transgressive segregation for resistance to yellow rust in wheat. Euphytica 33: 123–132.

Watson, I.A., 1970. Changes in virulence and population shifts in plant pathogens. Ann. Rev. Phytopathology 8: 209–230.

Watson, I.A., 1981. Wheat and its rust parasites in Australia. In: L.T. Evans & W.J. Peacock (Eds). Wheat Science Today and Tomorrow. pp. 129–147. Cambridge University Press Cambridge.

Watson, I.A. & C.N.A. de Sousa, 1982. Long distance transfer of spores of *Puccinia graminis tritici* in the southern hemisphere. Proc. Linn. Soc. N.S.W. 106: 311–321.

Wellings, C.R. & R.A. McIntosh, 1990. *Puccinia striiformis* f. sp. *tritici* in Australasia: pathogenic changes during the first 10 years. Plant Pathology 39: 316–325.

Wellings, C.R., R.A. McIntosh & J. Walker, 1987. *Puccinia striiformis* f. sp. *tritici* in Australia – possible means of entry and implications for plant quarantine. Plant Pathology 36: 239–241.

Worland, A.J., C.N. Law, T.W. Hollins, R.M.D. Koebner & A. Guira, 1988. Location of a gene for resistance to eyespot (*Pseudocercosporella herpotrichoides*) on chromosome 7D of bread wheat. Plant Breeding 101: 43–51.

Wright, R.G. & J.H. Lennard, 1980. Origin of a new race of *Puccinia striiformis*. Trans. Br. Mycol. Soc. 74: 283–287.

Zwer, P.K., R.F. Park & R.A. Mcintosh, 1992. Wheat stem rust in Australia 1969–1985. Aust. J. Agric. Res. 43: 399–431.

Euphytica **63**: 115–123, 1992.
R. Johnson and G.J. Jellis (eds), Breeding for Disease Resistance

Durable resistance to rice blast disease – environmental influences

J.M. Bonman
Division of Plant Pathology, International Rice Research Institute, P.O. Box 933, Manila, Philippines; present address: The Du Pont Co., Agricultural Products Department, Stine-Haskell Research Center, P.O. Box 30, Newark 19714 DE, USA

Key words: environmental influence, *Oryza sativa, Pyricularia grisea, P. oryzae,* rice, rice blast

Summary

Blast is one of the most serious diseases of rice worldwide. The pathogen, *Pyricularia grisea,* can infect nearly all parts of the shoot and is commonly found on the leaf blade and the panicle neck node. Host resistance is the most desirable means of managing blast, especially in developing countries. Rice cultivars with durable blast resistance have been recognized in several production systems. The durable resistance of these cultivars is associated with polygenic partial resistance that shows no evidence of race specificity. This partial resistance is expressed as fewer and smaller lesions on the leaf blade but latent period does not appear to be an important component. Partial resistance to leaf blast is positively correlated with partial resistance to panicle blast, although some cultivars have been found showing leaf-blast susceptibility and panicle-blast resistance. A diverse set of environmental factors can influence the expression of partial resistance, including temperature, duration of leaf-wetness, nitrogen fertilization, soil type, and water deficit. Because of the great diversity of rice-growing environments, resistance that proves durable in one system may or may not prove useful in another. In highly blast-conducive environments, other means of disease management must be applied to assist host-plant resistance.

Introduction

Blast, a primary disease of rice worldwide, is caused by the fungus *Pyricularia grisea* (= *P. oryzae*) (Rossman et al., 1990). *P. grisea* is one of most widely distributed pathogens of rice, being found in nearly all rice-growing environments. The teleomorph of the pathogen, *Magnaporthe grisea,* has not been found in nature but can be produced in culture and is now used in several laboratories in studies of the genetics of the host-pathogen interaction (Valent, 1990). Blast disease remains a serious production problem in temperate and sub-tropical rice production areas, at high elevation in the tropics, and in tropical upland rice.

Aside from causing direct losses, the disease is also a constraint to increased production in many marginal rice-growing environments because it is affected by crop management, particularly the use of nitrogen (N) fertilizer. In many parts of Asia where farmers practice intensive, high-input agriculture, blast frequently causes direct yield losses or increases production costs when fungicides are used for control. However, the disease is rarely found in areas where low-input, traditional rice culture is practiced because the farmers apply little or no N fertilizer. In many tropical upland areas where traditional rice cultivars are grown, farmers' fields show no disease. In yield trials, however, the same cultivars may show severe infection, even at relatively low fertilizer input levels. In such environments, the disease is a constraint to increased yields rather than a current production problem.

The epidemiological potential of blast is not only

associated with methods of crop management but is also linked to the rice production environment. There are five broad categories of rice environments: irrigated, rainfed lowland, upland, deepwater, and tidal wetland rice (Khush, 1984). These environments differ in their general conduciveness to blast. The disease has its greatest potential for causing severe epidemics in irrigated rice in temperate regions and in areas where rice is grown as an upland crop. It is not considered a major production problem in deepwater and tidal wetland rice.

Tropical lowland rice grown with good water control is the least prone to blast damage. In the Philippines and other parts of tropical Asia, the importance of the disease in lowland areas seems to have declined during the past 25 years. This decline was probably caused in part by the increase in the area of irrigated ricelands. Drought stress greatly increases the susceptibility of rice to blast; however, irrigation has lessened the probability of such stress. The change from traditional to modern cultivars may also be involved since indications are that some of the modern cultivars are more blast resistant than the traditional cultivars used in the past.

In spite of these improvements, the disease continues to be a problem in irrigated areas of tropical and sub-tropical countries, especially where blast-susceptible cultivars have been released (Loganathan & Ramaswamy, 1984). In fact, the disease re-emerged during the 1980s as an important production problem in both irrigated and rainfed regions of India (Reddy & Bonman, 1987), presumably because of the popularity of certain susceptible cultivars.

Developing resistant cultivars is the most desirable means of managing blast, particularly since small farmers can easily adopt technology packaged in the form of an improved cultivar. The importance of disease resistance in rice cultivars is often unrecognized. In some areas, blast disease is 'not a problem' because resistant cultivars grown prevent losses that would otherwise occur if susceptible cultivars were used.

Resistance types and durability

Durable resistance to blast has been recognized in certain rice cultivars in certain environments (Bonman & Mackill, 1988; Johnson & Bonman, 1990). This durable resistance is associated with some specific characteristics of the resistance, again depending on the environment in question. In this discussion, three aspects of the type of resistance to leaf blast will be considered: its effects on pathogen reproduction, its inheritance, and its race specificity (Fry, 1982). Resistance to the neck blast phase of the disease will also be examined.

Effects on pathogen reproduction: complete or partial. Complete resistance to blast occurs when the fungus is unable to cause sporulating lesions on the plant. Complete blast resistance has been associated with spectacular 'breakdowns' in cultivar resistance. In Korea, the complete resistance of the Tongil cultivars was effective for 5 years before fungus races able to overcome that resistance appeared in 1976 (Lee et al., 1976). The weather during 1977 in Korea did not favour the disease and little blast was observed in the rice crop. In 1978, though, weather conditions favoured blast and by that time the pathogen had become established throughout the country; a devastating blast epidemic resulted (Crill et al., 1981). In Japan, the lifetime of complete resistance (termed 'true resistance' by Japanese researchers) appears to be about 3 years (Kiyosawa, 1982). Reiho, released in 1969 in Japan as a blast-resistant cultivar, possessed the gene *Pi-ta*^{2} for complete resistance to Japanese races of *P. grisea.* Its area of cultivation increased until 1973, when it was severely damaged by blast (Matsumoto, 1974). Similarly, Reiho was completely resistant to races of *P. grisea* common in Egypt at its release in 1984. Resistance was lost during its first year in production, resulting in a serious epidemic (Reddy & Bonman, 1987). In Colombia, resistant cultivars have been released, but their resistance lasted only 1 or 2 years before being overcome by previously unidentified virulent races (Ahn & Mukelar, 1986). There do not seem to be any examples of complete blast resistance that have proved durable.

When a cultivar allows the pathogen to reproduce, yet not as much as a fully susceptible cultivar does, that cultivar has a form of resistance variously referred to as quantitative (Ahn & Ou, 1982), slow blasting (Villareal et al., 1981b), dilatory (Marchetti, 1983), field (Ezuka, 1979), or partial (Yeh & Bonman, 1986). The term partial resistance is used here for 'a form of incomplete resistance in which spore production is reduced even though the host plants are susceptible to infection (susceptible infection type)' (Parlevliet, 1979). A more recent definition of the term, 'quantitative resistance based on minor genes' (Parlevliet, 1988), is not adopted here because the genetic basis of blast resistance in most rice cultivars is unknown. Several examples of partial blast resistance are associated with durable resistance; one will be described in detail here.

In the Philippines and many other countries in Asia, the partially resistant cultivar IR36 has been cultivated in vast areas, but has rarely suffered damage due to blast. Its performance contrasts with that of IR50, which has often been blasted in wet season plantings (Loganathan & Ramaswamy, 1984). The complete resistance of IR36 and IR50 to some races appears to be identical because races that can infect IR36 can also infect IR50 and vice versa (Bonman et al., 1986). However, the two differ in level of partial resistance. When both are inoculated with compatible isolates, IR36 shows fewer and smaller lesions than does IR50 (Table 1). The difference is also evident in blast nursery mini-plot tests (Yeh & Bonman, 1986). The latent period has been reported to be an important component of partial resistance to blast (Villareal et al., 1981a; Castano et al., 1989). However, IR36 and IR50 do not differ in latent period (Table 1). Similarly, no differences in latent period were found when a range of germplasm was tested at the International Rice Research Institute (IRRI) (E. Roumen, personal communication). Thus, infection efficiency and lesion size are the main components of the partial resistance that has been durable under tropical lowland conditions.

Table 1. Components of partial resistance in two rice cultivars: susceptible IR50 and durably resistant IR36

Resistance component	Cultivar	
	IR36	IR50
Lesions per dm^2 fully extended fifth leaf	3	31
Lesions per dm^2 partially extended sixth leaf	31	345
Lesion size (mm^2)	1.6	4.0
Latent period (days)	5.7	5.7

Adapted from Yeh & Bonman (1986).

Partial resistance has also been described in many of the blast-resistant upland cultivars developed in West Africa and Brazil (Prabhu & Morais, 1986). At least one of these cultivars, Moroberekan, has shown durable resistance in the highly blast-conducive upland environments of West Africa (Bonman & Mackill, 1988). In most tropical upland areas, however, no cultivars with durable blast resistance have been identified. In Japan, traditional upland rice cultivars have a high level of partial resistance; the same cultivars have shown long-lasting resistance in upland culture, where blast is generally more severe (Toriyama, 1975). Although durable under Japanese conditions, the resistance of these cultivars was not sufficient against races present in Latin America (K. Toriyama, personal communication). Thus, the level of partial resistance that proves durable in one environment may not necessarily be useful in other, more blast-conducive environments.

Inheritance. Much of the work on inheritance of complete blast resistance has been done in Japan, where 13 major genes have been identified (Ezuka, 1979). Major resistance genes are common in rice germplasm, and even the most susceptible cultivars will show complete resistance to some isolates of the fungus. Few systematic studies have been conducted using tropical rices, but recent work at IRRI in the Philippines indicates that one or two dominant genes present in the cultivars studied confer complete resistance against each fungus isolate (Yu et al., 1987). Several of the genes identified at IRRI have been incorporated into a common susceptible background through backcrossing, their allelic relationships determined (Mackill & Bonman, 1992), and their chromosomal location mapped using molecular markers (Yu et al., 1991). The relationship between the major genes identified in Japan and at IRRI is now under investigation.

Partial resistance is generally inherited through an undetermined number of minor genes, as is the case with tropical lowland (Wang et al., 1989) and upland rice (Notteghem, 1985). However, there is at least one exception to this generalization – the single gene *Pi-f* controls partial resistance in Japanese cultivars St 1 and Chugoku 31 (Toriyama, 1975).

Recently, restriction fragment length polymorphism (RFLP) mapping has been applied to the study of partial resistance in rice (Wang, 1992). The durably resistant rice cultivar Moroberekan was crossed with a highly susceptible cultivar CO39. Recombinant inbred (RI) lines were produced by single seed descent and analyzed with 100 RFLP markers. Field and greenhouse tests indicated the presence of partial resistance in the RI lines. To characterize this resistance, a single isolate was used in polycyclic tests, and each line was scored for lesion number, lesion size, and diseased leaf area. Fourteen RFLP markers defining 10 chromosomal segments were associated with effects on lesion number. The six markers with the strongest effects accounted for 62% of the variance for lesion number. Most of the loci identified affected all three of the parameters measured. Thus, partial resistance in the traditional rice cultivar Moroberekan is of complex inheritance. Future work will determine the practical utility of marker-based selection for improving blast resistance in rice.

Race specificity. From the previous discussion, it is clear that complete resistance is race-specific in that a cultivar may show resistance to some races of the pathogen but may be infected by other races. Such specificity has also been found in partially resistant cultivars with the *Pi-f* gene. St 1 and Chugoko 31 were partially resistant to races of *P. grisea* in Japan at the time of their release, but their resistance later broke down because new races evolved (Toriyama, 1975). Race-specific partial resistance was also encountered when susceptible Korean japonica cultivars tested at IRRI showed partial resistance (Bonman et al., 1989). Inoculation experiments revealed that these cultivars' resistance to Philippine races was race-specific. The cultivars showed few sporulating lesions with Philippine isolates but many such lesions with Korean isolates (Fig. 1). Resistance against the Philippine isolates was not due to lower aggressiveness in the pathogen, since the susceptible control cultivar CO 39 was equally susceptible to both sets of isolates. It is likely that this race-specific partial resistance is simply inherited.

It is not known if polygenic partial resistance can be race-specific. Using different partially resistant lowland cultivars, Yeh & Bonman (1986) found no evidence of isolate-cultivar interaction. Villareal et al. (1981b) reported 'the occurrence of differential interaction' between rice cultivars and isolates of the pathogen for components of partial resistance. At least one of the cultivars they studied, IRAT 13, has polygenic partial resistance (Notteghem, 1985). Similarly, pathologists in Brazil empirically observed that resistance in IRAT 13 had 'eroded' after having been sown for several years at the same highly blast-conducive site (A. Prabhu, personal communication). This field observation should be substantiated with tests using isolates collected over several years.

Neck blast resistance. Most research on blast resistance has focused on leaf blast, which occurs at the vegetative stage of crop development. During this phase of the disease, the fungus infects the leaf blades and leaf collars thus reducing the photosynthetic capability of the plant directly by reducing leaf area and indirectly via physiological effects (Baastians, 1991). However, the pathogen can also infect the panicle after flowering. When the panicle neck node, branches, and spikelets are attacked and colonized, the flow of photosynthates to the developing grains is reduced or completely inhibited. Infection of the panicle neck node, called 'neck blast', is the most destructive stage of the disease because it causes direct yield loss (Bonman et al., 1991). For convenience, screening for blast resistance is often done at the early vegetative stage, with the assumption that lines found resist-

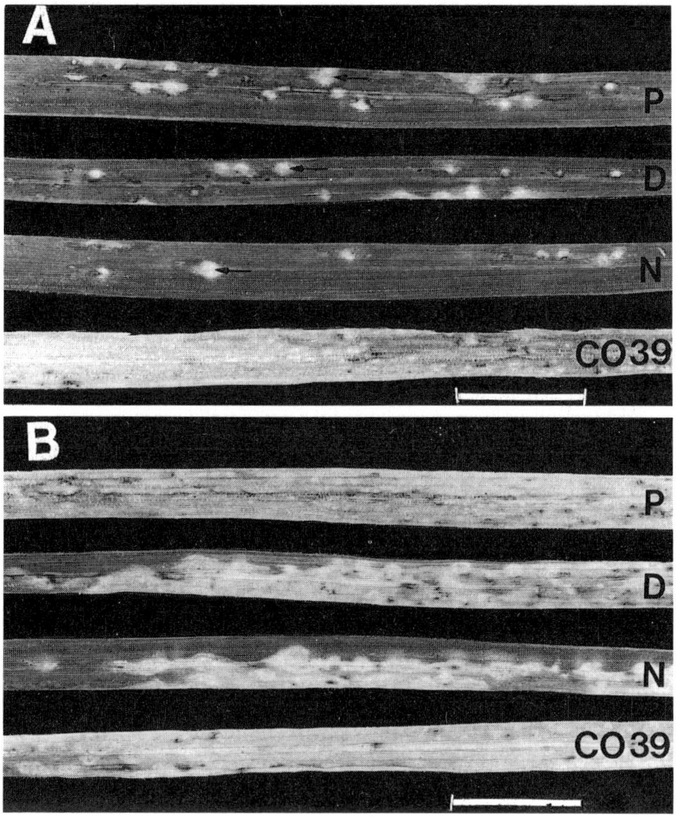

Fig 1. Reactions of Korean japonica rice cultivars Palgeum (P), Daechang (D), and Nagdong (N) and susceptible control CO39 inoculated with isolates of *Pyricularia grisea* from (A) the Philippines and (B) Korea. Arrows indicate typical individual lesions caused by the Philippine isolate. Scales bar = 1 cm (Adapted from Bonman et al., 1989a).

ant to leaf blast will also be resistant to neck blast (Ou, 1985). Similarly, resistance studies are also more convenient using leaves rather than panicles.

There is evidence that leaf and neck blast resistance are linked. In early work, cultivars completely resistant to a particular race at the seedling stage were also completely resistant to neck infection and those susceptible at the seedling stage were likewise susceptible to neck infection (Ou & Nuque, 1963). Researchers sometimes report a lack of correspondence between tests of leaf blast and neck blast resistance. Part of this discrepancy may be due to environmental differences at the different times various test cultivars flower. However, there is increasing evidence that relative differences in resistance exist among cultivars, especially when detailed measurements of partial resistance are made (Chung et al., 1980; Hwang et al., 1987; Bonman et al., 1989b).

The level of neck blast resistance was measured in 27 lowland rice lines in three field plantings at IRRI (Bonman et al., 1989b). The results for neck blast were compared with the leaf blast resistance of the same lines measured in several nursery experiments. The two sets of data showed positive correlation ($r^2 = 0.66$) (Fig. 2). The correlation was highest when the means of the three plantings were used. This may indicate errors due to differences in maturity of the entries in any single planting. Some lines appeared to be exceptions to this general relationship. For example, IR25604 was more susceptible to leaf blast but more resistant to neck blast than the partially resistant control IR36. Thus, although resistance to leaf and to neck blast

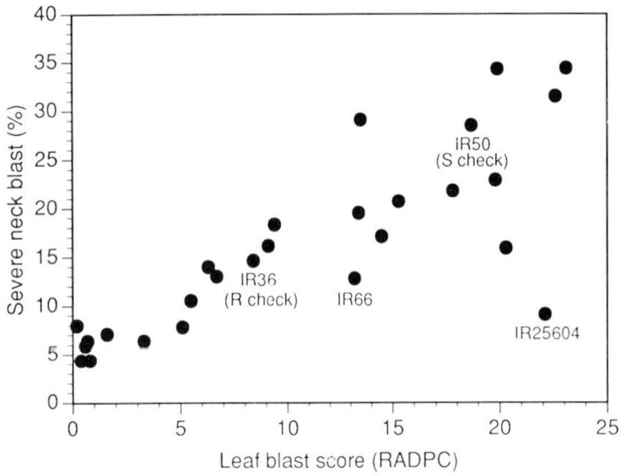

Fig. 2. Relationship between resistance to neck blast and resistance to leaf blast in 27 rice lines. Neck blast measurements are from five lowland field plantings and leaf blast measurements are from four upland miniplot trials. RADPC = the relative area under the disease progress curve. (Adapted from Bonman et al. 1989b).

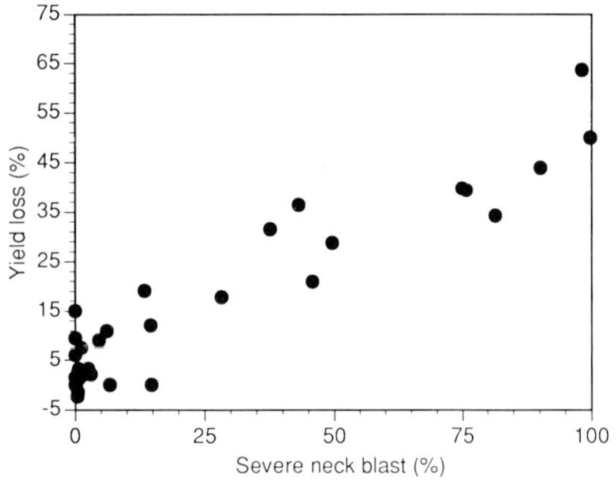

Fig. 3. Relationship between yield loss and percentage severe neck blast as measured in field trials in Korea and the Philippines. (Adapted from Bonman et al., 1991).

are usually positively correlated, some cultivars may be relatively resistant to one phase of the disease and relatively susceptible to another.

Neck blast is the most economically important phase of the disease. In some rice-growing areas, damage from leaf blast is a production problem, but as a general rule neck blast is responsible for most of the yield loss due to blast. The amount of loss in grain yield measured in field trials in the Philippines and Korea was positively correlated with the percentage of panicles with severe neck blast infection (Fig. 3). The Philippine trials also demonstrated the value of neck blast resistance in preventing yield losses. In the experiments cultivar IR66 was one of the test entries. This cultivar is susceptible to leaf blast but resistant to neck blast (Fig. 2) (Bonman et al., 1989a). Although its leaf blast infection was equal to that of the susceptible control IR50, IR66 had a low incidence of neck blast and no yield loss relative to plots protected with fungicide. Thus, neck blast resistance alone was sufficient to protect the crop from loss due to blast (Bonman et al., 1991).

Environment and durability of resistance

Blast disease is greatly affected by the environment. The key environmental factors favouring disease are 1) night temperatures between 17 and 23° C, 2) long duration of leaf wetness, 3) N fertilization, 4) aerobic soil, 5) water deficit, and 6) still air at night. Environmental influence is primarily through the effects on the physiology of the host, on the pathogen itself, or on the host-pathogen interaction.

The effect of aerobic soil is probably a direct effect on host susceptibility. Recently, Osuna-Canizalez et al. (1991) showed that plants given NO_3^--N in nutrient solution culture were much more susceptible than plants given NH_4^+-N. They concluded that N form may be the mechanism causing plants grown in aerobic soil to have higher susceptibility to blast, since NO_3^--N is the predominant N form under aerobic conditions.

Long duration of leaf wetness has a direct effect on the pathogen. At optimum temperature (about 25° C), the pathogen can infect the host after 6–8 hours of wetness. At about 16° C, though, infection occurs later (16–20 hours).

Temperature affects the interaction between the host and the pathogen. High temperatures (above 28° C) favour the growth of *P. grisea* but also stimulate host resistance. Plants are more susceptible at lower temperatures (about 20° C), thus favouring infection even though pathogen growth at these temperatures is slower.

As well as the six factors described, the soil type also affects the susceptibility of rice to blast disease. In the Philippines, for example, field observations indicated that upland rice grown at the Santo Tomas testing site in Batangas Province generally showed much less blast than rice grown at the Cavinti testing site in Laguna Province. E. Kurschner et al. (unpublished results) found that the site differences could be attributed to differences in the soil at the two sites. For two seasons at each site, plants were grown in pots containing soil from Cavinti and in pots containing soil from Santo Tomas. Disease was always highest in the Cavinti soil (Table 2).

Because of the many strong, complex, and interacting effects of environment on blast, there is great diversity in the 'disease potential' of various sites. Disease potential is a measure of the conduciveness of the environment to the disease and is affected by climatic, edaphic, and hydrologic conditions as well as the agronomic practices of farmers. There is at present no quantitative measure of blast disease potential for rice-growing environments, but blast simulation models may eventually prove useful in predicting the potential for the disease in regions where rice production practices are changing.

For blast disease, the durability of resistance not only is a function of the genotype of the host, but

Table 2. Effect of soil on leaf blast on potted plants grown at two upland rice testing sites for two years

	Site			
	Santo Tomas soil		Cavinti soil	
	Santo Tomas	Cavinti	Santo Tomas	Cavinti
1988	0.03 c	1.94 a	0.03 c	0.43 b
1989	0.65 b	2.26 a	0.28 c	0.56 b
Means	0.34 bc	2.10 a	0.15 c	0.50 b

Data are percentage leaf areas diseased 45 days after seeding; mean of two rice cultivars.
Means within a row followed by a common letter are not significantly different by LSD (p = 0.05).
Unpublished data from E. Kürschner, J.M. Bonman, I. Müller, J. Breithaupt & J. Kranz.

also is dependent upon the disease potential of the environment. In less blast-conducive environments, such as most tropical lowland areas, a level of partial resistance to leaf and neck blast similar to that of IR36 will probably be sufficient provided it is of the durable type. It may even be possible for the partial resistance to neck blast of cultivars like IR66 to show durability despite leaf blast susceptibility.

In environments with higher disease potential, such as temperate irrigated rice and probably some tropical rainfed lowland areas, a higher level of partial resistance will be required. For each target environment, field experiments using cultivars with various degrees of partial resistance should be conducted to ascertain the resistance level useful for blast management. Empirical studies are at present the only way of obtaining this information, since blast simulation models are not yet sophisticated enough to account for the many interacting plant and environment variables that affect the amount of disease that occurs. Once the required level of partial resistance is known, appropriate controls can be chosen for use in segregating populations generated by breeding programmes.

In environments with very high blast potential, such as many upland sites, it may not be possible to obtain high yields and at the same time manage blast disease solely with partial resistance. It is unlikely, for example, that the level of partial resistance in a cultivar such as IR36 would be of any direct use in most tropical upland environments. At the same time, it is important to recognize the value of such resistance in the appropriate target environment. For areas with very high blast potential, it will probably be necessary to incorporate major gene resistance into genotypes with high levels of partial resistance and to develop and use crop management practices that minimize disease development.

Certain cultivars show durable resistance because they '... remain resistant ... even though they are extensively cultivated in environments favourable to disease' (Johnson, 1981). Johnson's concept of durable resistance has been useful for scientists involved in rice improvement because it focuses on a practical issue: the utility and longevity of resistance in farmers' fields. The application of the concept in rice research has led to a much clearer understanding of how blast can be managed using resistant germplasm in the remarkably diverse environments in which the crop is produced.

References

Ahn, S.W. & A. Mukelar, 1986. Rice blast management under upland conditions. In: Progress in upland rice research, pp. 363–374. International Rice Research Institute, Manila, Philippines.

Ahn, S.W. & S.H. Ou, 1982. Quantitative resistance of rice to blast disease. Phytopathology 72: 279–282.

Baastians, L., 1991. Ratio between virtual and visual lesion size as a measure of describe reduction in leaf photosynthesis of rice due to leaf blast. Phytopathology 81: 611–615.

Bonman, J.M., J.M. Bandong, Y.H. Lee, E.J. Lee & B. Valent, 1989a. Race-specific partial resistance to blast in temperate japonica rice cultivars. Plant Dis. 73: 496–499.

Bonman, J.M., B.A. Estrada & J.M. Bandong, 1989b. Leaf and neck blast resistance in tropical lowland rice cultivars. Plant Dis. 73: 388–390.

Bonman, J.M., B.A. Estrada, C.K. Kim, D.S. Ra & E.J. Lee, 1991. Assessment of blast disease and yield loss in susceptible and partially resistant rice cultivars in two irrigated lowland environments. Plant Dis. 75: 462–466.

Bonman, J.M. & D.J. Mackill, 1988. Durable resistance to rice blast disease. Oryza 25: 103–110.

Bonman, J.M. & M.C. Rush, 1985. Report on rice blast in Egypt. National Rice Institute, Egypt. 15 pp.

Bonman, J.M., T.I. Vergel de Dios & M.M. Khin, 1986. Physiologic specialization of *Pyricularia oryzae* in the Philippines. Plant Dis. 70: 767–769.

Castano, J., D.R. MacKenzie & R.R. Nelson, 1989. Components analysis of race non-specific resistance to blast of rice caused by *Pyricularia oryzae*. Phytopathology 127: 89–99.

Chung, H.S., G.H. Choi & D. Shakya, 1980. Slow blasting of rice cultivars at leaf and heading stages in paddy field. Office of Rural Development, Suweon, Korea. 50 pp. (in Korean with English summary).

Crill, P., Y.S. Ham & H.M. Beachell, 1981. The rice blast disease in Korea and its control with rice prediction and gene rotation. Korean J. Breed. 13: 106–114.

Ezuka, A., 1979. Breeding for and genetics of resistance in Japan. In: Proc. Rice Blast Workshop, pp. 27–28. Int. Rice Res. Inst., Manila, Philippines.

Fry, W.E., 1982. Principles of Plant Disease Management. Academic Press, Inc., New York.

Hwang, B.K., Y.J. Koh & H.S. Chung, 1987. Effects of adult-plant resistance on blast severity and yield of rice. Plant Dis. 71: 1035–1038.

Johnson, R., 1981. Durable disease resistance. In: J.F. Jenkyn & R.T. Plumb (Eds.), Strategies for the control of cereal diseases, pp. 55–63. Blackwell Scientific Publications, Oxford.

Johnson, R. & J.M. Bonman, 1990. Durable resistance to blast disease in rice and to yellow rust in wheat. Abstract. In: Proc. Int. Symp. Rice Research: New Frontiers. Indian Council of Agricultural Research. November 1990.

Khush, G.S., 1984. Terminology for rice-growing environments. In: Terminology for Rice-Growing Environments, pp. 5–10. Int. Rice Res. Inst., Manila, Philippines.

Kiyosawa, S., 1982. Genetics and epidemiological modeling of breakdown of plant disease resistance. Annu. Rev. Phytopathol. 20: 93–117.

Lee, E.J., H.K. Kim & J.D. Ryu, 1976. Studies on the resistance of rice varieties to the blast fungus, *P. oryzae* Cav. Research Report for 1976. Institute of Agricultural Sciences, Office of Rural Development, Korea (in Korean).

Loganathan, M. & V. Ramaswamy, 1984. Effect of blast on IR50 in late samba. Int. Rice Res. Newsl. 9: 6.

Mackill, D.J. & J.M. Bonman, 1992. Inheritance of blast resistance in near-isogenic lines of rice. Phytopathology (in press).

Marchetti, M.A., 1983. Dilatory resistance to rice blast in USA rice. Phytopathology 73: 645–649.

Matsumoto, S., 1974. Pathogenic race occurred on a blast resistant variety Reiho. Proc. Assoc. Plant Prot. of Kyushu 20: 72–74 (in Japanese with English summary).

Notteghem, J.-L., 1985. Définition d'une stratégie d'utilisation de la résistance par analyse génétique des relations hôte-parasite Cas du couple riz – *Pyricularia oryzae.* L'Agron. Trop. 40: 129–147.

Osuna-Canizalez, F.J., S.K. De Datta, S.K. & J.M. Bonman, 1991. Nitrogen and silicon nutrition effects on resistance to blast disease of rice. Plant and Soil 135: 223–231.

Ou, S.H., 1985. Rice Diseases, 2nd ed. Commonwealth Mycological Institute, Kew, Surrey.

Ou, S.H. & F.L. Nuque, 1963. The relation between leaf and neck blast resistance to the rice blast disease. Int. Rice Comm. Newsl. 12: 30–34.

Parlevliet, J.E., 1979. Components of resistance that reduce the rate of epidemic development. Annu. Rev. Phytopathol. 17: 203–222.

Parlevliet, J.E., 1988. Identification and evaluation of quantitative resistance. In: K.J. Leonard & W.E. Fry (Eds.), Plant Disease Epidemiology, pp. 215–248. McGraw Hill, New York.

Prabhu, A.S. & O.P. Morais, 1986. Blast disease management in upland rice in Brazil. In: Proc. Symposium on Progress in Upland Rice Research, Int. Rice Res., pp. 383–392. Inst., Manila, Philippines.

Reddy, A.P.K. & J.M. Bonman, 1987. Recent epidemics of rice blast in India and Egypt. Plant Dis. 71: 850.

Rossman, A.Y., R.J. Howard & B. Valent, 1990. *Pyricularia grisea,* the correct name for the rice blast disease fungus. Mycologia 82: 509–512.

Toriyama, K., 1975. Recent progress of studies on horizontal resistance in rice breeding for blast resistance in Japan. In: Proc. Seminar on Horizontal Resistance to the Blast Disease of Rice, pp. 65–100. Series CE-No. 9, Centro Internacional de Agricultura Tropical.

Valent, B., 1990. APS Plenary Session Lecture (1989): Rice blast as a model system for plant pathology. Phytopathology 80: 33–36.

Villareal, R.L., D.R. MacKenzie, R.R. Nelson & W.R. Coffman, 1981a. The components of slow blasting of rice. Phytopathology 71: 263 (Abstr.).

Villareal, R.L., R.R. Nelson, D.R. MacKenzie & W.R. Coffman, 1981b. Some components of slow-blasting resistance in rice. Phytopathology 71: 608–611.

Wang, Z., D.J. Mackill & J.M. Bonman, 1989. Inheritance of partial resistance to blast in indica rice cultivars. Crop Sci. 29: 848–853.

Wang, G.L., 1992. RFLP mapping of major and minor genes for blast resistance in a durably resistant rice cultivar. PhD dissertation. University of the Philippines at Los Banos.

Yeh, W.H. & J.M. Bonman, 1986. Assessment of partial resistance to *Pyricularia oryzae* in six rice cultivars. Plant Pathol. 35: 319–323.

Yeh, W.H., J.M. Bonman & E.J. Lee, 1989. Effects of temperature, leaf wetness duration, and leaf age on partial resistance to rice blast. J. Plant Prot. Trop, 6 (3): 223–230.

Yu, Z.H., D.J. Mackill & J.M. Bonman, 1987. Inheritance of resistance to blast in some traditional and improved rice cultivars. Phytopathology 323–326.

Yu, Z.H., D.J. Mackill, J.M. Bonman & S.D. Tanksley, 1991. Tagging genes for blast resistance in rice via linkage to RFLP markers. Theor. Appl. Genet. 81: 471–476.

Euphytica **63**: 125–139, 1992.
R. Johnson and G.J. Jellis (eds), Breeding for Disease Resistance

Barley mildew in Europe: population biology and host resistance

M.S. Wolfe, U. Brändle, B. Koller, E. Limpert, J.M. McDermott, K. Müller & D. Schaffner
Phytopathology Group, Institute for Plant Sciences, Swiss Federal Institute of Technology, CH 8092 Zürich, Switzerland

Key words: barley mildew, DNA markers, *Erysiphe graminis* f. sp. *hordei,* fungicide resistance, *Hordeum vulgare,* population genetics, variety mixtures, virulence

Summary

Isolates of the barley mildew pathogen from the air spora over a large part of Europe and from fields of variety mixtures, were tested for virulence against 12 host resistance alleles. Subsamples were tested for their response to triadimenol fungicide and analyzed for 10 DNA loci using RAPD markers and PCR. There was a large range of haplotypes spread over Europe; irregularity in the distribution was probably due mainly to non-uniform use of the corresponding host resistances and fungicides. A large range of variation was also detectable within individual fields. Positive gametic disequilibria distorted the distribution of virulence alleles among haplotypes and reduced the number of haplotypes detectable in the sample. Analysis of the spread of the newly selected *Val3* allele into different European sub-populations indicated that gene flow throughout the population may be rapid for alleles that have a selective advantage.

Fungicide resistance was widespread in areas known for intensive use of fungicides for mildew control. Four classes of fungicide response were detectable and particular virulence haplotypes were found to be characteristic for each class.

Variety mixtures used in the former German Democratic Republic (GDR) reduced mildew infection, and thus fungicide use, during the years 1984–1991 despite the limited variation in host resistance among the mixtures. A tendency for complex pathogen races to increase in mixture crops was reversed by the large-scale re-introduction of fungicides for mildew control in 1991. The mixture strategy appeared to be more successful than using the same resistance alleles in pure monoculture or combining them in a single host genotype.

Introduction

Erysiphe graminis f. sp. *hordei* is a highly successful pathogen of barley, particularly in the intensive agricultural regions of Europe. It has overcome many host resistances (Wolfe & Schwarzbach, 1978a, b; Brückner, 1982, 1987) and adapted to various major fungicides (Wolfe et al., 1988) used in the monocultural system. We can presume that it does so through the effects of the major forces of evolution, mutation, selection, migration, recombination and drift, but we do not know the relative importance of these forces, although research has started to focus more on these questions in recent years (Brown & Wolfe, 1990; Burdon & Leather, 1990; Hovmoller & Ostergaard, 1991; Ostergaard & Hovmoller, 1991; Welz & Kranz, 1987; Wolfe & Caten, 1987).

There is a long history of national surveys of the pathogen populations, but only recently at the international level (Limpert & Schwarzbach, 1981; Limpert, 1987a), after Hermansen et al. (1978) had shown that viable spores of the pathogen can be spread over long distances by wind. On the basis of transnational virulence surveys, Limpert (1987a, b) concluded from the distribution of pathogen characters not under selection that long distance dispersal is important, and that migration of the pathogen population mainly occurs in Europe on prevailing winds from west to east. A novel approach of O'Dell et al. (1989) using RFLPs as unselected markers was expected to give further insight into quantification of the different forces involved. An early result was that a pathogen isolate found in England in 1986 with *Val3* appeared to be identical to a clone collected in Lower Austria, but probably originating from Czechoslovakia.

To try to understand the population biology of the pathogen further on an international scale and thus to define possible strategies for the improved use of host resistances and fungicides, a large European survey was started by screening for virulence characters and fungicide resistance. In addition, for unselected characters, a system of defining and monitoring RAPD (Random Amplified Polymorphic DNA) markers was also developed. The present paper is a summary of current progress with the survey.

One strategy of using host resistance that has been introduced and used on a large scale in recent years in Europe is the use of variety mixtures to control the disease. It was clearly important also to try to monitor the response of the pathogen to this new challenge.

For convenience, we refer to the overall genetic structure of this haploid pathogen as the haplotype. This is sub-divided into the virulence haplotype (V-haplotype), the fungicide haplotype (F-haplotype) and the DNA marker haplotype (D-haplotype). V-, D- and F-haplotypes can be aggregated in different combinations for population analysis. The term pathotype is equivalent to V-haplotype; race, or physiologic race, refers to a group of haplotypes with a defined range of identical virulence characters.

Materials and methods

a. Sampling

Road journeys were made throughout Europe from Scotland to the former Soviet Union and from Denmark to Italy with a Schwarzbach spore trap (Schwarzbach, 1979) mounted on a vehicle roof to sample the air spora. Healthy segments of the first leaves of the winter barley variety Igri (resistance *Ml-ra,* not effective in the area of Europe investigated) were loaded into the trap at roughly 100 km intervals, depending on the density of barley cropping in the region. Sampling runs did not cross borders between areas with different patterns of barley cultivation. Exposed segments were maintained in a refrigerator before returning to the laboratory. Approximately 3000 isolates per season were collected using this simple and rapid system.

For within-field sampling, used for analysing the pathogen populations on variety mixtures and on host plants at the local level, two different methods were used. In the first, leaf segments of the mildew susceptible variety Igri were loaded into a small mobile settling tower and then exposed to infection. This was repeated with fresh sets at different points in a field; new infections were incubated to obtain single colony isolates for subsequent testing. In the second method, infected segments were cut from plants selected at random. Existing spores were blown off the segments which were then maintained on agar for two days. Freshly developing spores were transferred to healthy leaf segments for multiplication and testing. More than 2500 isolates were collected and tested in this way each season.

b. Virulence testing

Conidia from single colonies of the pathogen were isolated from the incubated leaf segments and inoculated on to leaf segments of differential varieties laid on benzimidazole agar (30 ppm). After incubation for approximately 7 days at 17° C, the infections were assessed on simple scales (Limpert & Fischbeck, 1987). The differential varieties used were the near-isogenic lines developed by Kolster et al. (1986) in Denmark, containing the following range of resistances:

Group 1: P21 *Mlg,* P03 *Mla6,* P23 *Ml(La)*
Group 2: P04B *Mla7,* P16 *Mlk,* (Triumph) *Mla7* + *Ml(Ab)*

Group 3: P08B *Mla9,* P11 *Mla13,* (Alexis) *mlo*
Group 4: P10 *Mla12,* P01 *Mla1,* P02 *Mla3*

c. Fungicide testing

Ten-day-old plants of the variety Igri were sprayed with triadimenol at 0, 1 or 20 ppm. Eighteen hours later, segments from the first leaf were inserted in benzimidazole agar (30 ppm) and inoculated from a settling tower with four replications per treatment; segments with different fungicide doses were separated in disposable Petri dishes to avoid any interference from the vapour phase. Percentage sporulation of the fungus was estimated after 8 days. The limited set of fungicide treatments meant that it was not possible to determine ED50 values accurately or to carry out probit analysis of individual isolates. On the other hand, the system allowed a simple classification of the isolates into one of four classes of fungicide response. These classes were determined from a random subset of 106 isolates tested previously with a comprehensive set of fungicide treatments covering a finer scale. The classes were defined as follows:

Per cent sporulation	0 ppm	1 ppm	20 ppm
Class 1:	100	0	0
Class 2:	100	10–60	0
Class 3:	70–100	70–100	0–20
Class 4:	70–100	70–100	70–100

With this simple system, it was possible to analyse about 1200 isolates per annum taken as sub-samples of the isolate collection from the virulence survey.

d. DNA variation

Williams et al. (1990) introduced the use of synthetic oligonucleotide primers comprising 10 bases, in combination with the polymerase chain reaction (PCR), as a means of arbitrarily amplifying DNA sequences. DNA polymorphisms based on this technique are called RAPD markers for Randomly Amplified Polymorphic DNA. This technique was applied to the study of population differentiation of the powdery mildew pathogen in Europe.

DNA was extracted from freeze-dried conidia using a modified CTAB preparation (Murray & Thompson, 1980). PCR reactions were performed according to Williams et al. (1990) on a Perkin Elmer Cetus Gene Amp 9600 thermocycler programmed for 2 cycles of 30 sec at 94° C, 30 sec at 36° C and 120 sec at 72° C followed by 38 cycles of 20 sec at 94° C, 15 sec at 36° C, 15 sec at 45° C and 120 sec at 72° C followed by 10 min at 72° C. Reaction products were resolved by electrophoresis in a 1.5% agarose gel and 1X TPE buffer. The three primers used were PJ-2 5′-ACGAGGGACT, OE-3 5′-CCAGATGCAC and OE-7 5′-AGATGCAGCC.

With this technique, it was possible to determine the DNA structure of a sub-sample of about 600 isolates from the pathogen collection already tested for virulence and for fungicide response.

Results and discussion

1. General distribution of characters in the pathogen population

a. Virulence

The overall pattern for Europe for 1989–91 (Table 1) showed high frequencies for the six virulence alleles, *Vg, Va6, V(La), Vk, Va7* and *Va12,* for which the corresponding resistances have been in use for many years. There were moderate frequencies for the three alleles, *V(Ab), Va9* and *Va13,* for which the corresponding resistances were introduced more recently or used on a smaller scale. *Va1* and *Va3,* for which there has been little selection, occurred only at low frequencies. Virulence for the durable resistance *mlo,* was not detected.

The 11 alleles occurred in 743 of the possible 2048 (2^{11}) combinations among the 8893 isolates tested over the years 1989–1991, indicating a massive reservoir of pathogen variation in Europe as a whole. A consequence of the occurrence of so many V-haplotypes is that the most common had a low relative frequency, although the absolute frequencies were high. Thus, the ten most common of the V-haplotypes reached a frequency of only 1–5 per cent (Table 1), although all were significantly in excess of the expected values determined from the overall frequencies of the individual alleles. Only two of the most common V-haplotypes (3001 and 7301) occurred at frequencies similar to those expected.

Given the size of the survey, the number of V-haplotypes identified represents less than half of the number potentially available, calculated from the overall frequencies shown in Table 1. This discrepancy was mainly due to the occurrence of positive gametic disequilibria involving *Vk, Va7,* and *Va9,* often together with *V(Ab)* and *Val3* (Table 2). The reason for this was probably that many of the original varieties introduced with *Mla7* or *Mla9* also carried *Mlk* (Brown & Jorgensen, 1991), so that *Vk* was selected in combination with *Va7* or *Va9* and acted as a bridge between these sub-populations; these gametic disequilibria have been noted widely over many years. Concerning *V(Ab)* and *Val3*, the resistance of the variety Triumph, *Mla7, Ml(Ab),* used intensively in the former GDR, Britain and elsewhere, probably selected virulence from the *Va7-Vk* sub-population, though not exclusively, and *Val3* was first detected in a sub-population of V-haplotypes with *Vk, Va7* and *Va9* (see below). Over-sampling of the V-haplotypes in disequilibrium further decreased the probability of detecting rarer haplotypes. There was a tendency for *Va6* and *V(La),* which were in positive disequilibrium to be in negative disequilibrium with the virulences in the *Vk, Va7, Va9* group, though not consistently so.

From the overall frequencies of the virulence alleles (Table 1), the expected and observed distribution of the number of alleles among the V-haplotypes was approximately normal around a mean of four to five of the 11 alleles tested; this value varied among regions.

The more common virulence alleles and haplotypes were distributed relatively uniformly over the whole region (e.g. *Vg,* Fig. 1a), whereas the less common and more recently selected alleles (e.g.

Table 1. Virulence allele frequencies and the ten most common V-haplotypes of the pathogen over the survey period and their observed and expected frequencies (per cent)

Allele	*Vg*	*Va6*	*VLa*	*Va7*	*Vk*	*VAb*	*Va9*	*Val3*	*Val2*	*Val*	*Va3*		
Grp.	1	1	1	2	2	2	3	3	4	4	4		
Bin.	1	2	4	1	2	4	1	2	1	2	4		
Freq.*	70.5	55.3	44.1	54.5	50.0	20.7	23.2	18.6	53.9	9.5	5.7		
Race[1]												Freq. obs.	Freq. exp.
7000	+[2]	+	+	–	–	–	–	–	–	–	–	4.6	0.8
7001	+	+	+	–	–	–	–	–	+	–	–	3.3	0.9
3001	+	+	–	–	–	–	–	–	+	–	–	2.3	1.1
3121	+	+	–	+	–	–	–	+	+	–	–	2.3	0.3
1001	+	–	–	–	–	–	–	–	+	–	–	2.1	0.9
2000	–	+	–	–	–	–	–	–	–	–	–	1.6	0.4
5001	+	–	+	–	–	–	–	–	+	–	–	1.6	0.7
7301	+	+	+	+	+	–	–	–	+	–	–	1.5	1.1
5000	+	–	+	–	–	–	–	–	–	–	–	1.4	0.6
3000	+	+	–	–	–	–	–	–	–	–	–	1.3	1.0

* Overall frequencies of each allele in the total population of 8893 isolates.

[1] As an example, race 7001 is coded as virulent on Grp 1 varieties, *Mlg, Ml-a6* and *M-La,* giving the binary number 1 + 2 + 4, and avirulent on all others except the first variety of Group 4, *Mla12,* to give the final binary number, 1, in the race code. As with all ten races, its observed frequency, 3.3%, is significantly greater than expectation, 0.9%, at $P < 0.001$.

[2] + = virulent, – = avirulent.

Va13, Fig. 1b) and haplotypes were not. Irregularity in distribution of the virulence alleles was largely due to matching irregularity in the distribution of the corresponding host resistance alleles. Analysis of this aspect is not yet complete.

Since the overall data set represents a mixture of different sub-populations, detailed genetic analysis of the whole may lead to erroneous conclusions (Wolfe & Knott, 1982). Statistical tests were made therefore for homogeneity between sets of adjacent samples to try to estimate the limits of sub-populations. There were often several sub-populations within national boundaries, but some overlap did occur. There was a tendency for adjacent sub-populations to be more related to neighbouring than to distant sub-populations, but there were several exceptions (Fig. 2).

For within-field sampling, isolates were taken from unreplicated plots of the winter barley varieties Narcis, Nefta, Triton and their mixture grown in a farm crop near the Greifensee in Switzerland. A total of 652 isolates from this area was tested for virulence and RAPD markers.

The pathogen population in the mixture crop was dominated by haplotypes with *Va13* virulence because one-third of the crop area was occupied by the highly susceptible variety Triton (*Mla13*). Nevertheless, there was a large range of haplotypes, particularly on the pure stands, and the genetic richness was almost as high as in the European pathogen population as a whole. This indicates that, although sub-populations can be defined statistically on a regional basis, there appears to be little hierarchical diminution of variation from the regional to the local level. This agrees with previous observations of the maintenance of genetic

Table 2. Comparison of observed (O) and expected (E) values for all possible pairs of virulence alleles as percentages of the whole survey sample (8893 isolates)

	O = E			O > E				O > E	
Alleles	O	Alleles		O	E	Alleles		O	E
g-a1	7	*g-a6*	***[1]	42	39	*a1-a3*	*	0	1
g-a3	4	*g-La*	***	35	31	*a6-Tr*	**	10	11
g-k	36	*g-a7*	**	41	38	*a6-k*	***	25	28
g-Tr+	15	*g-a12*	***	41	38	*La-a7*	***	22	24
g-a9	17	*a6-a3*	**	4	3	*La-Tr*	***	4	9
g-a13	13	*a6-La*	*	26	24	*La-a13*	***	3	8
a6-a1	5	*La-a1*	***	6	4	*k-a12*	*	26	27
a6-a7	30	*a7-a3*	*	4	3	*Tr-a1*	***	1	2
a6-a9	13	*a7-k*	***	37	27	*a12-a9*	***	10	13
a6-a12	30	*a7-a9*	***	18	13	*a13-a1*	***	1	2
a6-a13	10	*a7-a13*	***	15	10	*a13-a3*	***	0	1
La-k	22	*a7-a12*	**	12	11				
La-a3	3	*Tr-a3*	*	2	1				
La-a9	11	*Tr-k*	***	15	10				
La-a12	24	*Tr-a13*	**	5	4				
a7-a1	5	*k-a1*	***	7	5				
k-a13	10	k-a3	*	2.8	3.4				
a9-a3	1	*a9-a1*	**	3	2				
a9-Tr	4	a9-k	***	20	12				
a12-a1	6	*a9-a13*	***	7	4				
a12-a3	3								
a12-Tr	12								
a12-a13	11								

+*Tr* = *a7, Ab*

[1] chi^2-test, levels of significance: * 0.005, ** 0.01, *** 0.001.

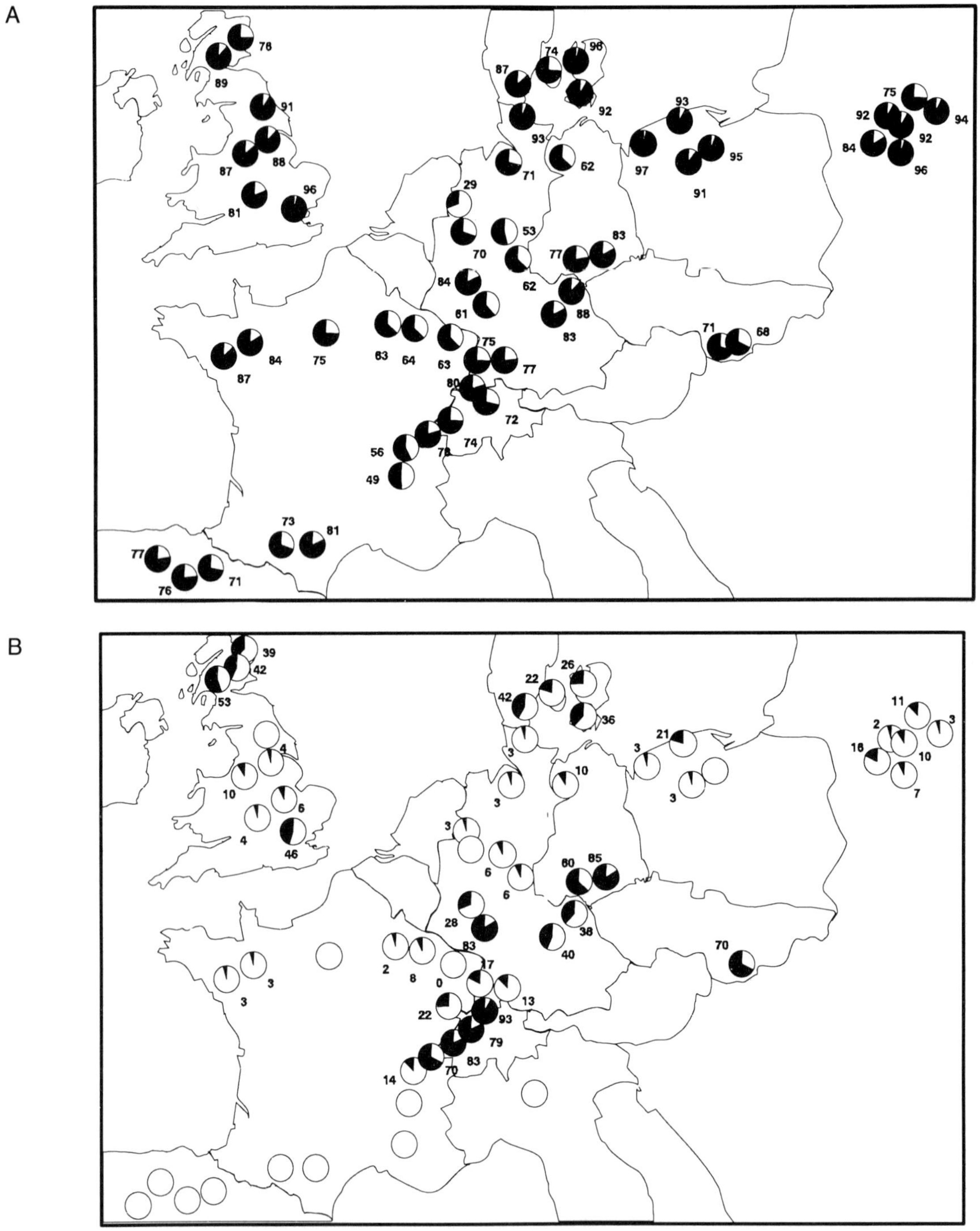

Fig. 1. Frequency distribution of *Vg* (Fig. 1a) and *Va13* (Fig. 1b) virulence alleles over Europe for 1991.

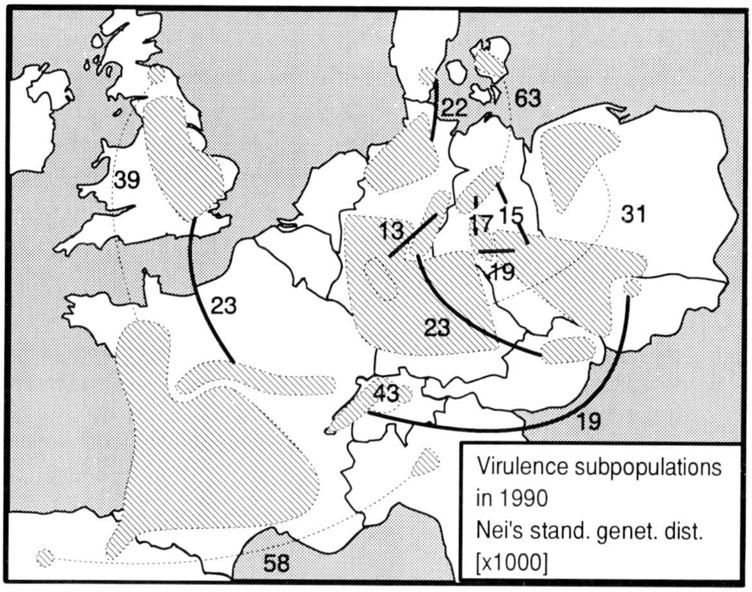

Fig. 2. National boundaries and the proposed distribution of the sub-populations of *Erysiphe graminis* f. sp. *hordei.*

variation in local populations of *Rhynchosporium secalis* (McDermott et al., 1989). On the other hand, with rusts such as *Puccinia striiformis,* single clones that are already virulent on common host varieties multiply and disperse over large distances and there is no detectable local variation (Newton et al., 1985).

Table 3. Distribution of fungicide response classes among countries in 1990

Group	Country	No. of isolates	Fungicide class			
			1	2	3	4
A	Italy	24	50	38	13	0
	Spain	47	53	28	19	0
B	Austria	46	24	33	43	0
	E. Germany	140	12	44	44	0
	Poland	157	5	30	65	0
	Switzerland	48	0	33	67	0
	France	113	13	12	75	0
C	Czechoslovakia	55	2	20	78	0
	Denmark	64	0	19	80	2
	W. Germany	335	1	18	81	1
D	Gt. Britain	179	2	23	63	12

The varieties Narcis and Nefta possess no known qualitative resistance genes so that it was remarkable to find that the pathogen populations on these varieties had relatively low frequencies of *Va13* and differed from each other in terms of the range and frequency of V-haplotypes that they supported. This was confirmed by the distribution of one of the DNA markers which was significantly less common on Narcis than on Nefta.

b. Fungicide resistance

The geographical distribution of fungicide response classes for 1990 (Table 3) shows four distinct groups (the data for 1991 showed a similar pattern). In Group A, Spain and Italy, where comparatively little fungicide has been used for mildew control, the pathogen populations were relatively sensitive, though less so than at the time of introduction of the DMI fungicide group. Group B was intermediate, comprising countries in which fungicide is used but not intensively. Within this Group, France could be sub-divided further into a southern region in which the populations resemble those of Spain and Italy, and a northern region which is more similar to the areas of intensive fungicide use in northern Europe.

In the Group C countries which have a relatively long history of intensive fungicide use, the majority of isolates tested fell into Class 3. The worst case was Group D, Great Britain, where a significant proportion of the isolates were classified in the highest resistance range; these isolates originated mainly from the south-east of the country. This corresponds with previous data (Limpert, 1987a; Wolfe et al., 1988) and with the intensive use of DMI fungicides since their introduction at the beginning of the 1980's. Other tests confirmed the occurrence of isolates in the east of Scotland that were resistant to DMIs and that also had reduced sensitivity to morpholine fungicides. They were not tested for ethirimol, although previous work showed that some isolates lacked sensitivity to all three fungicides (Wolfe et al., 1988).

The limited distribution of the highest level of fungicide response may indicate that Class 4 isolates carry some penalty in terms of fitness in the absence of fungicide and that selection in northern continental Europe towards a high frequency of Class 3 is adequate for the survival of the pathogen population. On the other hand, it is important to note that a change to a high level of fungicide resistance was a key characteristic in the evolution of *Va13* isolates selected in Britain (see below). The new haplotype carrying *Va13* and high fungicide resistance increased in frequency from 1988 to 1990–91, so that the British mildew population may now provide an inoculum source sufficiently large to carry these two important characters, in combination, back to continental Europe.

From the total of 1208 isolates tested for fungicide response in 1990, it was clear that the four fungicide response classes were heterogeneous for the distribution of the most common virulence alleles. Further analysis showed that certain V-haplotypes were associated with specific fungicide response classes, summarized in Table 4. V-haplotypes with *Vg* were associated with most fungicide classes, indicating the wide distribution of this allele. *Va9*, on the other hand, appeared not to be associated with a positive fungicide response indicating a limited distribution among crops treated less often than average. V-haplotypes containing the more recently selected virulence alleles, *VTr* (= *Va7, VAb*) and *Va13* were associated with the most fungicide resistant classes. In particular, the Class 4 V-F-haplotype was strongly selected in Great Britain because it was effective against the currently important host varieties as well as being poorly controlled by DMI fungicides.

A superficial view of Table 4 may suggest that there is a direct, functional, association between certain V-haplotypes and level of fungicide response. However, by taking into account the geographical distribution of the characters involved, it becomes clear that these associations in the pathogen probably indicate rather the effects of the history of local use of particular resistant varieties and the intensity of fungicide application.

With the exception of the GDR, where fungicide resistance increased considerably from 1990 to 1991, there was little difference in the distribution of fungicide response between these years, even though there were considerable changes in the pattern of distribution of some virulence alleles. Consequently, we may infer that the pathogen population as a whole is highly buffered for its fungicide response because of widespread selection for this character; it is unlikely that changing the range of varieties that are treated with fungicide will lead

Table 4. Distribution of virulence haplotypes among fungicide classes in 1990

	Vg	*Va6*	*VLa*	*Vk*	*Va7*	*VTr*	*Va9*	*Va13*	Cl.1	Cl.2	Cl.3	Cl.4
Cl.1	±	–	±	±	±	–	–	–	26	8	8	0
Cl.2	+	+	+	+	+	–	–	–	2	36	28	0
Cl.3	+	+	±	–	±	–	–	±	0	4	89	0
Cl.4	+	–	–	+	+	+	–	+	0	0	2	6

+ = virulence allele, – = avirulence allele, ± = either allele.

to any long-term change or improvement in the current performance of the fungicides.

Official figures from the former GDR showed that the uptake of variety mixtures for use in the spring barley crop led to a reduction in the national average level of mildew infection from about 50 per cent to 10 per cent, and a consequent reduction in the use of fungicide for mildew control, equivalent to non-treatment of about 100,000 ha (Wolfe et al., 1991). The effect of this reduction in use is likely to have been the cause of a discontinuity in the geographic distribution of fungicide response. From Table 3 it is clear that the fungicide resistance of the pathogen population in west Germany, Denmark, Poland and Czechoslovakia, which virtually surround the former GDR, was higher than in the former GDR itself.

c. DNA markers

Thus far, some 600 isolates have been analyzed with three oligonucleotide primers for the occurrence of ten putative loci. The geographical distribution of two of these loci are contrasted in Figs. 3a, b, and their distribution can be compared with two major virulence alleles in Fig. 1. The pattern of distribution of DNA polymorphisms indicates that there is a large reservoir of 'background' variation in the pathogen population, with, on the average, approximately 80% of the total allelic variation being present on the local scale. The distributions of some DNA loci indicated significant geographical heterogeneity, confirming again, that sub-structuring of the pathogen population does occur (Fig. 3b). On a global (i.e. European) scale there was little association between D-haplotypes and V-haplotypes. On the other hand, at the regional or local scale there is clear evidence of associations between DNA and virulence variation. The associations appear to be fortuitous and most likely the result of hitch-hiking given the strong selection for virulence on the local spatial temporal scale. In addition, a comparison of the populations occurring in Czechoslovakia and Switzerland demonstrated that identical genotypes, based on 10 DNA and 10 virulence loci, occurred commonly in both areas and at frequencies well above random expectations. This observation is consistent with recent, large-scale gene flow involving a particular pathogen haplotype.

d. Variety mixtures

For the analysis of the pathogen response to the variety mixtures grown in the former GDR, isolates were collected directly at different times during the growing season from known mixture crops and, as far as possible, from pure stands of the components grown for seed production in the neighbourhood. This proved difficult in practice either because of the paucity of pure stands in 1990, or because the pure stands had been treated with fungicide.

Isolates from the field were analyzed and their genotypic structure was compared with isolates from the air spora within the former GDR and in neighbouring countries. Two major differences observed were first, that there was a larger range of common virulence alleles in the mixture area than in surrounding countries, and second, partly as a consequence of this, the average number of virulence alleles per isolate was higher. From fungicide tests on isolates from the air spora, it was evident that the population in the GDR was more sensitive to DMI fungicides than the populations in the neighbouring countries (Table 3).

2. Dynamics of population change

a. The example of Va13

A major increase of *Va13* in the European pathogen population developed from an increase in Czechoslovakia in the early 1980's (Brückner, 1982, 1987). During the mid-1980s, the predominant V-haplotype in the Czech population was probably *Vg, Vk, Va7, Va9, Va13.* This predominated in a nearby sub-population found in Lower Austria (Andrivon & Limpert, 1992). An isolate with the same virulence pattern was also found in 1986 at Cambridge, England and this was tested for RFLP homology with an isolate, probably from Austria (E. Limpert, personal communication) with the same V-haplotype. The English and Austrian isolates were identical suggesting long-distance migra-

A

B

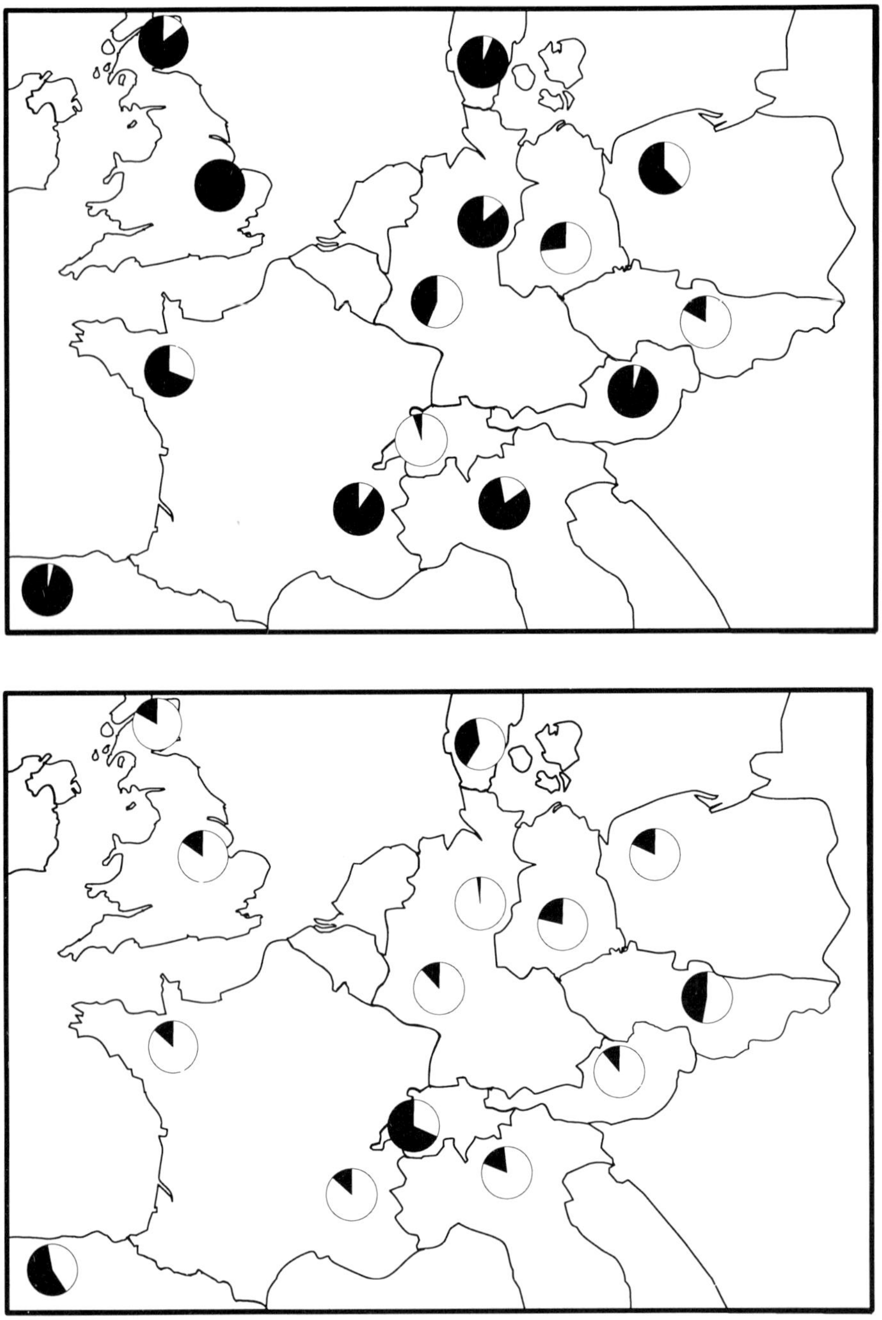

Fig. 3. Geographical distribution of DNA loci, P2-1 (Fig. 3a) and E3-1 (fig. 3b) determined from isolates of *Erysiphe graminis* f. sp. *hordei* collected in 1990.

tion of this haplotype from eastern Europe to the west, against the prevailing wind direction (unpublished data).

During the subsequent four to five years, however, a number of changes occurred. First, in Czechoslovakia, the virulence pattern of the common *Va13* haplotypes changed, principally by the loss of *Vk* and *Va9* and the gain of virulence *Va6* and

Va12. These changes also occurred more or less simultaneously on the winter barley variety Triton (*Mla13*) in Switzerland. From the RAPD tests, it was found that a number of the isolates from Czechoslovakia and Switzerland were identical, suggesting strongly that the epidemic of mildew that developed explosively on Triton in the late 1980's originated from the large inoculum source in Czechoslovakia. The majority of isolates in Switzerland still carry *Va13* despite the fact that the area of *Mla13* is already diminishing rapidly, illustrating the common 'over-run' phenomenon of continued high frequency of a matching virulence following the 'bust' phase of a 'boom and bust' cycle.

Meanwhile, in Great Britain, other changes occurred in the *Va13* population (Brown et al., 1991). From isolates sampled in 1988, it was found that a particular V-F-haplotype occurred commonly throughout the area (58% of *Va13* isolates). The haplotype corresponded with that found in the earlier samples from Czechoslovakia and carried moderately high resistance to DMI fungicides (probably equivalent to our Class 3: see Table 2); all samples had the same genetic fingerprint determined from RFLP tests. A second V-F-haplotype also occurred frequently though less so than the first (18% of *Va13* isolates). It was restricted to England and lacked *Va9* but did have high resistance to the DMIs, equivalent to the Class 4 described above; the RFLP fingerprint was distinct from that of the more common isolate. From our European survey samples obtained in England and Scotland from 1989 to 1991, it appeared that the second genotype, which was also found to be virulent on Triumph, became more common during this period. Furthermore, they had a DNA structure that was different from the contemporary V-F-haplotypes in Czechoslovakia, which is consistent with the RFLP tests of Brown et al. (1991).

These observations raise several questions. First, it seems clear that the origin of the *Va13* epidemic not only in Switzerland but also in Britain (and perhaps in other European countries), was the large inoculum source generated initially in Czechoslovakia. The start of the British epidemic was noted in Cambridge though it may, of course, have occurred elsewhere at the same time. Certainly it spread rapidly over the whole area. However, within only two years of this event, a second haplotype was found in Britain, widely distributed, presumably because of the additional selective advantage provided by the combination of virulence to Triumph with high resistance to DMI fungicides. We can only speculate about the origin of this second haplotype, but it is reasonable to suppose that it may have arisen locally as a recombinant from a hybrid between the Czech haplotype and the common local haplotypes that already combined Triumph virulence with fungicide resistance. We hope to test this hypothesis by DNA comparisons among representative isolates.

Second, the changes observed occurred rapidly but we do not know whether the long-distance transport from Czechoslovakia to Britain was a single event or whether intermediate steps were involved. Analysis of pathogen populations in the intermediate areas may now be difficult because of the possibility of re-export of the common *Va13* clones from Britain back to the putative intermediate areas, and of the possibility of the recently selected pathogen population in eastern Germany acting as a new source of *Va13*. The crucial observation, however, is that with the present agricultural system in Europe, a newly favoured virulence allele can apparently spread and become locally adapted in any direction very rapidly and over the whole region.

b. Variety mixtures

An analysis of the variety mixture strategy during the period 1984–1991 in the former GDR revealed that diversification within and among mixtures was very limited. First, from Table 5a, it is evident that, although the number of mixtures grown and the number of varieties used in the mixtures were large, relatively few different resistance factors were involved. The distribution of these factors among the mixtures (Table 5b) shows further that two of these, *Mla13* and *mlo,* were used in all mixtures in both years and a third, *Mla12,* was used on more than half of the mixture area. Second, because of the farm structure, based mainly on the collective principle, farm fields were, on average, larger than anywhere in western Europe, with single mixtures often occupying single stands of 80–100 hectares.

Because of these limitations to the potential diversification, there was a detectable increase in the relative frequency of haplotypes with combined *Va12* and *Va13* in the mixtures in 1990. However, when the change in relative frequency was related to the absolute size of the pathogen population, it was found that, because of the overall restriction in population growth in the mixtures, the absolute frequency of the complex pathogen genotypes did not increase in relation to the absolute frequency observed in the pure stands, neither in 1990 nor in 1991. Nevertheless, early in the 1991 epidemic, the relative frequency of the combined virulence was higher than in the previous season. Surprisingly, it then declined during 1991, being replaced to a significant extent by isolates with Va12 or *Va13* alone (Table 6).

Two factors changed in 1991 relative to the previous season. First, the mixture area was reduced by about half and, second, there was a large-scale re-introduction of the use of fungicides for mildew control. However, the change in the proportion of mixtures and pure stands did not significantly affect the distribution of varietal resistances and was unlikely, therefore, to have caused the sudden reduction in frequency of the *Va12* + *Va13* haplotypes. On the other hand, it was noted from the isolates collected in 1990, that the fungicide sensitivity overall was greater than in the populations from neighbouring countries and that, in particular, the sensitivity of the isolates with combined virulence was greater than that of the simpler haplotypes. Subsequently, the data on fungicide response (not shown) revealed a significant increase in fungicide resistance in eastern Germany from 1990 to 1991. Presumably, therefore, fungicide use in 1991 removed a large proportion of the complex, but fungicide sensitive, haplotypes. Fungicide resistant types that were selected would have been simple haplotypes either surviving from the earlier period of intensive fungicide use or immigrating from neighbouring countries. If this reasoning is correct then, by accident, it shows the effectiveness of a large-scale disruptive change in selection in reducing the frequency of selected, complex pathogen genotypes in the population, thus restoring, potentially, the efficiency of disease control of the mixtures. In this example, the disruption was caused by the re-introduction of fungicide use, but it could also be achieved by the intervention of varieties or variety mixtures containing different resistance genes.

Despite the increase in frequency of the complex haplotypes in 1990, the mixture strategy was effective during the period 1984–91, allowing a reduction in fungicide use equivalent to a single spray on more than 100,000 hectares, with no indication that the mixtures involved were losing effectiveness in practice, since yields were maintained at a high level during this period. This continued usefulness of a resistance gene combination over a large area during 7 years contrasts strongly with the rapid failure of *Mla13* used as a monoculture in Czechoslovakia (Brückner, 1982, 1987) and in Switzerland

Table 5. Structure of variety mixtures in eastern Germany in 1990 and 1991

a) Area of mixtures relative to the whole spring barley crop general composition

Year	Relative area (%)	Hectares (× 100000)	No. of mixtures	No. of varieties	No. of resistances
1990	97.0	3.6	8	19	7
1991	49.5	1.9	6	16	6

b) Relative distribution of the resistances among the mixtures

Year	*Mla13*	*mlo*	*Mla12*	*Mla7*	*MlLa*	H. 1613	H. 12939
1990	100	100	72	52	12	3	12
1991	100	100	52	11	8	33	0

(Müller et al., 1991), which suggests that if the *Mla12* and *Mla13* resistance alleles had been used in separate pure stands, then the rapid increase of autoinfection in such crops would have led to the failure of the relevant varieties probably within 3 years. The continued value of *Mla12* and *Mla13* in mixtures also contrasts with the rapid failures that occur when previously exposed resistance genes are combined in a single host variety and then introduced into farm use (Wolfe & Barrett, 1976; Brown et al., 1992). This consideration, of course, is largely theoretical for these two resistance genes since they are allelic or very closely linked.

Conclusions

a. Population biology

Under the current agricultural system, the mildew pathogen population in Europe is enormous and maintains a large range of variation, despite a relatively limited range of major selective characters, at both regional and local levels. At the European level, there is evidence of distinct sub-populations, generated mainly as a result of local use of host resistance and fungicides. The sub-populations tend to remain distinct and intact, but only because of local selection; they are not isolated, except where there is a physical barrier, such as the Alps or Pyrenees, between neighbouring populations. The illusion of isolation arises because the locally selected populations are large and normally little influenced by the much smaller populations of migrating spores. However, if a new niche is created, for example, by a newly introduced resistant variety, then, as the area increases, the variety is exposed to increasing numbers of immigrant spores from the whole area. Eventually, it is overcome by an immigrant haplotype that may have travelled any distance from a few hundred metres to a few hundred kilometres. The same consideration is true for fungicide response. The only exception so far, is for the *mlo* resistance, but we cannot forecast that it will continue in this way (Wolfe, 1992b). In other words, our impression is that gene flow in the European mildew population is potentially unrestricted, but it becomes apparent only when identifiable alleles have a selective advantage in a new area.

Table 6. Relative frequencies of isolates with the four possible combinations of *Va12* and *Va13* in samples from eastern Europe in 1990 and 1991

Haplotype a12	Haplotype a13	Czech. '90	East Ger. '90	East Ger. '91	Poland '90
V	V	53	80	33	5
V	A	10	6	17	84
A	V	22	12	29	5
A	A	15	1	21	6

b. Strategies for disease control

Because of the large potential for rapid gene flow, strategic options that involve diversification or planned planting of different host resistances, either geographically or over particular time periods, are unlikely to be useful in practice.

One of the features of the current system that encourages rapid evolution of the pathogen population is that the barley area is more or less equally divided between winter and spring crops. The winter crop is largely susceptible, which helps the pathogen population to avoid a bottleneck and encourages survival of newly selected genotypes of the pathogen. It has often been argued that limiting certain resistances strictly to the winter crop and others to the spring crop could have a significant epidemiological impact (Schwarzbach & Fischbeck, 1981). Unfortunately, experience shows that there is no practical possibility for implementing such a strategy on an appropriate scale.

The basis of a future strategy must depend on further diversification of resistance. Fortunately, there are indications that a number of new qualitative (Islam et al., 1991) and quantitative resistances will become available through breeding in the next few years, which increases the possibility of finding further durable resistance. In the longer term, re-cycling of used resistances may become feasible as some disappear from use and disruptive selection or drift leads to disappearance of the corresponding virulences. However, the effect of this strategy is unlikely to be as great as that of the initial introduction of the resistance, unless it is incorporated in new genetic backgrounds (Wolfe, 1992a). Moreover, the successful exploitation of such a policy would need continuous monitoring of the pathogen population over a large area.

The enormous range of variability available in the pathogen population helps to explain the rapid demise of varieties that have combinations of resistance genes that have been exposed previously as separate resistances (Wolfe & Barrett, 1976; Brown et al., 1992).

c. Variety mixtures

One effective strategy for using new resistances is in variety mixtures and it is encouraging to see that barley mixtures with acceptable malting quality can be developed and used on a large scale. However, to maximise the reliability of the system, it is essential to diversify as far as possible among mixtures within a region and from year to year.

d. Future development

Because the current agricultural system is so favourable to the pathogen, it seems that the only safe way forward is through a change in the system towards a more ecological approach in which different features are combined to protect host resistance (Wolfe, 1992a); this should include reduced input of inorganic fertilizer, improved rotations in space and time and mixed cropping at the species as well as at the variety level. Such changes can, of course, provide many benefits additional to improved disease control.

Acknowledgement

The authors gratefully acknowledge the support of the Research Commission of the Swiss Federal Institute of Technology.

References

Andrivon, D. & E. Limpert, 1992. Origin and proportions of components of populations of *Erysiphe graminis* f. sp. *hordei*. J. Phytopathol. 135: 6–19.

Brown, J.K.M. & J.H. Jorgensen, 1991. A catalogue of mildew resistance genes in European barley varieties. In: J.H. Jorgensen (Ed.), Integrated control of cereal mildews: virulence patterns and their change, 263–286. Riso National Laboratory, Roskilde.

Brown, J.K.M. & M.S. Wolfe, 1990. Structure and evolution of a population of *Erysiphe graminis* f. sp. *hordei*. Plant Pathol. 39: 376–390.

Brown, J.K.M., A.C. Jessop & H.N. Rezanoor, 1991. Genetic uniformity in barley and its powdery mildew pathogen. Proc. Roy. Soc. London B 246: 83–90.

Brown, J.K.M., C. Simpson & M.S. Wolfe, 1992. Adaptation of barley powdery mildew to varieties in which previously exposed resistance genes are combined. Plant Pathol. (submitted).

Brückner, F., 1982. Nalez rasi padli na jecmeni (*Erysiphe graminis* DC. *var. hordei* Marchal) virulentni pro geny rezistence *Mla9* a *Mla14*. Sbor. UVTIZ – Ochr. Rostl., 18: 101–105.

Brückner, F., 1987. Experiences with breeding of spring barley for resistance to powdery mildew in Czechoslovakia. In: M.S. Wolfe & E. Limpert (Eds.), Integrated Control of Cereal Mildews: Monitoring the Pathogen, pp. 39–41. Martinus Nijhoff Publishers, Dordrecht.

Burdon, J.J. & S.R. Leather (Eds.), 1990. Pests, Pathogens and Plant Communities. Blackwell Scientific Publications, Oxford.

Hermansen, J.E., U. Torp & L.P. Prahm, 1978. Studies of transport of live spores of cereal mildew and rust fungi across the north sea. Grana 17: 41–46.

Hovmoller, M. & H. Ostergaard, 1991. Gametic disequilibria between virulence genes in barley mildew populations in relation to selection and recombination. II. Danish populations. Plant Pathol. 40: 178–189.

Islam, M.R., A. Jahoor & G. Fischbeck, 1991. Reactions of differential barley lines carrying *Mla* alleles to European and Israeli powdery mildew isolates. Proc. 6th Int. Barley Gen. Symp. 589–592.

Kolster, P., L. Munk, O. Stolen & J. Lohde, 1986. Near-isogenic barley lines with genes for resistance to barley mildew. Crop Sci. 26: 903–907.

Limpert, E., 1987a. Frequencies of virulence and fungicide resistance in the European barley mildew population in 1985. J. Phytopath. 119: 298–311.

Limpert, E., 1987b. Spread of barley mildew by wind and its significance for phytopathology, aerobiology and for barley cultivation in Europe. In: G. Boehm & R.M. Leuschner (Eds.), Advances in Aerobiology. Birkhäuser, Basel.

Limpert, E. & G. Fischbeck, 1987. Distribution of virulence and of fungicide resistance in the European barley mildew population. In: M.S. Wolfe & E. Limpert (Eds.), Integrated

Control of Cereal Mildews; Monitoring the Pathogen, pp. 9–30. Martinus Nijhoff Publishers, Dordrecht.

Limpert, E. & E. Schwarzbach, 1981. Virulence analysis of powdery mildew of barley in different European regions in 1979 and 1980. Proc. 4th Int. Barley Genet. Symp. 458–465.

McDermott, J.M., B.A. McDonald, R.W. Allard & R.K. Webster, 1989. Genetic variability for pathogenicity, isozyme, ribosomal DNA and colony color variants in populations of *Rhynchosporium secalis*. Genetics 122: 561–565.

Müller, K., E. Limpert & M.S. Wolfe, 1991. Breakdown of barley powdery mildew resistance *Mla13* in Switzerland. Bull. Swiss Soc. Phytiatry 2: 14 (abstract).

Murray, M.G. & W.F. Thompson, 1980. Rapid isolation of high molecular weight plant DNA. Nucleic Acid Res. 8: 4321–4325.

Newton, A., C.E. Caten & R. Johnson, 1985. Variation for isozymes and double-stranded RNA among isolates of *Puccinia striiformis* and two other cereal rusts. Plant Pathol. 34: 235–247.

O'Dell, M., M.S. Wolfe, R.B. Flavell, C.G. Simpson & R.W. Summers, 1989. Molecular variation in populations of *Erysiphe graminis* on barley, oats and rye. Plant Pathol. 34: 340–351.

Ostergaard, H. & M. Hovmoller, 1991. Gametic disequilibria between virulence genes in barley mildew populations in relation to selection and recombination. I. Models. Plant Pathol. 40: 166–177.

Schwarzbach, E., 1979. A high throughput jet trap for collecting mildew spores on living leaves. Phytopath. Z. 94: 165–171.

Schwarzbach, E. & G. Fischbeck, 1981. Die Mehltauresistenzfaktoren von Sommer- und Wintergerstensorten in der Bundesrepublik Deutschland. Zeit. Pflanzenz. 87: 309–318.

Welz, G. & J. Kranz, 1987. Effects of recombination on races of a barley powdery mildew population. Plant Pathol. 36: 107–113.

Williams, J.G.K., A.R. Kubelik, K.J. Livak, J.A. Rafalski & S.V. Tingey, 1990. DNA polymorphisms amplified by arbitrary primers are useful as genetic markers. Nucleic Acids Res. 18 (22): 6531–6535.

Wolfe, M.S., 1992a. Barley diseases: maintaining the value of our varieties. Sixth International Barley Genetics Symposium, Vol. 2 1055–1067.

Wolfe, M.S., 1992b. Can the strategic use of disease resistant hosts protect their inherent durability? In: Durable Resistance. Wageningen: PUDOC (in press).

Wolfe, M.S. & J.A. Barrett, 1976. The influence and management of host resistance on control of powdery mildew on barley. Barley Genetics III. Proceedings of the 3rd International Barley Genetics Symposium, Garching, 433–439.

Wolfe, M.S. & C. Caten (Eds.), 1987. Populations of Plant Pathogens: their Dynamics and Genetics. Blackwell Scientific Publications, Oxford.

Wolfe, M.S. & D.R. Knott, 1982. Populations of plant pathogens: some constraints on analysis of variation in pathogenicity. Plant Pathol. 31: 79–90.

Wolfe, M.S. & E. Schwarzbach, 1978a. The recent history of the evolution of barley powdery mildew in Europe. In: D.M. Spencer (Ed.), Powdery Mildews, pp. 129–157. Academic Press, London.

Wolfe, M.S. & E. Schwarzbach, 1978b. Patterns of race changes in powdery mildews. Ann. Rev. Phytopathol. 16: 159–180.

Wolfe, M.S., S.E. Slater & P.N. Minchin, 1988. Mildew of barley. In: U.K. Cer. Path. Vir. Surv.: 1987 Ann. Rep. 22–28.

Wolfe, M.S., H. Hartleb, E. Sachs & H. Zimmermann, 1991. Sortenmischungen von Braugerste sind gesünder. Pflanzenschutz-Praxis 2: 33–35.

Euphytica **63**: 141–152, 1992.
R. Johnson and G.J. Jellis (eds), Breeding for Disease Resistance

Discovery, characterization and exploitation of Mlo powdery mildew resistance in barley

J. Helms Jørgensen
Plant Biology Section, Environmental Science and Technology Department, Risø National Laboratory, DK-4000 Roskilde, Denmark

Key words: barley, disease resistance, *Erysiphe graminis hordei, Hordeum vulgare*

Summary

Mlo resistance to barley powdery mildew is a relatively new kind of resistance. It was originally described in a powdery mildew resistant barley mutant in 1942 and has been mutagen-induced repeatedly since then. About 1970 it was also recognized in barley landraces collected in Ethiopia in the 1930s. It is unique in that 1) Mlo resistance does not conform to the gene-for-gene system; 2) *mlo* genes originating from different mutational events map as non-complementing recessive alleles in one locus; 3) all alleles confer the same phenotype, though with small quantitative differences; 4) it is effective against all isolates of the pathogen; and 5) the resistance is caused by rapid formation of large cell wall appositions at the encounter sites preventing penetration by the fungus. Powdery mildew isolates with elevated Mlo aggressiveness have been produced on barley in the laboratory, but have not been found in nature. Mlo resistance is considered very durable. The exploitation of Mlo resistance has been hampered by pleiotropic effects of the *mlo* genes, viz. necrotic leaf spotting and reduced grain yield, but they have been overcome by recent breeding work. During the 1980s Mlo-resistant spring barley varieties have become cultivated extensively in several European countries, in 1990 on about 700,000 ha.

Introduction

The Mlo resistance of barley (*Hordeum vulgare*) to the powdery mildew fungus (*Erysiphe graminis* f. sp. *hordei*) is a relatively new kind of resistance. It has become widely utilized recently in European barley breeding and production. It combines the main advantages of the other two kinds of mildew resistance amenable for barley breeding, viz. race-specific (gene-for-gene), and partial (horizontal) resistance, by being monogenic, non-race-specific and durable.

The main characteristics of Mlo resistance have been reviewed by Jørgensen (1984, 1987, 1991) based mainly on studies of mutagen-induced *mlo* resistance genes. In the present chapter, the current knowledge on Mlo resistance derived from spontaneously arisen and mutagen-induced *mlo* resistance genes, the characteristics of Mlo resistance, and the exploitation of this resistance in European barley breeding and production are reviewed. Selected references to recent literature only are given.

Mutagen-induced Mlo resistance

The first powdery mildew resistant mutant of barley, Mutante 66 (M66), was induced by X-rays in the German variety cv. Haisa and described in 1942. Subsequently, many powdery mildew resistant mutants were described. Ten of them (Table 1) were shown to possess independently induced mutant genes in one locus (Jørgensen, 1976). These ten alleles were designated *mlo1* to *mlo10*. In the 1970s and 1980s many more Mlo mutants were reported, e.g. from Japan (Yamaguchi & Yamashita, 1985), Sweden (Lundqvist, 1991) and Germany, East (Hentrich, 1979) as well as West (Röbbelen & Heun, 1991).

In total more than 150 *mlo* mutant genes have been reported to be induced in a variety of genetic backgrounds constituting mainly highly bred varieties of 2-rowed spring barley. Many of these mutants were described in the literature and made available for research and breeding purposes. In the three decades from the mid 1940s to the mid 1970s *mlo* powdery mildew resistance genes were known from induced mutations only (Jørgensen, 1971, 1976).

Spontaneously occurring Mlo resistance

German expeditions to Ethiopia in 1937 and 1938 collected numerous barley seed samples, which were described in detail in the subsequent years on bulk samples and single-plant progenies (designated E.P. (Einzelpflanzen nachkommenschaften)) (Table 2) at Halle-Hohenturm in eastern Germany. Some data and much of the plant material was lost during the second world war, but Giessen et al. (1956) published extensive data on 250 Ethiopian accessions. Some of them were resistant to powdery mildew in the field and to individual mildew isolates when tested in the greenhouse. Most of the resistant ones were 2-rowed, awnless or awnletted, naked spring barleys collected at locations Bulchi Gofa, Demhi Dollo and Ubamer Baco in southwest Ethiopia. The outstanding powdery mildew resistance of some of these accessions or plant progenies (Table 2) viz. 9 from the E.P. collection at Halle-Hohenturm and 30 from the Abyssinian collection at Köln-Vogelsang was described by Hoffmann & Nover (1959) and subsequently by Nover (1968, 1972) and Meyer & Lehmann (1979). A monogenic, recessive inheritance was shown for two of them (Nover, 1972). The allocation of these genes to locus *mlo*, previously known from the induced mutant genes only, was proven by Nover & Schwarzbach (1971) and Jørgensen (1971). The spontaneously occurring *mlo* allele in the Ethiopian barley line Grannenlose Zweizeilige was designated *mlo11* (Jørgensen 1976). The exact identity of several of the Ethiopian Mlo-resistant lines is dubious because many of them have been shuttled back and forth between breeding institutions and gene banks in Europe and USA, and assigned various designations over time (Table 2). One important line, a six-rowed black seeded line, designated L100 at the former Foundation for Agricultural Plant Breeding in The Netherlands (and in the European Barley Disease Nursery), is deposited at the Dutch Centre for Genetic Resources in Wageningen as accession number GGN00527. It apparently originates from Halle-Hohenturm or the gene

Table 1. Origin of ten barley mutants with Mlo powdery mildew resistance

Mother variety	Mutant	Country	Year	Mutagen	Allele
Haisa	M66	Germany	1942	X-rays	mlo1
Vollkorn	H3502	Austria	1953	X-rays	mlo2
Malteria Heda	MC20	Argentina	< 1960	γ-rays	mlo3
Foma	SR1	Sweden	1961	X-rays	mlo4
Carlsberg II	R5678	Denmark	1963	EMS	mlo5
Carlsberg II	R6018	Denmark	1963	EMS	mlo6
Carlsberg II	R7085	Denmark	1963	EMS	mlo7
Carlsberg II	R7372	Denmark	1963	EMS	mlo8
Diamant	SZ5139b	Czechoslovakia	1965	EMS	mlo9
Foma	SR7	Sweden	1966	γ-rays	mlo10

bank at Gatersleben (in the former GDR) under the name Abyssinian 1140 or HOR2556 which makes it likely that it originates from the German expedition in 1937–1938. The apparently same accession is also in the gene banks at Braunschweig (FRG) as BBA1621, and at Wageningen as CGN1121.

More recently, Mlo-resistant landrace barleys have been described by Negassa (1985a). Among 421 landrace populations, one collected in southern Ethiopia in the village of Jinka in the province of Gemugofa contained four Mlo-resistant lines, viz. 24 per cent of that population. Mlo resistance was not found at any other location. In total the *mlo* gene was found at a frequency of 0.24 per cent of the material.

Over the past 10 years the present author has screened about 4,100 spring barley accessions (*H. vulgare*) for powdery mildew resistance in the field (Jørgensen, 1988). Approximately 3,200 of them originated from Ethiopia. Twenty-four accessions were suspected to have Mlo resistance and 17 of them that were tested in the greenhouse produced the typical Mlo infection type. Among the 24 accessions, 20 originated from Ethiopia. Where specific locations are known, these were all in the southwest of Ethiopia at Bulchi Gofa, Dembi Dollo, Ubamer Baco, Sulto and Jimma. It thus appears that spontaneous *mlo* resistance gene(s) are found with a frequency of 0.2 to 0.6 per cent in Ethiopian landrace material, but predominantly at a few locations in southwest Ethiopia. This region comprises highland areas with high rainfall (up to 1,500 mm/year) and barley cultivation at altitudes from 1,600 m to above the timberline, where the diversity of barley is highest (Nagassa 1985b).

Characteristics of Mlo resistance

a. Genetics. The Mlo powdery mildew resistance in barley is conferred by a long series of recessive, non-complementing alleles in locus *mlo* of which 10 induced mutant genes are designated *mlo1* to *mlo10* (Table 1) and a spontaneously occurring gene *mlo11* (Jørgensen, 1976). There are, however, many more mutagen-induced alleles, and possibly more than one which has arisen spontaneously. Locus *mlo* is located distally on the long arm of barley chromosome 4 (Jørgensen, 1984, 1987).

The genetic fine-structure of locus *mlo* studied by interallelic recombination (Jørgensen & Jensen,

Table 2. Fourteen barley accessions from Ethiopia known to possess Mlo powdery mildew resistance

Accession designation	HOR number	Other designation	In EBDN[a] from	In IBDN[b] from	Reference[c]
Abyssinian 6	1677		1967		1, 4
Abyssinian 9	42				2
Abyssinian 1102	2551	HOR3036, L94, BBA1465	1970		1, 2
Abyssinian 1105	3280	HOR4257			1
Abyssinian 1126	2574	HOR3075			2
Abyssinian 1139	2558	L99	1970		2
Abyssinian 1140	2556	L100, BBA1621	1969		2, 5
Abyssinian 2193	4259				2
Abyssinian 2231	3210				1, 3
Abyssinian 2232	3540				1, 3
Abyssinian 6208	3025	BBA1400	1969		1, 3
E.P. 79	1504	HOR4408, L92, CI14017	1970	1970	1
Grannenlose Zweizeilige	2937	HOR3028, BBA1437	1967	1971	1, 3, 4
Duplialbum	–	CI11793		1965	4

[a] European Barley Disease Nursery.
[b] International Barley Disease Nursery.
[c] 1) Meyer & Lehmann 1979; 2) Nover 1968; 3) Nover 1972; 4) International Barley Disease Nursery Report 1965; 5) Jørgensen 1977, van Hintum pers. comm.

1979) revealed at least three mutational sites within a distance of about 0.05 per cent recombination. One site apparently comprises two X-ray induced alleles *mlo1* and *mlo4;* another site comprises three EMS-induced alleles *mlo5, mlo8* and *mlo9;* and the third site is represented by the spontaneously occurring allele *mlo11* only. Since the 11 named *mlo* alleles all confer an identical phenotypic expression, it is likely that the dominant wild-type Mlo^+ allele can be changed to the recessive state, *mlo*, by mutational events at one of at least three sites within the locus. Furthermore, the occurrence of recombination within the locus excludes the possibility that deletions, except at the molecular level, can be the cause of the mutations. Lastly, the recombination study proves that the necrotic leaf spotting is a true pleiotropic effect of the *mlo* genes because the susceptible recombinants did not show any sign of leaf necrosis. Other *mlo* mutant genes differ, however, in severity of necrotic leaf spotting and frequency of occasional mildew colonies formed on the leaves (Lundqvist, 1991; Hentrich & Habekuss, 1991; Röbbelen & Heun, 1991) suggesting that some mutant genes disrupt the plants' metabolic and defence activity more severely than others. This and related questions may be solved when the *mlo* gene is cloned and sequenced (Hinze et al., 1991) or otherwise identified at the molecular level (Yokoyama et al., 1991).

The *mlo* resistance genes interact with other genes in the genotype of barley. This interaction is seen mainly as modifications in the frequency of the occasional mildew colonies formed on Mlo-resistant barley, and as different levels of necrotic leaf spotting. The *mlo* genes do not, however, interact with other mildew resistance genes in the plant, and they can be combined with other resistance genes so that multi-resistant plants may be produced combining Mlo resistance, race-specific resistance and partial resistance.

b. Phenotype. The phenotype of Mlo-resistant barley lines is characterized mainly by 1) the occurrence of occasional mildew colonies on the leaves, 2) a more or less pronounced tendency to necrotic and/or chlorotic leaf spotting and 3) a reduced grain yield apparently through reduced grain size.

The resistance of Mlo barleys is universal in that an isolate of *E. graminis* which renders them susceptible has never been found (Jørgensen, 1977; Andersen, 1991). When Mlo-resistant barley seedlings are inoculated with *E. graminis,* the plants remain green except that occasionally there are a few small mildew colonies (Fig. 1). These arise mainly from primary infections in the subsidiary cells adjacent to the stomata (Jørgensen & Mortensen, 1977; Andersen & Jørgensen, 1991). This infection type is denoted by an '0' followed by a '4' in parenthesis i.e. 0/(4). If, however, the barley seedlings have an additional, effective resistance gene conferring infection type 2, the seedlings will exhibit infection type 0/(2). If the additional gene confers infection type 0, the seedlings will show infection type 0 because no colonies will be produced. The frequency of occasional mildew colonies may vary from very low to very high depending on several factors. One is density of inoculum; a high density results in a high frequency of colonies (Jørgensen & Mortensen, 1977). A second factor is the genotype of the powdery mildew isolate. All isolates collected in nature (Jørgensen, 1977) cause the formation of about the same low frequency of colonies (Andersen, 1991). Only an isolate selected for high aggressiveness in the laboratory causes many more colonies, i.e. about 10 per cent of the infection occurring in a compatible interaction between a virulent isolate and a susceptible barley (Andersen & Jørgensen, 1991). A third factor is the particular *mlo* allele. Studies in Germany (Hentrich & Habekuss, 1991; Röbbelen & Heun, 1991) and Sweden (Lundqvist, 1991) show that *mlo* alleles of different mutational origin may differ in the quantitative expression of the resistance i.e. the number of occasional mildew colonies. Fourthly, the host gene background may strongly affect the degree of expression of Mlo resistance in terms of the number of colonies (Jørgensen & Mortensen, 1977).

In the field a substantial amount of powdery mildew has occasionally been observed on Mlo-resistant barley. This appears to be the result of unusual environmental factors viz. large amounts of plant-available nitrogen and/or fast growth, for instance, after rainfall following upon a long dry

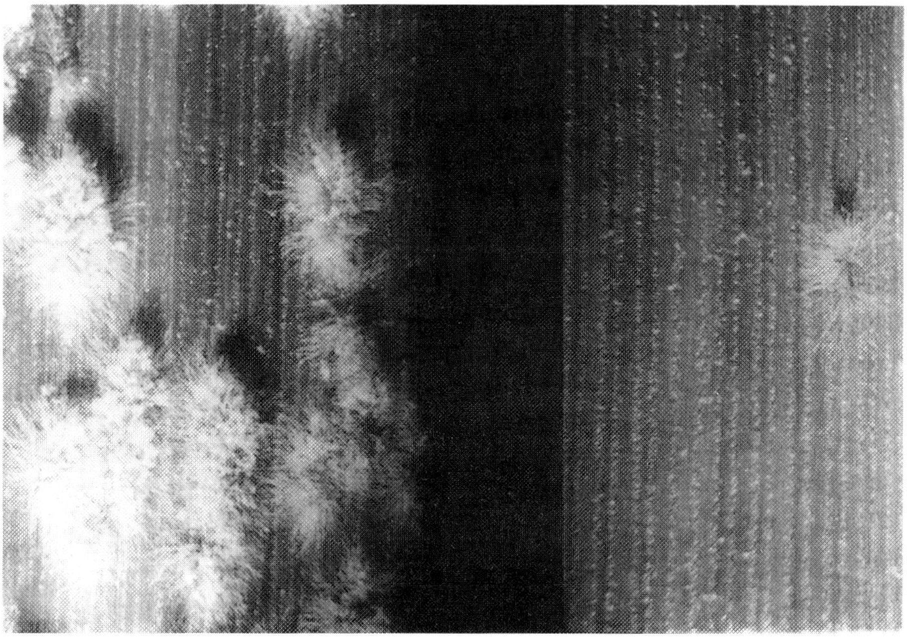

Fig. 1. Mildew colonies developed on non-Mlo-resistant (left) and Mlo-resistant (right) barley seedlings (eight days after inoculation).

period, or heavy and continuous inoculation from severely mildewed barley in neighbouring plots.

The second characteristic feature of Mlo-resistant barleys is a more or less pronounced tendency to develop necrotic (and/or chlorotic) leaf spotting (Schwarzbach, 1976). This is most easily seen at heading (Fig. 2). The necrosis is a pleiotropic effect of the *mlo* genes and is not an expression of hypersensitivity. It may be expressed to a varying degree by mutants with different *mlo* mutant genes (Hentrich & Habekuss, 1991; Röbbelen & Heun, 1991; Lundqvist, 1991). One of the most important factors affecting severity of necrosis is the overall genotype of the barley line with the *mlo* gene. Some genotypes, including lines of Mlo-resistant barley landraces from Ethiopia, may develop quite severe necroses (Fig. 2). Other genotypes, including most of the recent Mlo-resistant spring barley cultivars have only a very slight tendency to develop necrosis (Bjørnstad & Aastveit, 1990; Schwarzbach, 1976). The environment also plays a role in determining the severity and phenotype of necroses. This has been frequently observed but has not been analysed in detail.

The third characteristic trait of Mlo-resistant barley mutants is a reduced grain yield (Hänsel, 1971; Schwarzbach, 1976; Kjær et al., 1990). In many cases most of this effect may be ascribed to simultaneously induced mutant genes that generally reduce plant vigour. In a recent study, three alleles *mlo5, mlo6* and *mlo10* were evaluated in a common but heterogenous gene background (Kjær et al., 1990). All three alleles reduced grain yield and grain size equally, by about 4 per cent, whereas the remainder of the yield reduction was ascribed to other genes independent of locus *mlo*. The reduced grain yield was ascribed to the necrotic leaf spotting that reduces the effective photosynthetic leaf area and the translocation of products of photosynthesis from leaves to spikes during grain filling. There was, however, a considerable variation in grain yield and necrotic leaf spotting among the lines studied, and the best Mlo-resistant lines were equal to the two mildew-susceptible mother varieties Foma and Carlsberg II when tested in disease-free trials.

A considerable number of commercial spring barley varieties with Mlo resistance have been released in Europe. At Risø these barleys have very little or slight necrotic leaf spotting and they have

Fig. 2. Necrotic and chlorotic leaf spotting on Ethiopian Mlo-resistant barley. The leaf below the flag leaf of: Proctor (top), Abyssinian 6, Duplialbum, HOR2936, HOR2937, HOR2938 (bottom).

competed well on the European market with non-Mlo-resistant varieties with respect to agronomic traits and to quality traits such as malting quality. These observations indicate that high-yielding Mlo-resistant spring barley lines can be produced provided that appropriate adjustments are made to the genetic background. It also suggests that absence of necrotic leaf spotting may be an easy selection criterion for removing undesirable pleiotropic effects of the *mlo* resistance genes.

c. Histology. Time course studies on barley seedlings inoculated with powdery mildew have revealed that Mlo-resistant barley rapidly develop large cell wall appositions (papillae) below the encounter sites of the pathogen (Skou et al., 1984). The subsidiary cells at the stomata do not develop papillae (Skou, 1985), probably because these cells have a unique physiology and flexible cell walls enabling them to open and close stomata. It has also been shown that non-Mlo-resistant barley develops small-sized papillae when challenged by the infection peg of the germinated powdery mildew conidium, irrespective of the presence or absence of other (gene-for-gene) powdery mildew resistance genes. Other kinds of mechanical damage such as the stylet of an aphid, a microneedle or abrasion by carborundum treatment also induce a rapid formation of large papillae or papilla-like structures in epidermal cells of Mlo-resistant barley (Skou et al., 1984; Skou, 1985; Russo & Bushnell, 1989; Aist & Gold, 1987).

The papillae contain mainly callose but also basic staining material, carbohydrates, phenols and protein (Russo & Bushnell, 1989). An extensive series of studies on papillae and their formation in barley conducted at Cornell University, USA, have substantiated that early papilla formation is the mechanism of Mlo resistance, and that callose plays a decisive role (Bayles et al., 1990; Yokoyama et al., 1991). When Mlo-resistant barley tissue was treated with an inhibitor of callose formation in plants the tissue became susceptible. The papillae in a Mlo-resistant line are about 50 per cent thicker than those in a non-resistant line, 3.9 versus 2.6 μm (Aist et al., 1988) supporting the view that the large papillae constitute a barrier for the infection peg of the fungus. A light absorbing component and basic staining material in the papillae may also be molecular components of Mlo resistance (loc. cit.). Results of other experiments suggest that calcium ion level strongly affects the effectiveness of Mlo resistance. When Ca^{++} chelators were added to Mlo-resistant tissue, it became susceptible, suggesting that the *mlo* mutant gene affects calcium regulation (Bayles & Aist, 1987). Callose deposition in plants occurs within minutes of mechanical pertubation. This suggests that the enzyme for callose synthesis, the plasma membrane associated 1,3-β-glucan synthase, may be present in an inactive state at the cell membrane and is activated by the influx of Ca^{++} induced by the pathogen. The mutated *mlo* resistance gene may thus allow for an early, rapid increase in the Ca^{++} level in the host cell (Bayles et al., 1990). This is supported by the finding that the specific activity of 1,3-β-glucan synthase did not differ between inoculated non-Mlo-resistant and Mlo-resistant barley (Pedersen, 1990). It has recently been reported that an aqueous extract from barley leaves induced barley and wheat plants to form large papillae in response to powdery mildew infection (Yokoyama et al., 1991), i.e. made susceptible barley and wheat Mlo-resistant. This further supports evidence that Mlo resistance is conditioned by an early, rapid formation of large cell wall papillae. It also supports the notion that this papilla-based resistance mechanism is present in plant species other than barley; it can be induced in barley and wheat by extracts from barley, cauliflower, cucumber and onion (Yokoyama et al., 1991).

d. Function. The data summarized above can be used to establish the function of Mlo resistance. One firm conclusion is that the Mlo resistance is due to a rapid formation of enlarged cell wall appositions below the fungus' encounter sites. They constitute a physical and chemical barrier that the infection peg can rarely penetrate. The mutational origin and recessiveness of the *mlo* genes suggests that the wild type *Mlo*$^+$ gene is a functional one that can be inactivated by mutation. The inactivation may be more or less effective because of the differences seen in the phenotype conferred by some mutant genes. The involvement of several complex molecules such as callose, protein and carbohydrates, and their deposition in the papillae in a layered structure, must involve many genes in the plant. Many of these compounds and processes are common in the normal metabolism in plants. The presence of single, non-functional (mutant) *mlo* genes and the involvement of many functional genes may be explained by assuming that the wild type *Mlo*$^+$ gene has a regulatory function affecting the expression of some of the genes controlling normal metabolism (Skou et al., 1984).

The tendency of Mlo-resistant barleys to exhibit necrotic leaf spotting may be the result of a number of metabolic processes being upset in the resistant plants. Any mechanical damage to the barley leaves, and probably other stress factors, may activate these processes. The absence of the genetical control mechanism implies that these processes run out of control when triggered, and that they are terminated only when the affected cell (and neighbouring ones) is exhausted of substrate(s) for the biochemical processes. The reduction in the tendency to form leaf necrosis by manipulating the gene background indicates that some of the genetic regulation missing in Mlo-resistant barleys may be compensated for by that of other genetic factors.

The Mlo resistance mechanism is limited to pathogens that infect living epidermal cells. It does not affect diseases such as leaf rust, *Puccinia hordei,* because the germ tubes of rust spores infect through open stomata and penetrate the mesophyll cells. In the case of diseases such as net blotch, *Drechslera teres,* or scald, *Rhynchosporium secalis,* the germinating spores exude toxins, which kill the epidermal cells, and thus inactivate possible effects of gene *mlo*. In practice, Mlo-resistant spring barley varieties react to diseases other than powdery mildew in the same way as non-Mlo-resistant barley varieties.

One general defence mechanism of higher plants against mechanical pertubation of the epidermal cell walls is wound-sealing by formation of a callose-rich cell wall apposition below the encounter site. When the recessive *mlo* gene is present, these defence processes occur more rapidly and excessively. Thus the Mlo resistance to powdery mildew barley is related more directly to a physical wound healing than to defence against the powdery mildew fungus. Since many other plant species possess this fundamentally identical wound-healing mechanism, this kind of powdery mildew resistance

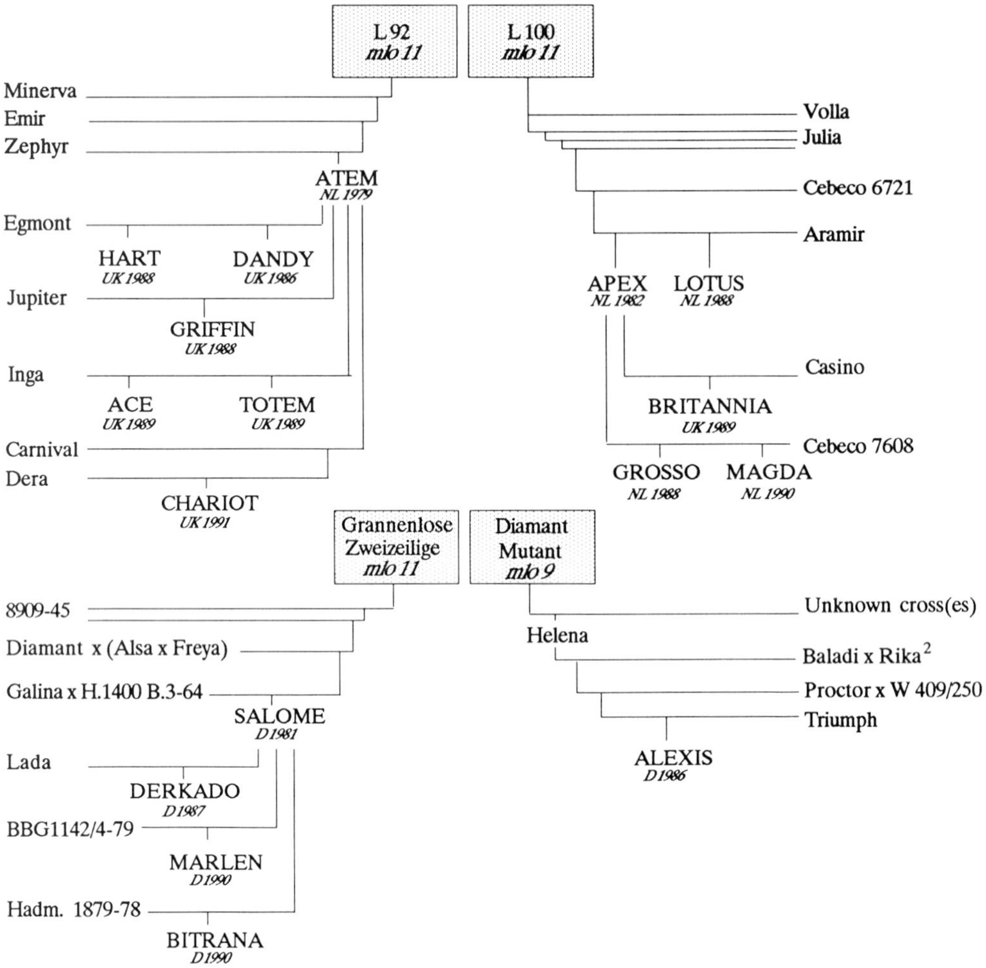

Fig. 3. Pedigree of Mlo-resistant spring barley varieties in Europe with their country and year of origin.

should also be present in other plant species. It should be borne in mind, however, that this resistance mechanism is effective in living epidermal cells only. It is not effective against pathogens producing toxins killing the host cells or against fungi that enter through the stomata and penetrate mesophyll cells.

Exploitation of Mlo resistance

Mutagen-induced powdery mildew resistance genes in locus *mlo* have been exploited in barley breeding for many years (Bjørnstad & Aastveit, 1990; Hänsel 1971; Schwarzbach, 1976; Czembor & Gacek, 1991). Due to the undesirable pleiotropic effects of the *mlo* genes viz. necrotic leaf spotting and reduced grain yield, and the availability of several other highly effective resistance genes such as *Mla7, Mla9* and *Mla12,* barley breeders tended to neglect Mlo resistance and concentrate on the other sources. During the 1970s and early 1980s, the ephemeral effectiveness of the race-specific genes became obvious and the possible durability of Mlo resistance became apparent. This probably led barley breeders to reassess the potential of Mlo resistance. In spite of the fact that the great majority of research was done on mutagen-derived *mlo* resistance genes, it was three Ethiopian sources of Mlo resistance, all probably with gene *mlo11,* that were first introduced in commercial varieties on the European market.

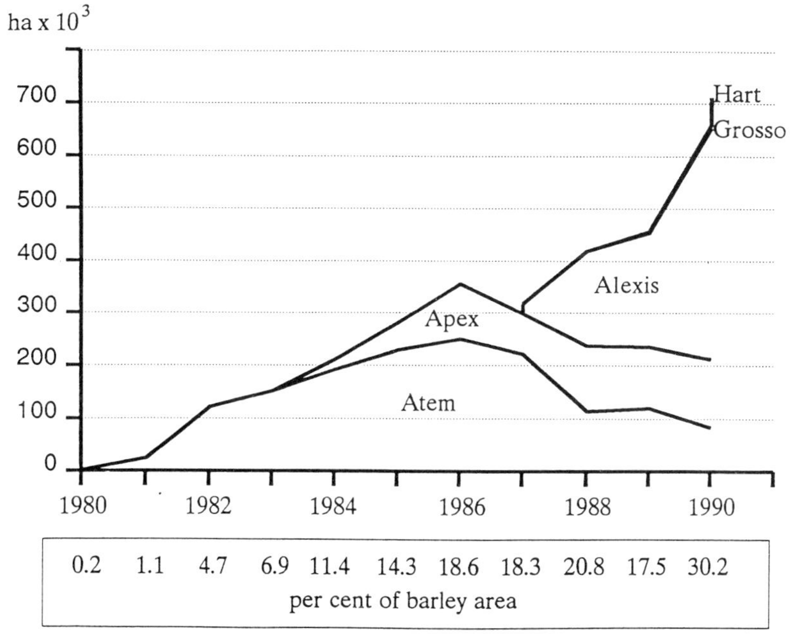

Fig. 4. Area (in ha × 10^3) cultivated with five Mlo powdery mildew resistant spring barley varieties and as percentage of the total spring barley area in the following European countries (from year): Austria (1981); Belgium (1980); West Germany (1985); Great Britain (1982); The Netherlands (1980); Denmark (1989); and Italy (1990).

The first Mlo-resistant barley variety in Europe derived its resistance from the Ethiopian accession L92 (= E.P.79 in Table 2) that gave rise to the variety Atem (released in The Netherlands in 1979) and Atem's descendants released in UK (Fig. 3). Shortly afterwards the variety Apex with Mlo resistance from the donor L100, (Table 2) was released by another breeder in The Netherlands in 1982. Another four varieties with this resistance were marketed subsequently (Fig. 3). The two donor lines L92 and L100 were selected for outstanding powdery mildew resistance in the European Barley Disease Nursery and incorporated into the prebreeding programme at the former Foundation for Agricultural Plant Breeding, The Netherlands, from where advanced breeding populations were released to Dutch barley breeders (L. Slootmaker, pers. comm.). In the former GDR, the variety Salome was marketed in 1981 with gene *mlo11* derived from Grannenlose Zweizeilige (Table 2). Salome was used extensively in variety mixtures in East Germany throughout the 1980s (Gabler & Fritsche, 1991; Wolfe, 1992). Its descendants Derkado and Krona, Bitrana and Marlen (Lau, pers. comm.) were released in 1987 and 1991, respectively (Fig. 3). The fourth source of Mlo resistance is distinctly different from the former three. It is line Helena derived from mutant SZ5139b (Table 1) carrying gene *mlo9* mutagen-induced in the variety Diamant (Schwarzbach, 1976). It gave rise to the malting barley variety Alexis (Fischbeck, 1992) marketed in West Germany in 1986.

Three of these varieties have been widely cultivated in western Europe (Fig. 4). Atem was popular in Austria, Belgium, The Netherlands and particularly in the UK. Apex was, and still is, widely grown in Austria and West Germany. During the last few years, Alexis has become very popular (Fig. 4), particularly in western Germany where it occupied around 35 per cent of the spring barley area in 1990, and in Denmark and Italy where it occupied about 20 per cent of the spring barley area in 1990. Over the past 5 years Mlo-resistant varieties have covered around 20 per cent of the spring barley area in western Europe (Fig. 4), rising to about 30 per cent or more than 700,000 ha. In 1990 two additional varieties, Grosso and Hart have become established in Austria and the UK, respectively.

Durability of Mlo resistance

The durability of Mlo resistance to powdery mildew in barley has been a subject of concern in recent years (Andersen, 1991; Andersen & Jørgensen, 1991; Jørgensen, 1984; Schwarzbach, 1979, 1987).

The rapid increase in the area of spring barley with Mlo resistance cultivated in Europe over the last decade (Fig. 4) may have served as a selective force favouring the emergence and multiplication of powdery mildew with elevated Mlo aggressiveness. Occasional reports have described finds of field samples of mildew with elevated Mlo aggressiveness (Schwarzbach, 1987), but the putative aggressiveness of some of them has not been confirmed (Schwarzbach, 1987; Andersen 1991). An analysis of the extent of cultivation of Mlo-resistant barley in Europe and its correlation with the severity of powdery mildew on Mlo-resistant and non-Mlo-resistant barley (data from the European Barley Disease Nursery 1976–1988) did not disclose any general trend towards an elevated level of Mlo aggressiveness, nor any individual location (local populations) with elevated aggressiveness (Andersen, 1991) in countries with extensive cultivation of Mlo-resistant barley.

Considering the data available, it appears safe to predict that Mlo resistance will be a very durable powdery mildew resistance of barley. If, however, Mlo-resistant spring and winter barley varieties are grown extensively, it is possible that the powdery mildew fungus will slowly but steadily evolve increased aggressiveness and gradually cause disease that may approach the threshold level for crop losses. Therefore, it has become highly relevant to initiate research on the potential of the fungus to 'overcome' Mlo resistance, to devise models and strategies that may aid in preserving the durability of Mlo resistance, and to develop and expand survey programmes for the occurrence of elevated levels of Mlo aggressiveness in European powdery mildew populations.

Conclusion

The Mlo resistance to powdery mildew in barley has been available to plant breeders for about 50 years as induced mutant genes in several high-yielding European spring barley varities, and since about 1970 from Ethiopian landrace barleys. Extensive studies mainly on induced mutants with *mlo* resistance genes disclosed the mechanism of the resistance, its world-wide effectiveness, its possible durability and its undesirable pleiotropic effects. The latter caused plant breeders to emphasize other promising sources of resistance rather than Mlo resistance. More by chance than upon deliberation Mlo resistance from three Ethiopian landraces, probably all with the spontaneously occurring gene *mlo11,* were introduced in commercial varieties around 1980. These varieties rapidly became progenitors for many new varieties. The other source of Mlo resistance is the induced mutant gene *mlo9*. During the 1980s Mlo-resistant varieties have become widely cultivated in many European countries; in 1990 on more than 700,000 ha or close to 30 per cent of the spring barley area.

Acknowledgements

The author is indebted to many colleagues for encouragement and constructive criticism over the years, and to those who have provided unpublished information for the present paper.

References

Aist, J.R. & R.E. Gold, 1987. Prevention of fungal ingress: The role of papillae and calcium. Japan Sci. Soc. Press, Tokyo/Springer-Verlag, Berlin, pp. 47–58.

Aist, J.R., R.E. Gold, C.J. Bayles, G.H. Morrison, S. Chandra & H.W. Israel, 1988. Evidence that molecular components of papillae may be involved in Ml-o resistance to barley powdery mildew. Physiol. Mol. Plant Pathol. 33: 17–32.

Andersen, L., 1991. Mlo aggressiveness in European barley

powdery mildew. In: J. Helms Jørgensen (Ed.), Integrated Control of Cereal Mildews: Virulence Patterns and Their Change. Risø National Laboratory, Roskilde, pp. 187–196.

Andersen, L. & J.H. Jørgensen, 1992. Mlo aggressiveness of barley powdery mildew. Norwegian J. Agric. Sci. Suppl. No. 7: 77–87.

Bayles, C.J. & J.R. Aist, 1987. Apparent calcium mediation of resistance of an *ml-o* barley mutant to powdery mildew. Physiol. Mol. Plant Pathol. 30: 337–345.

Bayles, C.J., M.S. Ghemawat & J.R. Aist, 1990. Inhibition by 2-deoxy-D-glucose of callose formation, papilla deposition, and resistance to powdery mildew in an *ml-o* barley mutant. Physiol. Mol. Plant Pathol. 36: 63–72.

Bjørnstad, Å. & K. Aastveit, 1990. Pleiotropic effects on the *ml-o* mildew resistance gene in barley in different genetical backgrounds. Euphytica 46: 217–226.

Czembor, H.J. & E. Gacek, 1991. Development of high-yielding and disease-resistant barley cultivars through combination of mutagenesis with conventional cross-breeding. Cereal Res. Comm. 19: 43–49.

Fischbeck, G., 1992. Barley cultivar development in Europe – success in the past and possible changes in the future. In: L. Munck (Ed.), Barley Genetics VI, vol. 2, Munksgård Intern Publ., Copenhagen, pp. 885–901.

Gabler, J. & H. Fritsche, 1991. Ergebnisse der Virulenzanalyse 1987–1990 bei Gerstenmehltau auf dem Teritorium der östlichen Bundesländer. Vortr. Pflanzenzüchtg. 19: 317–318.

Giessen, J.E., W. Hoffmann & R. Schottenloher, 1956. Die Gersten Äthiopiens und Erythräas. Z. Pflanzenzüchtg. 35: 377–440.

Hänsel, H., 1971. Experience with a mildew-resistant mutant (mut. 3502) of 'Volkorn' barley induced in 1952. In: Mutation Breeding for Disease Resistance, IAEA-PL-412/13, pp. 125–129.

Hentrich, W., 1979. Multiple Allelie, Pleiotropie und züchterische Nutzung mehltauresistenter Mutanten des mlo-Locus der Gerste. Tag-Ber., Akad. Landwirtsch.-Wiss., DDR, Berlin 175: 191–202.

Hentrich, W. & A. Habekuss, 1991. Untersuchungen an heteroallelen mehltauresistenten Mutanten des mlo-Locus der Sommergerste. Vortr. Pflanzenzüchtg. 19: 311–312.

Hinze, K., R.D. Thompson, E. Ritter, F. Salamini & P. Schulze-Lefert, 1991. RFLP-mediated targeting of the *ml-o* resistance locus in barley (*Hordeum vulgare*). Proc. Nat. Acad. Sci. USA 88: 3691–3695.

Hoffmann, W. & I. Nover, 1959. Ausgangsmaterial für die Züchtung mehltauresistenter Gersten. Z. Pflanzenzüchtg. 42: 68–78.

Jørgensen, J.H., 1971. Comparison of induced mutant genes with spontaneous genes in barley conditioning resistance to powdery mildew. In: Mutation Breeding for Disease Resistance, IAEA-PL-412/12, pp. 117–124.

Jørgensen, J.H., 1976. Identification of powdery mildew resistant barley mutants and their allelic relationship. In: Barley Genetics III, Karl Thiemig, München, pp. 446–455.

Jørgensen, J.H., 1977. Spectrum of resistance conferred by *ml-o* powdery mildew resistance genes in barley. Euphytica 26: 55–62.

Jørgensen, J.H., 1984. Durability of the *ml-o* powdery mildew resistance genes in barley. Vortr. Pflanzenzüchtg. 6: 22–31.

Jørgensen, J.H., 1987. Three kinds of powdery mildew resistance in barley. In: Barley Genetics V, Okayama Univ. Press, 583–592.

Jørgensen, J.H., 1988. Screening of *Hordeum vulgare* for powdery mildew resistance. Nordisk Jordbrugsforsk. 70: 529.

Jørgensen, J.H., 1991. Mechanism of Mlo resistance to barley powdery mildew. Sveriges Utsädesförenings Tidsskrift 2: 45–50.

Jørgensen, J.H. & H.P. Jensen, 1979. Inter-allelic recombination in the *ml-o* locus in barley. Barley Genet. Newsl. 9: 37–39.

Jørgensen, J.H. & K. Mortensen, 1977. Primary infection by *Erysiphe graminis* f. sp. *hordei* of barley mutants with resistance genes in the ml-o locus. Phytopathol. 67: 678–685.

Kjær, B., H.P. Jensen, J. Jensen & J.H. Jørgensen, 1990. Associations between three *ml-o* powdery mildew resistance genes and agronomic traits in barley. Euphytica 46: 185–193.

Lundqvist, U., 1991. Swedish mutation research in barley with plant breeding aspects. A historical review. In: Plant Mutation Breeding for Crop Improvement, IAEA-SM-311/25, pp. 135–147.

Meyer, H. & C.O. Lehmann, 1979. Resistenzeigenschaften im Gersten- und Weizensortiment Gatersleben. 22. Prüfung von Sommergersten auf ihr Verhalten gegen zwei neue Rassen von Mehltau (*Erysiphe graminis* DC. f. sp. *hordei* Marchal). Kulturpflanze 27: 181–188.

Negassa, M., 1985a. Geographic distribution and genotypic diversity of resistance to powdery mildew of barley in Ethiopia. Hereditas 102: 113–121.

Negassa, M., 1985b. Patterns of phenotypic diversity in an Ethiopian barley collection, and the Arussi-Bale Highland as a center of origin of barley. Hereditas 102: 139–150.

Nover, I., 1968. Eine neue, für die Resistenzzüchtung bedeutungsvolle Rasse von *Erysiphe graminis* DC. f. sp. *hordei* Marchal. Phytopath. Z. 62: 199–201.

Nover, I., 1972. Untersuchungen mit einer für den Resistenzträger 'Lyallpur 3645' virulenten Rasse von *Erysiphe graminis* DC. f. sp. *hordei* Marchal. Arch. Pflanzenschutz 8: 439–445.

Nover, I. & E. Schwarzbach, 1971. Inheritance studies with a mildew resistant barley mutant. Barley Genet. Newsl. 1: 36–37.

Pedersen, L.H., 1990. 1,3-β-glucansynthetase activity and callose synthesis in barley *mlo* mutants and mother varieties. (Abstract no 582) Plant Physiol. 79: 102.

Russo, V.M. & W.R. Bushnell, 1989. Responses of barley cells to puncture by microneedles and to attempted penetration by *Erysiphe graminis* f. sp. *hordei*. Can. J. Bot. 67: 2912–2921.

Röbbelen, G. & M. Heun, 1991. Genetic analysis of partial resistance against powdery mildew in induced mutants of barley. In: Plant Mutation Breeding for Crop Improvement, IAEA-SM-311/157, pp. 93–105.

Schwarzbach, E., 1976. The pleiotropic effects of the *ml-o* gene

and their implications in breeding. In: Barley Genetics III, Karl Thiemig, München, pp. 440–445.
Schwarzbach, E., 1979. Response to selection for virulence against the *ml-o* based mildew resistance in barley, not fitting the gene-for-gene hypothesis. Barley Genet. Newsl. 9: 85–88.
Schwarzbach, E., 1987. Shifts to increased pathogenecity on *ml-o* varieties. In: M.S. Wolfe & E. Limpert (Eds.), Integrated Control of Cereal Mildews: Monitoring the Pathogen. Martinus Nijhoff Publishers, Dordrecht, pp. 5–7.
Skou, J.P., 1985. On the enhanced callose deposition in barley with *mlo* powdery mildew resistance genes. Phytopath. Z. 112: 207–216.
Skou, J.P., J.H. Jørgensen & U. Lilholt, 1984. Comparative studies on callose formation in powdery mildew compatible and incompatible barley. Phytopath. Z. 109: 147–168.
Wolfe, M.S., 1992. Barley diseases: maintaining the value of our varieties. In: L. Munck (Ed.), Barley Genetics VI, vol. 2. Munksgård Intern. Publ., Copenhagen, pp. 1055–1067.
Yamaguchi, I. & A. Yamashita, 1985. Induction of mutation for powdery mildew resistance in two-rowed barley. JARQ 18: 171–175.
Yokoyama, K., J.R. Aist & C.J. Bayles, 1991. A papilla-regulating extract that induces resistance to barley powdery mildew. Physiol. and Mol. Plant Pathol. 39: 71–78.

Euphytica **63**: 153–167, 1992.
R. Johnson and G.J. Jellis (eds), Breeding for Disease Resistance

Analysis of durable resistance to stem rust in barley

Brian J. Steffenson
Department of Plant Pathology, North Dakota State University, Fargo, ND 58105, USA

Key words: barley, doubled haploids, durable resistance, gene pyramiding, genetic diversity, *Hordeum vulgare,* molecular marker assisted selection, *Puccinia graminis* f. sp. *secalis, Puccinia graminis* f. sp. *tritici,* resistance genes, rye stem rust, wheat stem rust

Summary

Since the mid-1940's, barley cultivars grown in the northern Great Plains of the USA and Canada have been resistant to stem rust caused by *Puccinia graminis* f. sp. *tritici.* This durable resistance is largely conferred by a single gene, *Rpg*1, derived from a single plant selection of the cultivar Wisconsin 37 and an unimproved Swiss cultivar. At the seedling stage, barley genotypes with *Rpg*1 generally exhibit low mesothetic reactions at 16–20° C and slightly higher mesothetic reactions at 24–28° C to many stem rust pathotypes. This resistance is manifested by a low level of rust infection and mostly incompatible type uredia on adult plants. *Rpg*1 reacts in a pathotype-specific manner since some genotypes of *P. g.* f. sp. *tritici* are virulent on cultivars carrying this gene in the field. Several factors may have contributed to the longevity of stem rust resistance in barley, a) since barley is planted early and matures early, it can sometimes escape damage from stem rust inoculum carried from the south; b) one or more minor genes may augment the level of resistance already provided by *Rpg*1; c) the cultivation of resistant wheat cultivars and eradication of barberry have reduced the effective population size and number of potential new pathotypes of *P. g.* f. sp. *tritici,* respectively; and d) virulent pathotypes of *P. g.* f. sp. *tritici* and *P. g.* f. sp. *secalis* have not become established. This situation changed in 1989 when a virulent pathotype (Pgt-QCC) of *P. g.* f. sp. *tritici* became widely distributed over the Great Plains. However, *Rpg*1 may still confer some degree of resistance to pathotype QCC because stem rust severities have been low to moderate and yield losses light on barley cultivars carrying the gene during the last four seasons (1989–1992). Several sources of incomplete resistance to pathotype QCC have been identified in barley. To facilitate the transfer of resistance genes from these sources into advanced breeding lines, molecular marker assisted selection is being employed.

Introduction

The stem rust (or black rust) fungus, *Puccinia graminis,* is one of the most devastating pathogens of cereals worldwide. In the northern Great Plains of the USA and Canada, the wheat stem rust pathogen (*P. g.* f. sp. *tritici*) has caused a number of spectacular epidemics on both wheat (*Triticum aestivum* and *T. turgidum* var. *durum*) and barley (*Hordeum vulgare*) resulting in catastrophic yield losses (Stakman & Harrar, 1957). These epidemics were the impetus for the initiation of cereal breeding programmes. With wheat, breeding programmes for stem rust resistance began in earnest after the widespread and destructive epidemic of 1904 (Stakman & Harrar, 1957). Before any resistant cultivars were developed, the Canadian cultivar, Marquis, was widely grown because it matured early and largely escaped damage by stem rust. Marquis was heavily rusted during the 1916 epidemic, but was still grown over large areas until 1936. The release of Ceres in 1926 marked the beginning of the 'boom and bust' cycles of plant breeding since it was the first major cultivar bred for stem rust resistance in the region (Dyck & Kerber, 1985). Ceres was heavily damaged in the 1935 and 1937 epidemics caused by pathotype (race) 56 of *P.g.*f.sp. *tritici* (Stakman & Harrar, 1957). Cultivars with other genes for stem rust resistance supplanted Ceres from 1938 to 1950. During this period, only traces of stem rust were observed on the durum and bread wheats. However, in the early 1950's, a dangerous pathotype (15B) with virulence on most of the durum and bread wheats became widely distributed. Pathotype 15B was responsible for the destruction of the durum wheats and nearly one quarter of the bread wheats during the epidemics of 1953 and 1954 (Stakman & Harrar, 1957). Since the mid-1950's, there has not been a serious outbreak of stem rust on wheat in the northern Great Plains. This success is due primarily to the development of wheat cultivars that possess a number of stem rust resistance genes (Dyck & Kerber, 1985; Roelfs, 1985), the diligent monitoring of pathotypes of *P. g.* f. sp. *tritici* in the USA and Canada, and the eradication of the barberry (*Berberis vulgaris*) (Roelfs, 1982).

A different sequence of events has taken place with regard to stem rust on barley during the same era (1904 to the present) and in the same area of production (the northern Great Plains). The Red River Valley of northwestern Minnesota, eastern North Dakota, northeastern South Dakota, and southern Manitoba is one of the most productive cereal-growing regions in the northern Great Plains (Fig. 1). However, stem rust is present every year in this area and is capable of causing serious damage on barley and wheat. Prior to 1940, yield losses in barley from stem rust were nearly as common as they were in wheat during epidemic years, although never as severe (Table 1). In 1942, the first barley cultivar with stem rust resistance (Kindred) was released to US farmers in the southern Red River Valley region. Since this time, there have been no significant losses due to stem rust in barley, even during the 1953 and 1954 epidemics when the wheat crop was devastated (Table 1). During the 1954 epidemic, barley in the northern Red River Valley region of Manitoba, Canada suffered serious damage (total losses exceeded $9 million) because over 90% of the hectarage was still being sown to stem rust susceptible cultivars (McDonald, 1970). The hectarage of resistant barley cultivars increased in Manitoba after the 1954 epidemic, and losses to stem rust have been minimal ever since.

The resistance of barley to stem rust is durable (Steffenson, 1989) according to the definition of Johnson (1984). This resistance has remained effective for 50 years in different cultivars that have been widely grown. The hectarage and range of these resistant cultivars is immense; in the USA, between 1.4 and 1.8 m ha are planted annually across the northern tier of states from Wisconsin in the east to Washington in the west and south to Wyoming and Colorado. The plantings of resistant cultivars are concentrated in the Red River Valley states of Minnesota, North Dakota, and South Dakota. Large areas of barley with the same resistance are also cultivated in Canada, mainly in the province of Manitoba. In the Red River Valley region alone, over 1 m ha of resistant barley are planted each year. Stem rust is particularly severe on cereal crops grown in the Red River Valley due, primarily, to the region's location along the main

Table 1. Estimated percent yield loss of barley and wheat to stem rust during several epidemics in North Dakota and Minnesota from 1916–55[a]

Year of epidemic	North Dakota		Minnesota	
	Barley	Wheat	Barley	Wheat
1916	–[b]	70.0	–[b]	61.0
1923	3.0	12.0	1.5	15.0
1925	0.8	5.0	0.5	11.0
1935	15.0	56.5	15.0	51.6
1937	8.0	25.0	5.0	10.0
1952	Trace	5.6	Trace	2.3
1953	Trace	37.7	Trace	13.4
1954	Trace	42.9	Trace	18.0
1955	Trace	8.0	Trace	2.4

[a] From Roelfs, A.P., 1978. Estimated losses caused by rust in small grain cereals in the United States – 1918–76. U.S. Dept. Agri., Misc. Pub. # 1363.

[b] Data not available for barley. The data for wheat in 1916 were provided by M.E. Hughes, USDA-ARS Cereal Rust Laboratory, St. Paul.

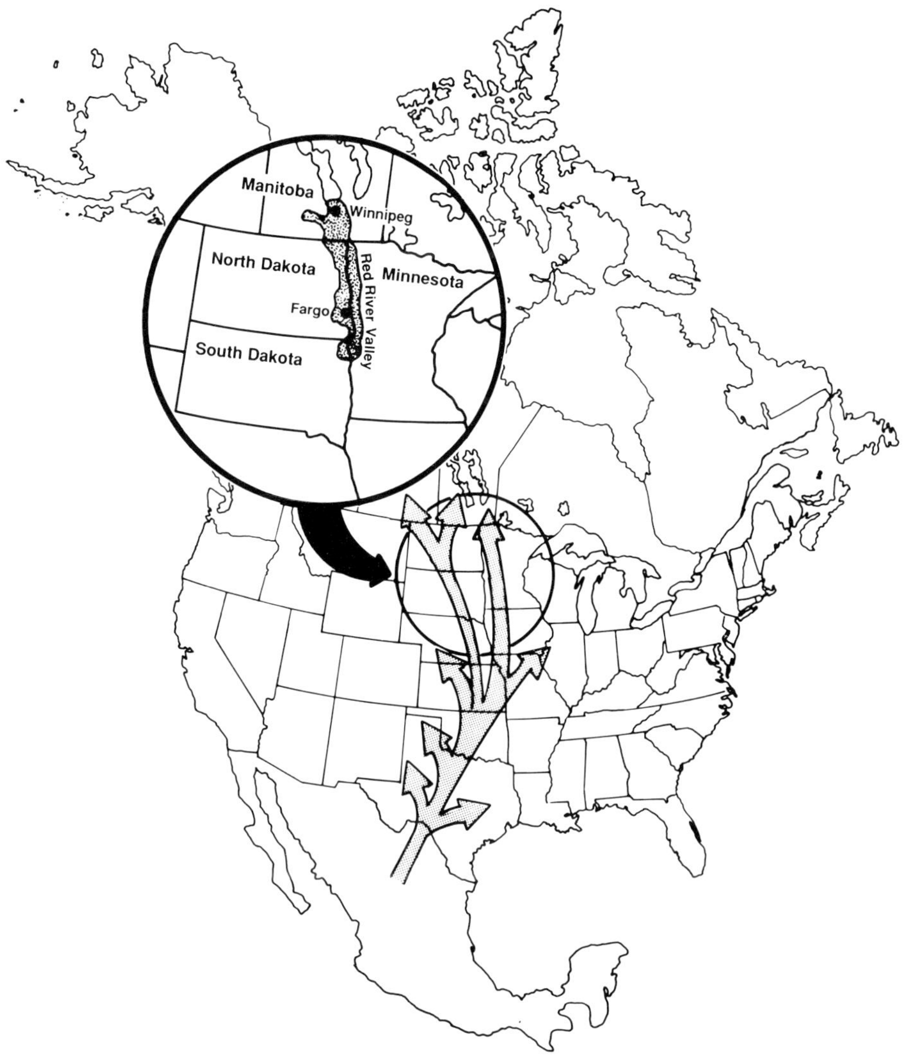

Fig. 1. The Red River Valley of Minnesota, North Dakota, and South Dakota in the USA and Manitoba in Canada. Stippled arrows represent the generalized movement of stem rust urediospores (*Puccinia graminis*) during the spring season.

pathway of stem rust inoculum (Fig. 1) and a favourable environment for disease development. The severity of stem rust in a given year will depend on the date of urediospore arrival, the amount of inoculum transported from the south, and the weather conditions. In this paper, only the Red River Valley and surrounding regions will be considered since stem rust is a continual threat to barley in this area.

Durable resistance was defined by Johnson (1984) as a type of resistance that remains effective for a long time in cultivars that are widely grown. This definition was made without reference to any underlying genetic mechanisms or associated characters of resistance. The unrestrictive nature of this definition is appropriate since some researchers have presumptuously assumed certain unifying mechanisms and principles for the terms they have used to describe host resistance. These underlying mechanisms or principles have not proven universal. In order to gain an understanding as to the range of characteristics and possible mechanisms of durable resistance, we must dissect and analyse each example of resistance that proves durable.

Table 2. CI or PI number, use, origin, year of introduction or release, pedigree, reaction to stem rust, probable source(s) of stem rust resistance, and period of peak hectarage of major barley cultivars grown in the tri-state region of Minnesota, North Dakota, and South Dakota in the USA and Manitoba in Canada from 1945 to the present[a]

Cultivar	CI or PI number	Use	Origin	Year introduced or released	Pedigree	Reaction to stem rust	Probable source(s) of stem rust resistance	Period of peak hectarage
Tri-state region of Minnesota, North Dakota, and South Dakota								
Kindred	6969	Malting	S. Lykken, Kindred, ND	1942	Single plant selection from Wisc. 37	Resistant	unknown	1945–58
Traill	9538	Malting	N. Dakota Agr. Expt. Station, Fargo, ND	1956	Kindred/Titan	Resistant	Kindred	1959–62
Trophy	10647	Malting	N. Dakota Agr. Expt. Station, Fargo, ND	1961	Traill/White aleurone UM570	Resistant	Kindred + Peatland	1963–64
Larker	10648	Malting	N. Dakota Agr. Expt. Station, Fargo, ND	1961	Traill/White aleurone UM570	Resistant	Kindred + Peatland	1965–79
Dickson	10968	Malting	N. Dakota Agr. Expt. Station, Fargo, ND	1964	Traill/Kindred/CI 7117–77	Resistant	Kindred	1967–72
Beacon	15480	Malting	N. Dakota Agr. Expt. Station, Fargo, ND	1973	Conquest/Dickson	Resistant	Kindred + Peatland	1975–78
Morex	15773	Malting	Minnesota Agr. Expt. Station, St. Paul, MN	1978	Cree/Bonanza	Resistant	Kindred + Peatland	1980–84
Glenn	15769	Malting	N. Dakota Agr. Expt. Station, Fargo, ND	1978	Br5755–3/Trophy//ND B138	Resistant	Kindred + Peatland	1981–83
Azure	15865	Malting	N. Dakota Agr. Expt. Station, Fargo, ND	1982	Bonanza/Nordic/ND B130	Resistant	Kindred + Peatland	1984–89
Robust	476976	Malting	Minnesota Agr. Expt. Station, St. Paul, MN	1983	Morex/Manker	Resistant	Kindred + Peatland	1985–present
Excel	542047	Malting	Minnesota Agr. Expt. Station, St. Paul, MN	1990	Cree/Bonanza//Manker/Robust	Resistant	Kindred + Peatland	released 1990
Manitoba, Canada								
O.A.C. 21[b]	1470	Malting	Ontario Agricultural College, Guelph, Ont.	1910	Single plant selection from Manscheuri	Susceptible	–	pre-1945–49
Gartons	7016	Feed	Farmer, Portage la Prairie, Man.	1930	Selection of barley from John Garton Company, England	Moderately resistant	unknown	pre-1945–49 and 1955
Plush	6093	Feed	Dominion Experimental Farm, Brandon, Man.	1939	Lion/Bearer	Susceptible	–	pre-1945–48
Montcalm	7149	Malting	MacDonald College Quebec, Que.	1945	Michigan 31604/Common 6-rowed 4307 M.C./Mandscheuri 1807 M.C.	Susceptible	–	1950–57
Vantage	7324	Feed	Dominion Experimental Farm, Brandon, Man.	1947	Newal/Peatland/Plush	Resistant	Peatland	1954–55
Husky	9537	Feed	University of Saskatchewan, Saskatoon, Sask.	1953	Peatland/Regal/O.A.C. 21/Newal	Resistant	Peatland	1955–57
Herta	8097	Feed	Plant Breeding Inst. Weibullsholm, Landskrona, Sweden	1956	Kenia/Isaria	Moderately Susceptible	–	1969–71
Parkland	10001	Malting	Dominion Experimental Farm, Brandon, Man.	1956	Newal/Peatland/O.A.C. 21/Olli/Montcalm	Resistant	Peatland	1958–1966
Keystone	10877	Feed	Dominion Experimental Farm, Brandon, Man.	1961	Vantage/Jet/Vantmore	Resistant	Peatland	1964–65
Conquest	11638	Malting	Canada Dept. of Agriculture Expt. Farm, Brandon, Man.	1965	Vantage/Jet/Vantmore/Br4635/Swan/ Parkland	Resistant	Peatland	1967–72
Fergus	13797	Feed	University of Guelph, Guelph, Ont.	1968	Selection from Firlbecks III	Susceptible	–	1972–77
Bonanza	14003	Malting	Canada Dept. of Agriculture Expt. Farm, Brandon, Man.	1970	Vantage/Jet/Vantmore/Parkland/Conquest	Resistant	Peatland	1975–87
Bedford	15774	Feed	Agriculture Canada Res. Sta., Brandon, Man.	1979	Keystone/Vantage/Jet/Vantmore/Husky/ Cree	Resistant	Peatland	1984–present
Norbert	452125	Malting	Agriculture Canada Res. Sta., Winnipeg, Man.	1980	CI 5791/Parkland/Betzes/Piroline/Akka/ Centennial/Klages	Resistant	Peatland	1984–86
Argyle	496255	Malting	University of Manitoba, Winnipeg, Man.	1981	Bonanza/UM67–907	Resistant	Peatland	1988–present

[a] Data on the CI or PI number, use, origin, year of introduction or release, pedigree, and reaction to stem rust of most entries were obtained from the Barley Variety Dictionary published by the American Malting Barley Association, Inc., Milwaukee, Wisconsin (Anonymous, 1990). Hectarages of major barley cultivars grown in the USA and Canada were provided by Mr. S.E. Heisel, American Malting Barley Association, Inc., and Dr. N.T. Kendall, Brewing and Malting Barley Research Institute, Winnipeg, Manitoba, respectively. The probable source(s) of stem rust resistance in cultivars is based on published pedigrees.
[b] O.A.C. 21 was reported to possess some resistance to stem rust (Jedel et al., 1989).

Some of the features that should be considered with regard to resistance are the sources, genetics, pathotype-specificity, ontogenetic expression, components, and effect of environmental factors. The objectives of this paper are to describe and analyse what is known about the durable resistance to stem rust in barley and to discuss possible breeding strategies for the lasting control of this disease.

Important features of stem rust resistance in barley

Sources of resistance. The origin of the first stem rust resistant barley cultivar is unusual since it was produced by an astute farmer rather than a scientific breeding programme. During the severe epidemic of 1935, a farmer named Sam Lykken of Kindred, North Dakota identified a single green (rust-free) plant in his heavily infected field of Wisconsin 37 barley (Lejeune, 1951). This plant was probably the result of an admixture rather than a mutation because it was distinctly different from Wisconsin 37 for a number of characters (Wiebe & Reid, 1961). Thinking this plant may possess some useful resistance to stem rust, Lykken harvested all 18 seeds and increased them for six consecutive seasons. The stem rust resistance of this genotype was later confirmed by researchers at the North Dakota Agricultural Experiment Station. In 1942, Lykken's single plant selection was released as the commercial cultivar, Kindred (Lejeune, 1951). In addition to being resistant to stem rust, Kindred also passed the stringent industry requirements for classification as a malting barley cultivar. Kindred and cultivars derived from it have dominated the barley hectarage in the north central region of the USA from 1942 to the present (Table 2). Indeed, Lykken's single plant selection was probably the most important factor contributing to stable barley production in the southern Red River Valley area during the early post-World War II era.

Another source of stem rust resistance in barley was derived from an umimproved cultivar from the Canton of Lucerne in Switzerland. This germplasm was brought to the United States Department of Agriculture (USDA) in 1914. Two sclections from this bulk seed lot, Chevron and Peatland, proved to be highly resistant during the pathotype 56 epidemics of 1935 and 1937 (Shands, 1939). Peatland was the source of stem rust resistance in most Canadian barleys starting with the release of Vantage in 1947 (Table 2). Larker and Trophy, released to growers in 1961, were the first major US malting barley cultivars with Peatland in their pedigree.

Selections from the cultivar Gartons were reported to possess resistance to stem rust in Canada (Lejeune, 1947). The original seed lot of Gartons traces to a Manitoba farmer who procured a barley sample from the John Garton Company in England during the early 1930's (Anonymous, 1990). The stem rust resistant selections of Gartons were cultivated under the same name from the 1930's to the early 1960's, mainly in Manitoba (N.T. Kendall, personal communication). No studies have been advanced to determine the genetics of stem rust resistance in Gartons or the relation of this resistance to other known sources.

Genetics of resistance. The inheritance of resistance in barley to *P. g.* f. sp. *tritici* was first studied by Powers & Hines (1933) who found that resistance in Peatland was governed by a single dominant gene. This gene was designated 'T' since it conferred resistance to the *tritici* forma specialis of *P. graminis* (Powers & Hines, 1933). The current locus designation for the *T* gene is *Rpg*1 (Søgaard & von Wettstein-Knowles, 1987). A number of other studies have been completed on the genetics of resistance in Peatland, Chevron, or their derivatives to several different pathotypes of the wheat stem rust fungus. Some of these studies corroborated the original findings of Powers & Hines (1933) that a single dominant gene confers stem rust resistance (Andrews, 1956; Brookins, 1940; Jedel et al., 1989; Shands, 1939). However, other studies documented the presence of genes with minor effects in addition to *Rpg*1 (Lejeune, 1946; Miller & Lambert, 1955; Patterson, 1950).

In a cross between Kindred and Minn. 615 (a derivative of Peatland), no obvious segregation was observed to pathotype 15B of *P. g.* f. sp. *tritici* (Miller & Lambert, 1955). Similar results were obtained with a cross between Kindred and Peatland to pathotype 56 (S. Fox, personal communication).

These data indicate that the resistance factor in Kindred is allelic to that found in Peatland, namely *Rpg*1. Thus, the durable resistance of North American barley cultivars to *P. g.* f. sp. *tritici* is oligogenic.

Specificity to pathotypes of Puccinia graminis. When the cultivar Kindred and those derived from Peatland became widely grown in the late 1940's, it appeared that their resistance was effective against a number of pathotypes of the wheat stem rust pathogen, including 15B, a pathotype known to be virulent on many wheat cultivars (Lejeune, 1947). However, in 1950, large, compatible type uredia were observed on barley breeding lines known to carry *Rpg*1 at St. Paul, Minnesota (Miller & Lambert, 1956). Subsequent experiments confirmed the virulence of this rust, identified as pathotype 59A, on Peatland, Chevron, Kindred and their derivatives (Ali, 1954; Miller & Lambert, 1956). This was not the only report of stem rust on barley genotypes carrying *Rpg*1 since Canadian researchers found that Peatland and its derivatives were completely susceptible to strains of pathotype 11 identified in Winnipeg in 1952 (Johnson, 1954; Johnson, 1961). In 1959 and again in 1981, compatible type uredia were isolated from barley cultivars carrying *Rpg*1 in the Red River Valley region. This rust was identified as pathotype 23 (E.E. Banttari, personal communication) and 151 (Franckowiak & Miller, 1983), respectively.

The resistance of selected barley genotypes was studied to a composite of four pathotypes, Pgt-RTQ[a] (113-RTQ); -QTH (151-QSH); -TPL (15-TNM); and -HTC (29-HJC), and to the single pathotype -QFC (151-QFB) in isolated rust nurseries in 1989. The four pathotypes used in the composite were previously considered the most common or virulent pathotypes occurring on barley in the northern Great Plains (Steffenson et al., 1985). Pathotype QFC was studied to determine its virulence on adult plants carrying *Rpg*1 in the field.

[a] Pathotype (race) designations for *P.g.*f.sp. *tritici* are based on the nomenclatural system of Roelfs & Martens (1988). The number and letter codes in parentheses represent the previous pathotype designations under the Stakman (Stakman et al., 1962) and Cereal Rust Laboratory (CRL) (Roelfs & McVey, 1973) nomenclatural systems, respectively.

Genotypes with *Rpg*1 could be easily distinguished from those without the gene to the pathotype composite of *P. g.* f. sp. *tritici:* genotypes with *Rpg*1 had low to moderate levels of rust infection (7.5–20%) and low to moderately low infection responses (primarily resistant [R] to moderately resistant [MR] responses), whereas those without the gene had high levels of rust infection (52.5–90%) and high to moderately high infection responses (primarily susceptible [S] to moderately susceptible [MS] responses) (Table 3). Some differences also were noted among genotypes with *Rpg*1 as the level of rust infection on Chevron was significantly lower than on Azure, 80-TT-29, and Morex. When infected with pathotype QFC, genotypes with *Rpg*1 exhibited high terminal rust severities (77.5–90%) and high to moderately high infection responses (Table 3). Thus, it was not possible to differentiate genotypes with and without *Rpg*1 to pathotype QFC.

The specificity of *Rpg*1 to various pathotypes of *P. g.* f. sp. *tritici* also was documented in a number of greenhouse studies (Ali, 1954; Immer et al., 1943; Jedel et al., 1989; Johnson, 1954; Patterson et al., 1957; Steffenson et al., 1985).

In addition to the wheat stem rust pathogen, barley also can be attacked by the rye stem rust pathogen, *P. graminis* f. sp. *secalis.* Some pathotypes of the rye stem rust fungus are highly virulent on adult barley genotypes possessing *Rpg*1 at the field (Johnson, 1954; Johnson & Buchannon, 1954; Steffenson et al., 1985); however, in the seedling stage, many of these same genotypes exhibit low infection types (Immer et al., 1943; Johnson & Buchannon, 1954; Patterson et al., 1957; Steffenson et al., 1985).

Expression of resistance at the seedling stage. At the seedling stage, most barley genotypes, including those known to be very susceptible in the field, exhibit uredia that are associated with some degree of chlorosis (Patterson et al., 1957; Miller & Lambert, 1955; Waterhouse, 1948). This type of reaction makes it difficult to assess infection types using the conventional stem rust scale developed for wheat by Stakman et al. (1962). To overcome this difficulty, Miller & Lambert (1955) devised a scale for barley that was based primarily on uredium size.

Barley seedlings also commonly display mesothetic reactions (a mixture of different infection types on the same plant) to many pathotypes of the stem rust pathogen, especially at lower temperatures (Jedel et al., 1989; Miller & Lambert, 1955; Patterson et al., 1957; Steffenson et al., 1985). It is difficult to differentiate barley genotypes on the basis of the infection type when most exhibit a mesothetic response. A weighted infection type, based on the relative frequency of each infection type present (Steffenson et al., 1985), has been useful in separating some genotypes (e.g. with or without *Rpg*1) that display mesothetic responses (Jedel et al., 1989; Steffenson et al., 1985); however, this analysis is cumbersome and time-consuming.

Recently, Steffenson et al. (1991) found that pathotypes MCC (56-MBC) and HPH (29-HNH) could clearly differentiate genotypes with *Rpg*1 from those without the gene at the seedling stage. When infected with these pathotypes, barley genotypes with *Rpg*1 commonly exhibited infection types ranging from 0; to 10;, whereas those without the gene gave 23− to 33+ (Table 4). The reactions exhibited by the barley genotypes to pathotype TPM (15-TNM) are typical of those exhibited to other pathotypes in that similar infection types are sometimes observed on genotypes with (80-TT-29 and Hector) and without (80-tt-30, Steptoe, Harrington, Hietpas 5, and PI 382313) the gene, *Rpg*1. Some of these genotypes can be differentiated by the frequency of individual infection types (e.g. 23- vs 3-2) which was the basis of the weighted infection type described earlier.

Expression of resistance at the adult plant stage.

Table 3. Terminal rust severity (in percent) and infection response of selected barley genotypes inoculated with a composite of pathotypes (Pgt-HTC, QTH, RTQ, and TPL) and with the single pathotype Pgt-QFC of *Puccinia graminis* f. sp. *tritici* in isolated nurseries at Fargo and Casselton, North Dakota, respectively, in 1989

Genotype	Recognized allele for stem rust reaction[a]	Pathotype composite (HTC, QTH, RTQ, TPL)		Pathotype QFC	
		Terminal stem rust severity[b]	Infection response[c]	Terminal stem rust severity	Infection response
Chevron	*Rpg*1	7.5 a[d]	R-MR	77.5 ab	MS-S(MR)
Peatland	*Rpg*1	13.0 ab	R-MR	80.0 abc	MS-S(MR)
Kindred	*Rpg*1	14.3 ab	MR-R(MS)	80.0 abc	MS-S
Azure	*Rpg*1	18.8 b	MR-R(MS)	90.0 cd	S-MS
80-TT-29	*Rpg*1	18.8 b	R-MR	80.0 abc	S-MS
Morex	*Rpg*1	20.0 b	R-MR	85.0 bc	S-MS
Hietpas 5	*Rpg*2	52.5 c	MS-S	75.0 ab	S-MS
Black Hulless	*rpg*BH	57.5 cd	S-MS	80.0 abc	MS-S
80-tt-30	*rpg*1	62.5 d	MS-S(MR)	72.5 a	S-MS(MR)
Steptoe	*rpg*1	65.0 d	S-MS	80.0 abc	S-MS
Hiproly	*rpg*1	90.0 e	S	95.8 d	S-MS

[a] *Rpg*1 denotes the dominant homozygous resistant condition for the *T* gene and *Rpg*2 for *T*2. *rpg*1 denotes the recessive homozygous susceptible condition for the *T* gene. *rpg*BH denotes the recessive homozygous resistant condition for a gene (previously designated, *s*; Steffenson et al., 1984) that confers resistance to *Puccinia graminis* f. sp. *secalis*. Genotypes 80-TT-29 (*Rpg*1) and 80-tt-30 (*rpg*1) are near-isogenic (Steffenson et al., 1985).

[b] Terminal rust severity was assessed using the modified Cobb scale (Stubbs et al., 1986) when the plants were in the mid-dough stage of development.

[c] Infection responses were assessed using the criteria of Stubbs et al. (1986) where R = resistant, MR = moderately resistant, MS = moderately susceptible, and S = susceptible. Infection responses given in parenthesis were observed infrequently.

[d] Values are the means of four replicates. Means with different letters within a column are significantly different according to Duncan's Multiple Range Test (P = 0.05).

Adult plants with *Rpg*1 generally exhibit small (incompatible type) uredia and low percentages of stem rust infection to many of the common pathotypes of *P. g.* f. sp. *tritici* (Jedel et al., 1989; Lejeune, 1947; Shands, 1939; Steffenson & Wilcoxson, 1987; Steffenson et al., 1985) (Table 3). In some cases, a small percentage of compatible type uredia will develop on the stems or leaf sheaths to these same pathotypes (Brookins, 1940; Jedel et al., 1989; Steffenson et al., 1985). Jedel et al. (1989) found distinct differences with regard to the expression of *Rpg*1 in several two- versus six-rowed barleys; the former commonly exhibited higher infection levels than the latter. Virulent pathotypes of *P. graminis* have been isolated from barley in the past as described previously. When these virulent pathotypes (e.g. QFC) are inoculated onto genotypes with *Rpg*1, fully susceptible reactions are expressed (Table 3). The degree of association between the stem rust reaction at the seedling stage and that expressed at the adult plant stage is an important factor to consider, especially in the development of greenhouse screening techniques. In general, the seedling reaction to *P. g.* f. sp. *tritici* corresponds to that observed on adult plants in the field (Brookins, 1940; Immer et al., 1943; Miller & Lambert, 1955; Shands, 1939; Steffenson et al., 1985).

Components of resistance. Quantitative or rate-reducing components of resistance have been described in a number of cereal rust pathosystems (Parlevliet, 1979). In the barley:stem rust pathosystem, few studies have been made to determine if *Rpg*1 is involved in conferring any quantitative components of resistance to *P. g.* f. sp. *tritici*. Sellam & Wilcoxson (1976) studied the development of the stem rust fungus in barley cultivars with and without *Rpg*1. Urediospore germination, appressorium formation, and penetration by the fungus were similar in the two groups of barley; however, the number of uredia forming on leaves was lower and the growth of the pathogen restricted in genotypes with *Rpg*1 (Sellam & Wilcoxson, 1976). Low receptivity was closely associated with the presence of *Rpg*1 in barley genotypes infected with pathotypes RTQ and QTH in the field and also with the latter pathotype in the greenhouse (Steffenson & Wilcoxson, 1987). This was not the case with pathotype RTQ in the greenhouse, since the near-isogenic lines for *Rpg*1 did not differ significantly in numbers of uredia/cm^2 in either the seedling or adult plant stage. Thus, in addition to qualitative characters of resistance, *Rpg*1 also may be involved in reducing the number of successful infections that occur on barley; however, this quantitative component of resistance may vary with pathotype.

Effect of the environment. The effect of various environmental factors on the resistance conferred by *Rpg*1 has not adequately been studied. Temperature is known to affect the stem rust reaction in barley. At the seedling stage, barley genotypes with *Rpg*1 generally exhibit low mesothetic reac-

Table 4. Infection types of barley seedlings to pathotypes Pgt-MCC, HPH, and TPM of *Puccinia graminis* f. sp. *tritici* at 25–28° C

Genotype	Recognized allele for stem rust reaction[a]	Seedling infection type — Pathotypes MCC	HPH	TPM
Chevron	*Rpg*1	0;1[b]	0;1	21
Kindred	*Rpg*1	0;1	0;1	12
Glenn	*Rpg*1	0;1	0;1	21
Excel	*Rpg*1	0;	0;1	21
Bowman	*Rpg*1	0;1	0;1	21
Hector	*Rpg*1	0;1	10;	23−
80-TT-29	*Rpg*1	0;1	0;1	23−
80-tt−30	*rpg*1	3−3	33+	3-2
Steptoe	*rpg*1	3−3	3−3	3-2
Harrington	*rpg*1	3−3	3−3	3-2
Hietpas 5	*Rpg*2	23−	3−3	23−
PI 382313	*Rpg*3	23−	3−3	23−
Black Hulless	*rpg*BH	3−2	3−3	3−3

[a] See Table 3. *Rpg*3 denotes the dominant homozygous condition for the *T*3 gene.

[b] Data are the two most common infection types (most prevalent type listed first) observed on individual genotypes from five replicates. Infection types were assessed 12 days after inoculation using the system of Miller & Lambert (1955). Infection types 0;, 1, and 2 are indicative of resistance (low infection response), whereas infection types 3 and 4 are indicative of susceptibility (high infection response). The '+' and '−' symbols were used to denote more or less sporulation, respectively.

tions at low temperature (16–20° C) and slightly higher mesothetic reactions at high temperature (24–28° C) (Miller & Lambert, 1955; Steffenson et al., 1985). Infection type differences between genotypes known to be resistant or susceptible in the field are most distinct at high incubation temperatures because genotypes that lack *Rpg*1 generally exhibit high infection types (Andrews, 1956; Jedel et al., 1989; Miller & Lambert, 1955; Patterson et al., 1957; Steffenson et al., 1985). For this reason, most researchers have used incubation temperatures above 24° C when evaluating barley seedlings to *P. g.* f. sp. *tritici.* It is possible that higher ambient temperatures during the growing season could reduce the effectiveness of *Rpg*1 in adult barley plants in the field, but this aspect has not been investigated.

In another study, Patterson et al. (1957) examined the effect of pre-inoculation temperature and different combinations of post-inoculation daylength and light intensity treatments on the infection type of cultivar Valentine (possesses *Rpg*1). The infection type of this cultivar did not change appreciably in any of the different treatments.

Factors contributing to the longevity of stem rust resistance in barley.

The durability of the *Rpg*1 resistance in barley is remarkable considering the potential variability of *P. g.* f. sp. *tritici* and the vast hectarages of cultivars that are genetically uniform for stem rust resistance. Few single resistance genes have performed as well and as long as *Rpg*1 in a cereal rust pathosystem. However, the longevity of stem rust resistance in barley was undoubtedly aided by a number of factors.

First, because most of the barley is planted early in the spring and matures in a relatively short time, stem rust inoculum carried from the south will sometimes reach the crop in the north (Fig. 1) when it is nearly mature and cannot be seriously damaged. In the northern Great Plains, barley is a 88–90 day crop and matures approximately 7–12 days before many of the common wheat cultivars. Thus, like the early-maturing wheat cultivar, Marquis, barley can sometimes escape severe stem rust infection (Green, 1971). Barley will not escape rust infection by this means in all years. Cool, wet weather during the spring season can delay the seeding of barley providing additional time for the stem rust pathogen to reach and reproduce on the crop.

Second, one or more minor genes may augment the level of resistance provided by *Rpg*1 in barley cultivars. Additional genes for stem rust resistance have been documented in several genetic studies (Miller & Lambert, 1955; Lejeune, 1946; Patterson, 1950) as previously described, but their effect in suppressing stem rust development is not fully understood. Barley genotypes commonly exhibit chlorosis around uredia at the seedling stage. The ubiquitousness of this chlorotic reaction led Patterson et al. (1957) to conclude that many barley genotypes possess at least some level of resistance to *P. graminis.* It is not known whether this seedling resistance is effective in the later stages of plant development; however, in recent greenhouse and field experiments, we have not been able to achieve high levels of rust infection on some barley genotypes (with or without *Rpg*1) with virulent pathotypes of *P. g.* f. sp. *tritici* prior to anthesis (B.J. Steffenson & Y. Jin, unpublished data). Thus, some barleys may possess resistance genes that protect them from severe stem rust infection up to the stage of anthesis. According to Green (1971), the *H. vulgare:P. graminis* pathosystem exhibits the primitive host:parasite characteristics of low specialization and low parasite aggressiveness. He suggested that slow rusting resistance may have protected barley from stem rust damage in western Canada. If this is true, the genetic basis of this resistance is unknown.

Third, the cultivation of stem rust resistant wheats in the northern Great Plains has prevented local outbreaks of stem rust (Steffenson et al., 1985) and reduced the effective population size of *P. g.* f. sp. *tritici* in North America (Schafer & Roelfs, 1985). The reduced population size of *P. g.* f. sp. *tritici* may not be highly significant in contributing to the longevity of stem rust resistance in barley since the potential number of mutant spores produced on wheat that possess virulence on single resistance genes is great (Schafer & Roelfs, 1985).

Fourth, the eradication of the barberry in the Great Plains has greatly contributed to the stability of stem rust resistance in wheat (Roelfs, 1982) and certainly in barley as well even though this alternate host was essentially eliminated before cultivars with *Rpg*1 were released to growers.

Fifth, virulent pathotypes of *P. g.* f. sp. *secalis* have not increased to damaging levels in the Great Plains (Green, 1971). The reason for this may be due to the limited hectarage of the pathogen's primary host, rye, and eradication of the alternate host, barberry, especially in the USA. However, a number of wild grass species can harbour *P. g.* f. sp. *secalis* and are common in the region (Roelfs, 1985).

Finally, although virulent pathotypes of *P. g.* f. sp. *tritici* have been detected on barley periodically in the past, they have not increased and become established. The reasons for this are unclear. In the northern Great Plains, stem rust infections are initiated every spring by inoculum from the south (Fig. 1) since the fungus rarely overwinters north of Oklahoma. If virulent pathotypes are to survive each year, they must be capable of attacking winter wheats or wild grass species because the hectarage of barley is very small in the southern and central Great Plains. It is possible that virulent pathotypes were simply 'filtered-out' on different wheat genotypes in the south during the winter and decreased in frequency (Vanderplank, 1968) – that is, until the appearance of pathotype Pgt-QCC.

Pathotype QCC. In 1989, compatible stem rust uredia were observed on barley cultivars carrying *Rpg*1 in many fields throughout the northern areas of production in the USA (B.J. Steffenson, unpublished data). Subsequent pathotype determinations of these field collections by the Cereal Rust Laboratory in St. Paul revealed that the rust was pathotype QCC of *P. g.* f. sp. *tritici* – a pathotype new to the Great Plains of the USA (Roelfs et al., 1991a). Pathotype QCC was first detected in the Great Plains region one year earlier by Martens et al. (1989) in Canada. In identifying this pathotype in Saskatchewan and Manitoba, Martens et al. (1989) noted that, 'it is unusual to obtain virulence in the prairie region that is characteristic of the Pacific region.' This early discovery was significant because pathotype QCC was detected in nearly every barley field surveyed in the southern Red River Valley region during the following three seasons (1989–1991). It is likely that pathotype QCC originated in the Pacific Northwest because isolates with the same virulence pattern were described from that region prior to 1989 (Martens et al., 1989; Roelfs et al., 1990a). Additionally, it is possible that this pathotype arose from barberry since this sexual host is known to occur in the Pacific Northwest region. Regardless of whether this pathotype originated from a sexual or asexual host, this incident demonstrates the importance of periodic inoculum exchanges from the western to the central stem rust population.

Minor stem rust epidemics were recorded on barley in the northern Great Plains in 1990 and 1991. The severity of stem rust on barley in eastern North Dakota, northeastern South Dakota, and northwestern Minnesota ranged from trace to 60% (Roelfs et al., 1990b). Not since the late 1930's has barley stem rust been as prevalent and severe in this region. Fortunately, overall yield losses in barley cultivars with *Rpg*1 were light, ranging from 1–3% (Roelfs et al., 1990b; Roelfs et al., 1991b). Late planted crops, however, suffered higher yield losses (Harder & Dunsmore, 1991). The actual dollar loss from stem rust was greater than the yield loss estimates would predict since many growers did not receive the premium afforded malting barley due to a low percentage of plump kernels in their grain. This reduction in kernel size was likely due to stem rust infection (Dill-Macky et al., 1990).

Pathotype QCC is virulent on several winter wheat cultivars and wild grass species in the south and is now well established in the Great Plains. During the past four years, QCC has become one of the most prevalent pathotypes of *P. g.* f. sp. *tritici* on barley and wheat in North America (D.E. Harder & A.P. Roelfs, personal communication). This recent change in the frequency of wheat stem rust pathotypes is due primarily to the susceptibility of cultivated barley. The effect of barley on the pathotype composition of the wheat stem rust population should not be understated as indicated by Luig (1985). Indeed, Fox & Harder (1991) detected significant changes in the frequency of *P. g.* f. sp. *tritici* pathotypes on different barley genotypes grown in the field. It is possible that pathotypes with virulence on northern spring wheat cultivars (both bread and durum types) could arise from stem rust infected barley plants. This could result in stem rust epidemics on wheat because barley would provide an early and local source of inoculum. The potential vulnerability of spring wheat highlights the urgency of identifying and incorporating additional stem rust resistance genes into new barley cultivars.

Table 5. Reaction of selected barley genotypes to pathotype QCC of *Puccinia graminis* f. sp. *tritici* as seedlings in the greenhouse and adult plants in the field[a]

Genotype[b]	Seedling infection type[c]		Field reaction	
	Mode	Range	Average terminal rust severity[d]	Infection response[e]
Q21861	0;1	0;/23−	6.1 a[f]	MR-MS
Diamond	21	12/23−	7.1 a	MR
PI 452421	23−	12/3−2	9.4 ab	MS-S
PC11	123−	12/33−	10.0 ab	MR-R-MS
PC84	23−1	12/23−	10.1 ab	MR-MS
Hietpas 5	23−	23/3−2	13.9 b	MS-S
Robust	21	12/23−	25.0 c	S-MS

[a] Data from Jin et al., 1992.
[b] Diamond, PI 452421, and Hietpas 5 were selected as being resistant to pathotype QCC by B.J. Steffenson and coworkers at North Dakota State University; PC11 and PC84 by S. Fox at the University of Manitoba; and Q21861 by Dr. A.P. Roelfs, USDA-ARS Cereal Rust Laboratory, St. Paul. The resistance of Q21861 and Diamond was independently verified by B.J. Steffenson and S. Fox, respectively. Robust is a widely grown cultivar that is susceptible to pathotype QCC.
[c] Infection types were assessed 12 days after inoculation using the system of Miller & Lambert (1955). The seedlings were incubated in a greenhouse at 21–25° C. Mode = most commonly observed infection types and range = lowest/highest observed infection types.
[d] Terminal rust severity was assessed using the modified Cobb scale (Stubbs et al., 1986) when the plants were in the mid-dough stage of development. Average terminal severity is based on four replications at each of the following 1991 rust nurseries: Fargo, North Dakota and St. Paul, Minnesota.
[e] See Table 3.
[f] Values are the means of eight replicates. Means with different letters within a column are significantly different according to Duncan's Multiple Range Test (P = 0.05).

Identification of resistance to pathotype QCC. In 1989, a cooperative project was initiated among researchers at North Dakota State University, University of Minnesota, USDA, Agriculture Canada, and Busch Agricultural Resources, Inc. to identify resistance in barley to pathotype QCC. Disease evaluations were made mainly on accessions from the USDA National Small Grains Collection (NSGC) in the greenhouse and at several field nurseries in the USA and Canada. Resistance to pathotype QCC is rare in barley as fewer than 10 genotypes were selected as possessing adequate resistance from over 17,000 entries evaluated.

At North Dakota State University, detailed studies are being conducted on six selected genotypes: Q21861, PC11, and PC84, genotypes originating from the CIMMYT barley breeding programme; Hietpas 5, a selection from Oderbrucker that possesses the *Rpg2* (*T2*) gene for stem rust resistance (Patterson et al., 1957); and PI 452421, and PI 491573, selections from the USDA NSGC (Table 5). PI 491573 is the cultivar Diamond which was produced by the Agriculture Canada Research Station at Lacombe. None of these six genotypes possesses complete resistance to pathotype QCC. At the seedling stage, the infection types on these lines can be quite variable, ranging from 0; to 33−. Differences in resistance between the selected lines and common cultivars like Robust are detectable in the field after heading; the former exhibit relatively low terminal rust severities and numbers of compatible type uredia (moderately susceptible or susceptible types) and the latter higher terminal rust severities and numbers of compatible type uredia (Table 5). The incomplete resistance of Diamond may be due to a reduction in receptivity and uredium size (Steffenson et al., 1990).

Q21861 possesses the highest level of resistance among the six selections under study. This line was one of several hundred originally selected by Dr. W.J.R. Boyd from a CIMMYT barley nursery established by Dr. H. Vivar (R.G. Rees, personal communication). In Australia, Q21861 possesses resistance to pathotype Pgt-LSH in the field (Dill-Macky et al., 1992). The resistance of this line to pathotype QCC was first described by Dr. A.P. Roelfs at St. Paul and was independently confirmed by our research group at Fargo. At the seedling stage, Q21861 commonly exhibits infection types ranging from 0; to 21, although higher reactions are sometimes observed at elevated temperatures (> 24° C) (J.D. Miller, personal communication). Adult plants exhibit small to minute uredia with chlorosis; however, large compatible uredia are often present near the top node and on the peduncle – a common occurrence on many other barley genotypes. Preliminary genetic studies indicate that the resistance of Q21861 to pathotype QCC is governed by one or possibly two recessive genes. Additionally, the reaction of PC11, PC84, Diamond, and Hietpas 5 to pathotype QCC appears to be simply inherited with resistance dominant (Y. Jin & B.J. Steffenson, unpublished data).

Breeding for resistance to stem rust in barley

The appearance of pathotype QCC in the Great Plains has forced barley pathologists and breeders to reevaluate their strategies for the control of stem rust. The resistance gene, *Rpg*1, was effective for nearly 50 years in barley cultivars that were widely grown, but it is not realistic to expect such longevity from another resistance gene. Several gene deployment strategies have proven effective in a number of cereal pathosystems over the past three decades. Two of these strategies, the deployment of genetically diverse cultivars and gene pyramiding, may prove useful in the barley production areas of the northern Great Plains.

Deployment of cultivars that are genetically diverse for stem rust resistance. Strategies that increase the diversity of disease resistance have been used with some success in several cereal pathosystems (Mundt & Browning 1985; Wolfe & Barrett, 1980). These schemes may vary according to the level of diversity within a unit of area. For example, the cultivation of a genotype mixture (with the individual components of the mixture differing in the resistance genes they possess) within a single field comprises an intra-field diversity scheme. Inter-field diversity can be achieved by the planned deployment of genotypes over different fields or farms (Mundt & Browning, 1985). The exploitation of an intra-field diversity scheme may not be possible for malting barley in the Red River Valley region due to the rigid cultivar identity requirements of the malting and brewing industries in the USA and Canada. Furthermore, it is unlikely that a planned inter-field diversity scheme would be acceptable to most producers in the area. However, some action should be taken to reduce the extreme genetic uniformity that exists for stem rust resistance in barley today. There are six major breeding programmes that develop barley cultivars for the Red River Valley and surrounding areas. A modest level of inter-field diversity could be achieved if the future cultivars from these programmes were bred with different resistance genes. Unfortunately, the effectiveness of this strategy may be limited since a single cultivar usually becomes dominant in the area and remains so for several years (Wych & Rasmusson, 1983).

Pyramiding stem rust resistance genes in barley. The strategy of gene pyramiding has been effective against stem rust in wheat for over 37 years in North America and Australia (Dyck & Kerber, 1985). This same strategy will likely provide control of *P. g.* f. sp. *tritici* in barley for a long time. The malting and brewing industries dictate, to some extent, the strategy used in barley breeding programmes. The decision by industry to accept a barley cultivar for malting and brewing is based on about 25 different quality traits (Wych & Rasmusson, 1983). These stringent requirements have forced breeders to narrow their germplasm base. To obtain acceptable malting types, it is often necessary to cross closely related parents that already possess superior malting characteristics (Wych & Rasmusson, 1983). Prior to the appearance of pathotype QCC in 1989, breeding for stem rust resistance was easy because it required the incorporation of just one dominant gene (*Rpg*1) which was already present in many superior malting types. To control stem rust in the future, breeders may have to combine *Rpg*1 and gene(s) for resistance to pathotype QCC in new barley cultivars. The retention of *Rpg*1 in advanced breeding lines is essential since this gene has proven durable to many pathotypes of *P. g.* f. sp. *tritici* in the Great Plains.

Molecular marker assisted selection of genes for stem rust resistance. The incorporation of multiple stem rust resistance genes in barley will significantly increase the complexity of the breeding process. To aid breeders in this task, a project has been initiated for the molecular marker assisted selection of genes that confer stem rust resistance in barley. This technique is based on the identification of molecular markers (isozymes, restriction fragment length polymorphisms [RFLPs], or random amplified polymorphic DNAs [RAPDs]) that are closely linked to the genes of interest. If a tight linkage is found between a molecular marker and a disease resistance gene, the former can be exploited for the indirect selection of the latter without the need for laborious and sometimes variable pathogen inoculations (Melchinger, 1990). Genome maps of two doubled haploid populations (one derived from Steptoe/Morex and the other from Harrington/TR306) are under construction as part of the North American Barley Genome Mapping Project. These populations will serve as useful models for the molecular marker assisted selection of stem rust resistance in barley since both are polymorphic for *Rpg*1 (Steffenson & Dahleen, 1991; B.J. Steffenson, unpublished data). Two flanking RFLP markers (*Tel*1S [subtelomeric] and *Plc* [barley plastocyanin precursor]) to *Rpg*1 were recently identified in the Steptoe/Morex doubled haploid population (Steffenson et al., 1992). The DNA probe for *Plc* will best facilitate the transfer of *Rpg*1 in breeding programmes because it is closely linked (ca. 3 cM) to the gene and gives a strong, unambiguous hybridization signal.

The bulked segregant analysis procedure (Michelmore et al., 1991) is being used to identify molecular markers (RAPDs) that are closely linked to genes conferring resistance to pathotype QCC in a doubled haploid population derived from the cross, Q21861/SM89010. This population was developed by Drs. B.G. Rossnagel and K. Kao at the University of Saskatchewan and is polymorphic for reaction to the leaf (brown) rust (*Puccinia hordei*) and powdery mildew (*Blumeria* = *Erysiphe graminis* f. sp. *hordei*) pathogens in addition to pathotype QCC of *P. g.* f. sp. *tritici* (B.J. Steffenson & Y. Jin, unpublished data). Bulked segregant analysis is a simple and elegant technique that was first validated with genes for resistance (*Dm* genes) to the downy mildew pathogen (*Bremia lactucae*) in lettuce. Using this procedure, Michelmore et al. (1991) identified three markers that were 6, 8, and 12 cM away from the *Dm*5/8 locus. The use of this procedure on doubled haploid populations offers several advantages over the conventional analysis on F_2 progeny: a) there is no need to verify homozygous resistant plants in the F_3 generation since all the alleles are fixed, and b) replicated inoculations to determine the infection phenotype of individuals can be repeated as often as necessary and with several different pathogens over time since the doubled haploids represent a permanent or 'immortal' population. This latter attribute is valuable when the expression of a resistance gene is significantly altered by environmental factors or if the goal of a molecular marker assisted selection programme is to tag and transfer genes for resistance to more than one pathogen.

Concluding remarks

The durable resistance conferred by *Rpg*1 in barley to stem rust has faltered because a virulent pathotype of *P. g.* f. sp. *tritici* became widely distributed in the Great Plains. However, *Rpg*1 may still confer some degree of resistance to pathotype QCC because stem rust severities have been low and yield losses light on barley cultivars carrying the gene during the last four seasons (1989–1992) in the USA. In Canada, Harder & Dunsmore (1991) reported only light to moderate yield losses on cultivars carrying *Rpg*1 (e.g. Argyle, Leduc, and Bonanza) and heavy losses on the cultivar Tupper which lacks the resistance gene. Genotypes with *Rpg*1 (and some without the gene) appear to possess a level of resistance that prevents severe rust infection by pathotype QCC prior to anthesis in both greenhouse and field experiments (B.J. Steffenson & Y. Jin, unpublished data). This 'pre-anthesis' resistance, whether conferred by *Rpg*1 or other genes, may protect barley from serious damage by stem rust. Additionally, it is possible that pathotype QCC will decrease to innocuous levels in the wheat stem rust population. This situation apparently occurred four times during the past 42 years with the virulent pathotypes 59A, 11, 23, and 151. Whether pathotype QCC continues to predominate in the Great Plains population of *P. g.* f. sp. *tritici* or not, there are grounds for optimism that durable resistance will continue to be achieved in future barley cultivars by the incorporation of multiple resistance genes using molecular marker assisted selection.

Acknowledgements

I thank Drs. Y. Jin, J.D. Miller, R.G. Rees, S. Fox, D.E. Harder, A.P. Roelfs, and J.W. Martens for reviewing this manuscript and making valuable suggestions. I appreciate the assistance of S.E. Steffenson and C. Burkhart in proofreading the text and preparing the figure, respectively.

References

Anonymous, 1990. Barley Variety Dictionary. American Malting Barley Assoc., Inc., Milwaukee.

Ali, S.B., 1954. Behavior of isolates of race 59 of *Puccinia graminis tritici* on certain varieties of wheat and barley. (Abstr.) Phytopathology 44: 481.

Andrews, J.E., 1956. Inheritance of reaction to loose smut, *Ustilago nuda,* and to stem rust, *Puccinia graminis tritici,* in barley. Can. J. Agric. Sci. 36: 356–370.

Brookins, W.W., 1940. Linkage relationship of the genes differentiating stem rust reaction in barley. Ph.D. Thesis. University of Minnesota, St. Paul.

Dill-Macky, R., R.G. Rees & W.J.R. Boyd, 1992. Sources of resistance to stem rust in barley. Plant Dis. 76: 212.

Dill-Macky, R., R.G. Rees & G.J. Platz, 1990. Stem rust epidemics and their effects on grain yield and quality in Australian barley cultivars. Aust. J. Agric. Res. 41: 1057–1063.

Dyck, P.L. & E.R. Kerber, 1985. Resistance of the race-specific type. In: A.P. Roelfs & W.R. Bushnell (Eds.). The Cereal Rusts. Vol. II. Diseases, distribution, epidemiology, and control, pp. 469–500. Academic Press, Inc., New York.

Fox, S. & D. Harder, 1991. Selection pressure of barley genotypes on a stem rust population. (Abstr.) Can. J. Plant Pathol. 13: 276.

Franckowiak, J.D. & J.D. Miller, 1983. Reaction of barley to a culture of wheat stem rust, race 151. Barley Newsl. 27: 52.

Green, G.J., 1971. Hybridization between *Puccinia graminis tritici* and *Puccinia graminis secalis* and its evolutionary implications. Can. J. Bot. 49: 2089–2095.

Harder, D.E. & K.M. Dunsmore, 1991. Incidence and virulence of *Puccinia graminis* f. sp. *tritici* on wheat and barley in Canada in 1990. Can. J. Plant Pathol. 13: 361–364.

Immer, F.R., J.J. Christensen & W.Q. Loegering, 1943. Reaction of strains and varieties of barley to many physiologic races of stem rust. Phytopathology 33: 253–254.

Jedel, P.E., D.R. Metcalfe & J.W. Martens, 1989. Assessment of barley accessions PI 382313, PI 382474, PI 382915, and PI 382976 for stem rust resistance. Crop Sci. 29: 1473–1477.

Jin, Y., B.J. Steffenson & T.G. Fetch, Jr., 1992. Evaluation of barley for resistance to stem rust, 1991. Biol. & Cult. Tests 7: 73.

Johnson, R., 1984. A critical analysis of durable resistance. Ann. Rev. Phytopathol. 22: 309–330.

Johnson, T., 1954. The reaction of barley varieties to certain races of wheat stem rust. Div. of Botany & Plant Pathol., Sci. Serv., Dept. of Agr., Winnipeg. Rept. # 7.

Johnson, T., 1961. Rust research in Canada and related plant-disease investigations. Res. Branch, Canada Dept. Agr., Winnipeg. Publ. # 1098.

Johnson, T. & K.W. Buchannon, 1954. The reaction of barley varieties to rye stem rust, *Puccinia graminis* var. *secalis.* Can. J. Agr. Sci. 34: 473–482.

Lejeune, A.J., 1946. Correlated inheritance of stem rust reaction, nitrogen content of grain and kernel weight in a barley cross. Sci. Agr. 26: 198–211.

Lejeune, A.J., 1947. A note on the reaction of certain barley varieties to race 15B of stem rust (*Puccinia graminis tritici* Erikss. and Henn.). Sci. Agr. 27: 183–185.

Lejeune, A.J., 1951. The story of Kindred (L) barley. Barley Improv. Conf. Rept., January 23, 1951, Minneapolis.

Luig, N.H., 1985. Epidemiology in Australia and New Zealand. In: A.P. Roelfs & W.R. Bushnell (Eds.). The Cereal Rusts. Vol. II. Diseases, distribution, epidemiology, and control, pp. 301–328. Academic Press, Inc., New York.

McDonald, W.C., 1970. Diseases and other factors affecting average yields of barley in Manitoba, 1954–1968. Can. Plant Dis. Surv. 50: 113–117.

Martens, J.W., K.M. Dunsmore & D.E. Harder, 1989. Incidence and virulence of *Puccinia graminis* in Canada on wheat and barley in 1988. Can. J. Plant Pathol. 11: 424–430.

Melchinger, A.E., 1990. Use of molecular markers in breeding for oligogenic disease resistance. Plant Breed. 104: 1–19.

Michelmore, R.W., I. Paran & R.V. Kesseli, 1991. Identification of markers linked to disease-resistance genes by bulked segregant analysis: A rapid method to detect markers in specific genomic regions by using segregating populations. Proc. Natl. Acad. Sci. USA 88: 9828–9832.

Miller, J.D. & J.W. Lambert, 1955. Variability and inheritance of reaction of barley to race 15B of stem rust. Agron. J. 47: 373–377.

Miller, J.D. & J.W. Lambert, 1956. Reactions of certain spring barley lines to race 59A of stem rust. Plant Dis. Reptr. 40: 340–346.

Mundt, C.C. & J.A. Browning, 1985. Genetic diversity and cereal rust management. In: A.P. Roelfs & W.R. Bushnell (Eds.). The Cereal Rusts. Vol. II. Diseases, distribution, epidemiology, and control, pp. 527–560. Academic Press, Inc., New York.

Parlevliet, J.E., 1979. Components of resistance that reduce the rate of epidemic development. Ann. Rev. Phytopathol. 17: 203–222.

Patterson, F.L., 1950. Adult plant and seedling reactions of barley varieties and hybrids to three races of *Puccinia graminis tritici.* Ph.D. Thesis. University of Wisconsin, Madison.

Patterson, F.L., R.G. Shands & J.G. Dickson, 1957. Temperature and seasonal effects on seedling reactions of barley varieties to three races of *Puccinia graminis* f. sp. *tritici.* Phytopathology 47: 395–402.

Powers, L. & L. Hines, 1933. Inheritance of reaction to stem rust and barbing of awns in barley crosses. J. Agr. Res. 46: 1121–1129.

Roelfs, A.P., 1978. Estimated losses caused by rust in small grain cereals in the United States – 1918–76. U.S. Dept. Agri. Misc. Pub. # 1363.

Roelfs, A.P., 1982. Effects of barberry eradication on stem rust in the United States. Plant Dis. 66: 177–181.

Roelfs, A.P., 1985. Wheat and rye stem rust. In: A.P. Roelfs & W.R. Bushnell (Eds.). The Cereal Rusts. Vol. II. Diseases, distribution, epidemiology, and control, pp. 3–37. Academic Press, Inc., New York.

Roelfs, A.P., D.H. Casper, D.L. Long & J.J. Roberts, 1990a. Races of *Puccinia graminis* in the United States in 1988. Plant Dis. 74: 555–557.

Roelfs, A.P., D.H. Casper, D.L. Long & J.J. Roberts, 1991a. Races of *Puccinia graminis* in the United States in 1989. Plant Dis. 75: 1127–1130.

Roelfs, A.P., D.L. Long, B.J. Steffenson, Y. Jin, M.E. Hughes & D.H. Casper, 1990b. Barley rusts in the United States in 1990. Barley Newsl. 34: 73–76.

Roelfs, A.P. & D.V. McVey, 1973. Races of *Puccinia graminis* f. sp. *tritici* in the USA during 1972. Plant Dis. Reptr. 57: 880–884.

Roelfs, A.P. & J.W. Martens, 1988. An international system of nomenclature for *Puccinia gramminis* f. sp. *tritici.* Phytopathology 78: 526–533.

Roelfs, A.P., B.J. Steffenson, M.E. Hughes, D.L. Long, D.H. Casper, Y. Jin & J. Huerta, 1991b. Barley rust in the United States in 1991. Barley Newsl. 35: 71–77.

Schafer, J.F. & A.P. Roelfs, 1985. Estimated relation between numbers of urediniospores of *Puccinia graminis* f. sp. *tritici* and rates of occurrence of virulence. Phytopathology 75: 749–750.

Sellam, M.A. & R.D. Wilcoxson, 1976. Development of *Puccinia graminis* f. sp. *tritici* on resistant and susceptible barley cultivars. Phytopathology 66: 667–668.

Shands, R.G., 1939. Chevron, a barley variety resistant to stem rust and other diseases. Phytopathology 29: 209–211.

Søgaard, B. & P. von Wettstein-Knowles, 1987. Barley: Genes and chromosomes. Carlsberg Res. Commun. 52: 123–196.

Stakman, E.C. & J.G. Harrar, 1957. Principles of Plant Pathology. Ronald Press Co., New York.

Stakman, E.C., D.M. Stewart & W.Q. Loegering, 1962. Identification of physiologic races of *Puccinia graminis* var. *tritici.* U.S. Dept. Agric., Agric. Res. Serv. E-617.

Steffenson, B.J., 1989. Durable resistance to wheat stem rust in barley. Barley Newsl. 33: 119.

Steffenson, B.J. & L.S. Dahleen, 1991. The potential use of molecular markers for mapping disease resistance genes in barley. In: L. Munck (Ed.). Barley Genetics VI, pp. 644–646. Munksgaard Int. Pub. Ltd., Copenhagen.

Steffenson, B.J., Y. Jin, T.G. Fetch, Jr. & J.D. Miller, 1990. Reaction of five barley cultivars to infection by race QCC of *Puccinia graminis* f. sp. *tritici.* Barley Newsl. 34: 99–100.

Steffenson, B.J., A. Kilian & A. Kleinhofs, 1992. Identification of molecular markers linked with the *Rpg*1 gene for stem rust resistance in barley. (Abstr.) Phytopathology 82: 1083.

Steffenson, B.J., J.D. Miller & Y. Jin, 1991. Detection of the T gene for resistance to *Puccinia graminis* f. sp. *tritici* in barley seedlings. (Abstr.) Phytopathology 81: 1229.

Steffenson, B.J. & R.D. Wilcoxson, 1987. Receptivity of barley to *Puccinia graminis* f. sp. *tritici.* Can. J. Plant Pathol. 9: 36–40.

Steffenson, B.J., R.D. Wilcoxson & A.P. Roelfs, 1985. Resistance of barley to *Puccinia graminis* f. sp. *tritici* and *Puccinia graminis* f. sp. *secalis.* Phytopathology 75: 1108–1111.

Steffenson, B.J., R.D. Wilcoxson & A.P. Roelfs, 1984. Inheritance of resistance to *Puccinia graminis* f. sp. *secalis* in barley. Plant Dis. 68: 762–763.

Stubbs, R.W., J.M. Prescott, E.E. Saari & H.J. Dubin, 1986. Cereal Disease Methodology Manual. Centro Internacional de Mejoramiento de Maiz Y Trigo (CIMMYT), Mexico.

Vanderplank, J.E., 1968. Disease Resistance in Plants. Academic Press, New York.

Waterhouse, W.L., 1948. Studies in the inheritance of resistance to rust of barley. Part II. J. Roy. Soc. N. S. W. 81: 198–205.

Wiebe, G.A. & D.A. Reid, 1961. Classification of barley varieties grown in the United States and Canada in 1958. U.S. Dept. Agr. Tech. Bull. # 1224.

Wolfe, M.S. & J.A. Barrett, 1980. Can we lead the pathogen astray? Plant Dis. 64: 148–155.

Wych, R.D. & D.C. Rasmusson, 1983. Genetic improvement in malting barley cultivars since 1920. Crop Sci. 23: 1037–1040.

Euphytica **63**: 169–174, 1992.
R. Johnson and G.J. Jellis (eds), Breeding for Disease Resistance

Novel pathotypes of lettuce mosaic virus – breakdown of a durable resistance?

D.A.C. Pink[1], H. Lot[2] & R. Johnson[3]
[1] *Horticulture Research International, Wellesbourne, Warwick CV35 9EF, UK;* [2] *INRA Station de Pathologie Vegetale, Domaine St Maurice, B.P. 94-84143 Montfavet, Cedex, France;* [3] *Cambridge Laboratory, AFRC Institute of Plant Science, John Innes Centre, Colney, Norwich NR4 7UJ, UK*

Key words: Lettuce Mosaic Virus, resistance genes, gene-expression, pathotypes, durability of resistance, *Lactuca sativa,* lettuce

Summary

Resistance to lettuce mosaic virus (LMV) is derived either from cv. Gallega (*g* gene) or the wild accession PI251245 (*mo* gene). Previous studies indicated that these two genes were identical. Breeders in Europe produced numerous resistant cultivars utilising *g* while in the USA *mo* was used. The resistance has been effective for over 20 years. However, recently there have been reports of LMV isolates causing unusually severe and sometimes necrotic symptoms on cultivars with these resistance genes. Investigations of these 'severe' isolates have distinguished three new pathotypes in addition to the common pathotype (II) and identified a novel dominant gene for resistance. The *mo/g* genes confer resistance to pathotypes I and II but pathotype III possesses virulence for cultivars with *g* but not for those with *mo*. These two genes are therefore not identical but are probably either closely linked genes or alleles. Pathotype IV possesses virulence for all lettuce lines so far tested. Some isolates of this pathotype are seed transmitted in cultivars possessing *mo* or *g* and have caused severe crop losses in southern France. The durability of the resistance conditioned by these two genes is discussed.

Introduction

Lettuce Mosaic Virus (LMV) is a major disease of commercial lettuce crops in all lettuce growing areas of the world (Tomlinson, 1962; Ryder, 1968). Infected plants show mosaic and vein-clearing symptoms on the leaves which often have a frilly appearance and the plants are stunted, making them unmarketable.

The virus is seed-transmitted in susceptible cultivars and is also transmitted by aphids in a non-persistent manner (Tomlinson, 1970; Ryder, 1973). This combination of transmission characters can lead to epidemics of the disease, particularly if successional crops are grown. By the mid 1970's the disease was starting to be controlled by stringent testing of seed stocks to ensure that infection levels were below 0.1%. Use of mosaic free seed gave good control in the UK by 1984 (Walkey et al., 1985). However the production of virus free seed was expensive and the cheaper alternative of using genetic resistance was exploited.

Sources of resistance

Resistance was originally identified in Argentina in the cultivar Gallega by von der Pahlen & Crnko (1965) and later in the USA in PI accessions of *Lactuca sativa* from Egypt by Ryder (1968, 1970)

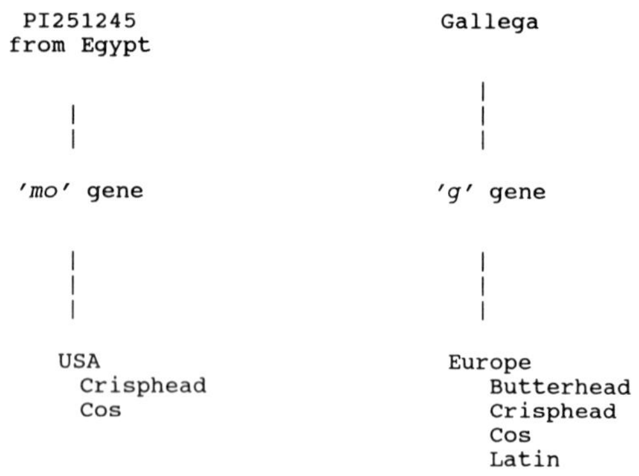

Fig. 1. Sources of resistance to LMV.

(originally wrongly identified as *L. serriola*). It was identified as resistance to the virus not to the aphid vector. Resistance was shown to be controlled by a single recessive gene in these lines, designated as *g* in Gallega (Bannerot et al., 1969) and as *mo* in the PI lines (Ryder, 1970). Allelism tests gave no susceptible segregants and it was assumed that these two genes were identical. The *mo* gene was also identified in two Spanish cultivars, Madrilenos and Mataro de los Tris Osos (Ryder, 1976).

Lettuce Mosaic Virus resistance in modern cultivars

Numerous resistant cultivars were produced in the USA using the *mo* gene and in Europe using the *g* gene (Fig. 1). These cultivars were widely used for at least 20 years and the resistance remained effective.

It was reported that seed transmission of the virus did not occur in cultivars possessing the *mo/g* gene but that the plants possessing the gene are not immune to the virus. They become infected but the virus multiplies much more slowly than in susceptible cultivars and symptom expression is delayed. Furthermore, the level of virus multiplication is significantly affected by the background genotype of cultivars possessing the recessive resistance gene (Table 1). Associated with this variation in the level of virus multiplication are variations in expression of the resistance. Thus resistant butterhead cultivars do often eventually develop mild symptoms associated with the greater multiplication of virus in them than in resistant crisphead cultivars.

Pink et al. (1992) suggested that this ability of the virus to multiply in such cultivars might provide selection pressure for the evolution of more pathogenic forms of the virus. In fact, Zink et al. (1973) recorded the occurrence of an isolate of LMV able to infect cv. Gallega systemically. In recent years there have been several reports of isolates of LMV causing unusually severe symptoms in lettuce crops from Spain (H. Lot, INRA, unpublished), USA (J.E. Duffus, USDA, personal communication) Greece, (Kyriakopoulou, 1985) and the Yemen Arab Republic (Walkey et al., 1990).

Variation in LMV virus isolates

In order to investigate the cause of these severe infections isolates of the virus were collected from various countries under UK Ministry of Agriculture Licence PHF 1227/52 (49). These comprised an isolate from the Yemen Arab Republic (LMV-YAR) collected by D.G.A. Walkey, the 'Fire-

Table 1. Symptom expression of LMV (Scale 0 to 10) and amount of virus detected by ELISA in lettuce cultivars with *mo/g* resistance

Cultivars	Symptom severity (35 days)	ELISA log dilution factor (39 days)
Crisphead cvs.		
Scala	0.0	− 0.10
Malika	0.0	− 0.07
Vanguard 75	0.0	0.61
Astral	0.0	0.85
Calona	0.1	0.54
Ithaca	3.2	3.38
Butterhead cvs.		
Tuba	1.0	0.70
Mantilla	1.5	1.25
Mondian	1.9	1.21
Capitan	1.9	1.22
Benita	2.0	1.33
Sabine	5.2	3.40

Data from Walkey et al. (1985).

stone' isolate from the USA (LMV-F) supplied by J.E. Duffus, the 'Greek' strain (LMV-G) supplied by P.E. Kyriakopoulou and the 'Spanish' strain (LMV-E) supplied by H. Lot. These were compared with a UK isolate (LMV-W) obtained from seed of naturally infected lettuce plants grown at Wellesbourne (Walkey et al., 1985).

The isolates were maintained on the susceptible butterhead cultivar Sabine and leaves of infected plants were used as a source of inoculum. Full experimental details are in Pink et al. (1992).

Because the response of butterhead cultivars to infection with LMV is more variable than those of crisphead cultivars only the latter were used in the tests. This allowed the inclusion of cultivars with the *mo* gene and also some with the *g* gene. Two susceptible cultivars, both of which possess no known genes for resistance to LMV, were included: Saladin (syn. Salinas) because it is the most widely grown cultivar in the UK and Ithaca. The cultivars Vanguard 75 and Salinas 88 possess the *mo* gene, the latter produced by backcrossing the resistance gene into Salinas. The cultivars Malika and Calona possess the *g* gene.

The lettuce seedlings were inoculated at the three to four leaf stage and were assessed for infection on a scale of 0 (no symptoms) to 10 (intensive symptoms, severe stunting and death of the plants) at 35, 52 and 60 days after inoculation. Plants were harvested 62 days after inoculation and fresh weight and marketability were recorded.

Saladin was susceptible to all the isolates but the isolates differed in severity (aggressiveness) on it (Table 2) in the order LMV-YAR > LMV-E > LMV-F > LMV-W. Although the symptoms on Saladin with LMV-W were intermediate at the recording date shown they were more severe at the earlier recording dates (Pink et al., 1992). The cultivar Ithaca, used as a susceptible control, was resistant to the isolates from the Yemen Arab Republic (LMV-YAR) and Greece (LMV-G). In fact, apart from Saladin, all the cultivars were resistant to LMV-YAR.

All the cultivars were susceptible to the isolate LMV-E, including those with the *mo* and *g* genes. However, the cultivar Salinas 88 showed a relatively lower final score (Table 2) and also a delay in the development of symptoms (Pink et al., 1992). Evidently this delay and reduction in symptoms were not due to the *mo* gene because Vanguard 75 also possesses this gene but was highly susceptible to LMV-E.

On the evidence of Table 2 three different pathotypes of LMV could be distinguished. The common type in the UK and the USA is represented by LMV-W and LMV-F. These are virulent on Ithaca and Saladin but not on cultivars possessing the *mo* or the *g* gene. A previously undescribed pathotype, represented by LMV-YAR and LMV-G, lacks virulence for Ithaca and for the cultivars with *mo*/*g*. Another previously undescribed pathotype is represented by LMV-E from Spain which possesses virulence for all the cultivars and caused severe loss in yield and marketability to all the cultivars, although the delay in symptom development of Salinas 88 was associated with somewhat less severe damage than in the other cultivars as indicated by its ability to produce a marketable head, albeit with a 50% reduction in yield (Pink et al., 1992). Unfortunately this effect of delayed disease development was not observed in a trial in France.

Subsequent to the preparation of the data of Pink et al. (1992), a further variation in virulence was detected among isolates of LMV. This variant, isolated from endive and named as LMV-9, possessed virulence for cultivars with the *g* gene but not for those with the *mo* gene (Dinant & Lot, 1992). This indicated for the first time that the genes *mo* and *g* are not identical.

Table 2. Means symptom scores (Scale 0 to 10) at 60 days after inoculation of lettuce cultivars infected with five isolates of LMV (Data from Pink et al., 1992)

	Gene	Isolates				
		LMV-W	LMV-F	LMV-G	LMV-YAR	LMV-E
Saladin		5.1	8.0	10.0	10.0	9.8
Ithaca		7.1	8.7	0.0	0.0	9.8
Salinas 88	*mo*	2.7	3.4	5.0	0.3	7.4
Vanguard 75	*mo*	1.8	1.1	0.0	0.0	9.3
Malika	*g*	0.2	2.0	4.6	0.5	9.6
Calona	*g*	1.0	2.7	0.9	0.4	9.6

Genetics of resistance

The discovery of this variation in the virulence of LMV isolates led to further studies on the genetic control of resistance. Ithaca was crossed with Saladin and F2 plants were inoculated with isolate LMV-YAR. The segregation was 91 resistant (R): 29 susceptible (S) which fitted a 3 resistant to 1 susceptible ratio ($\chi^2_{(1)} = 0.04$, ($P > 0.9$)) consistent with the operation of a single dominant gene. This contrasted with the known recessive action of both the *mo* and *g* genes.

Ithaca was also crossed with Vanguard 75 and unexpectedly 120 F2 plants were all resistant to isolate LMV-YAR indicating that they possessed a gene in common or closely linked. There is evidence in their pedigrees of some common ancestors. Vanguard 75 was also crossed with Saladin and F2 plants were tested with either LMV-F or LMV-YAR (Table 3). The segregation with LMV-F fitted a ratio of 1 resistant to three susceptible ($P > 0.25$), consistent with a single recessive gene, presumably *mo*. Segregation with LMV-YAR was 97 resistant : 33 susceptible which fitted a 13R : 3S ratio ($P > 0.05$). This indicates the combined action of the dominant gene in Vanguard 75 allelic with the Ithaca gene and the recessive gene *mo*. The F2 from the cross of Ithaca with Salinas 88 was also tested with LMV-YAR (Table 4) and the segregation fitted a 13R : 3S ratio ($P > 0.5$). This is consistent with the combined action of the dominant gene in Ithaca and the recessive *mo* gene in Salinas 88.

Further evidence of the difference between the *mo* and *g* genes was obtained from a cross of Malika (*g* gene, susceptible to LMV-9) with Vanguard 75 (*mo* gene, resistant to LMV-9). When tested with the LMV-9 isolate, three classes were identified in the F2 population as 27 symptomless : 55 large chlorotic lesions : 24 mosaic. This was good fit to a 1 : 2 : 1 ratio ($\chi^2_{(2)} = 0.32$, $P > 0.75$) in which *momo* homozygotes gave the symptomless plants, *mo, g* heterozygotes gave chlorotic lesions and *gg* homozygotes gave mosaic. This supports the evidence that the genes have different expression to this isolate although there is much evidence to indicate that they are either allelic or very closely linked.

The information derived from these investigations is summarized in Table 5. The dominant gene first identified in the cultivar Ithaca is shown as Mo_2 and the corresponding recessive allele that does not give resistance as Mo^+_2. The two recessive genes *g* and *mo* are shown respectively as mo^1_1 and mo^2_1 and the genotypes of several cultivars are indicated together with the resistance conditioned toward four different pathotypes of the LMV virus.

The numbering of the four pathotypes of LMV in Table 5 varies from that given by Pink et al. (1992). The difference is that the LMV-E isolate is here listed as pathotype IV whereas it was listed as III in Pink et al. (1992). This arose because of the construction of the Table in which the pathotypes differ in a stepwise fashion from the type attacking the fewest to the type attacking the most cultivars and the pathotype LMV-9 therefore fits better at position III than the LMV-E isolate.

Expression of the genes for resistance with the LMV pathotypes

As noted above, there is variation in expression of resistance conditioned by the genes at the mo_1 locus

Table 3. Segregation for resistance (R) and susceptibility (S) of F2 plants from the cross (Vanguard 75 × Saladin) inoculated with LMV-F of LMV-YAR isolates of LMV

F2 (Vanguard 75 × Saladin)		R	:	S	$\chi^2_{(1)}$
LMV Isolates:	LMV-F	37	:	89	1 : 3 = 1.28
	LMV-YAR	97	:	33	13 : 3 = 3.76

Table 4. Segregation for resistance (R) and susceptibility (S) among F2 plants from the cross (Ithaca × Salinas 88) inoculated with LMV-YAR isolate of LMV

Parents and control	R	:	S	$\chi^2_{(1)}$
Saladin (S control)	0	:	10	–
Ithaca	5	:	0	–
Salinas 88	5	:	0	–
F_2 (Ithaca × Salinas 88)	240	:	59	13 : 3 = 0.19

in different cultivars. This includes the amount of virus found in plants using the ELISA technique (Enzyme Linked Immuno-Sorbent Assay) and the expression of symptoms. In the further studies and the identification of the dominant Mo_2 gene it was shown that it apparently confers immunity to LMV-YAR since no virus can be detected in the cultivar Ithaca. In the cross of Ithaca with Salinas 88 the 13 resistant F2 plants shown in Table 4 could be further subdivided. Those with no virus comprised plants homozygous or heterozygous for Mo_2 (12 out of the 13) and the thirteenth plant homozygous for mo^2_1 which was resistant but carried virus detected by ELISA.

It was also noted above that plants with the mo_1 genes did not transmit virus via the seed. Recently, some pathotypes of the type IV class, although not LMV-E, were shown to be transmitted through the seed of cultivars with the mo_1 gene at high frequencies (Dinant & Lot, 1992). This increases the possible threat of these pathotypes to the lettuce industry and may require the re-emphasis of use of virus free seed stocks.

Durability of resistance to LMV

As noted above, the resistance conditioned by the genes at the mo_1 locus was effective in widespread use for over twenty years and only recently have severe infections been recorded on some crops due to the newly identified LMV pathotypes. Thus, the resistance could be classified as durable. As noted elsewhere (see Chapter 1), this does not imply that resistance must be permanent. When resistance becomes ineffective after a prolonged period of usefulness possible reasons for the change of effectiveness should be considered. It may be noted that pathotypes of the type IV have been first identified in areas where cropping is extremely heavy and continuous, such as Southern France and Spain, and also where there may be a continuity of host material in the form of other plants such as endive. The ability of the virus to multiply in cultivars with mo_1 resistance may have contributed to this ability to evolve. The demonstration of the possible damaging nature of these newly identified pathotypes in crops in some areas and a glasshouse experiments does not necessarily imply a rapid spread of such pathotypes to other areas, particularly where less intensive production occurs, such as the UK. Nevertheless their existence intensifies the need to search for further sources of resistance, and to reassess the role of seed hygiene where problems are already occurring.

Table 5. Resistance and susceptibility of lettuce cultivars to four pathotypes of the LMV virus showing the proposed resistance genotype of the cultivars

Cultivar	Genotype	LMV Pathotypes			
		I LMV-YAR	II LMV-W	III LMV-9	IV LMV-E
Saladin	$mo^+_1Mo^+_2$	S	S	S	S
Ithaca	$mo^+_1Mo_2$	R	S	S	S
Malika	$mo^1_1Mo^+_2$	R	R	S	S
Salinas 88	$mo^2_1Mo^+_2$	R	R	R	S
Vanguard 75	$mo^2_1Mo_2$	R	R	R	S

* Cultivars would be homozygous for the given genes, genotypes are given as single genes for clarity.

Conclusions

Breeding for resistance to Lettuce Mosaic Virus has been highly successful and has made a valuable contribution to protection of the lettuce crop, particularly in Europe. Recent changes in the pathogen present new challenges to identify new possible sources of resistance, and to obtain a deeper understanding of the pathogen and its variability. Present efforts for future exploitation include the investigation of serological relationships between virus strains in the UK and development of cDNA libraries and the study of coat protein gene sequences in France.

References

Dinant, S. & H. Lot, 1992. Lettuce mosaic virus: a review. Plant Pathol. 41: 528–542.

Kyriakopoulou, P.E., 1985. A lethal strain of lettuce mosaic virus in Greece. Phytoparasitica 13: 271.

Pink, D.A.C., D. Kostova & D.G.A. Walkey, 1992. Differentiation of pathotypes of lettuce mosaic virus. Plant Pathol. 41: 5–12.

Ryder, E.J., 1968. Evaluation of lettuce varieties and breeding lines for resistance to common lettuce mosaic. USDA Tech. Bull. 1391, 8 pp

Ryder, E.J., 1970. Inheritance of resistance to common lettuce mosaic. J. Amer. Soc. Hortic. Sci. 95: 378–379.

Ryder, E.J., 1973. Seed transmission of LMV in mosaic resistant lettuce. J. Amer. Soc. Hortic. Sci. 98: 610–614.

Ryder, E.J., 1976. The nature of resistance to LMV. Eucarpia Meeting of Leafy Vegetables, Wageningen, Holland 15–18 March.

Tomlinson, J.A., 1962. Control of lettuce mosaic by the use of healthy seed. Plant. Pathol. 11: 61–64.

Tomlinson, J.A., 1970. Lettuce mosaic virus. CMI/AAB Descriptions of Plant Viruses No. 9.

Von der Pahlen, A. & J. Crnko, 1965. El virus del mosaico de la lechuga (*Marmer lactucae*) en Menoza y Buenos Aires. Rev. Invest. Agropecuarias 11: 25–31.

Walkey, D.G.A., C.M. Ward & K. Phelps, 1985. Studies on lettuce mosaic virus resistance in commercial lettuce cultivars. Plant Pathol. 34: 545–551.

Walkey, D.G.A., A.A. Allubaishi & J.W. Webb, 1990. Plant virus diseases in the Yemen Arab Republic. Tropical Disease Management 36: 195–206.

Zink, F.W., J.E. Duffus & K.A. Kimble, 1973. Relationship of a non-lethal reaction to a virulent isolate of lettuce mosaic virus and turnip mosaic susceptibility in lettuce. J. Amer. Soc. Hortic. Sci. 98: 41–45.

Euphytica **63**: 175–185, 1992.
R. Johnson and G.J. Jellis (eds), Breeding for Disease Resistance

The genetics of plant-virus interactions: implications for plant breeding

R.S.S. Fraser
Horticulture Research International, Worthing Road, Littlehampton, West Sussex BN17 6LP, UK

Key words: resistance, virulence, gene-for-gene relationships, pathogenic fitness

Summary

Host resistance is the main means of control of plant virus diseases. This paper reviews the genetics of resistance and matching virulence. Theoretical models of basic compatibility between plant species and their viruses, and of resistance, are described and used to predict features of resistance genetics, and mechanisms. These predictions are compared with a survey of known examples of resistance.

Resistance is mainly controlled at a single genetic locus, although more complex systems are known. About half of the resistance alleles studied were dominant, the remainder were either incompletely dominant or recessive. Doubt is cast on the reliability of assessing resistance genotypes (numbers of loci and dominance relationships) from 'distant' phenotypic measurements such as symptom severity or plant growth. A model is proposed to reconcile apparent inconsistencies between genotype and phenotype.

Dominant resistance alleles are strongly associated with virus localising mechanisms normally involving local lesions. Incompletely dominant and recessive alleles allow spread of the virus, but inhibit multiplication or symptom development. Fully recessive alleles may be associated with complete immunity.

Most resistance genes in the survey had been overcome by virulent virus isolates with dominant localising resistance alleles especially vulnerable. Comparatively few resistance genes have proved exceptionally durable. Acquisition of virulence can be associated with loss of general pathogenic fitness, but in some cases this can be restored by further selection of the virus in resistant hosts.

Virulence/avirulence determinants have been mapped to individual base changes in different functional regions of the viral genome. A virus may contain several virulence determinants and may develop a stable gene-for-gene relationship with a host having several resistance genes. It may be possible to design robust, oligogenic resistance systems which will be difficult for the virus to overcome.

Introduction

Viruses can cause serious losses of yield and quality in many crops grown in agriculture, horticulture and forestry. Methods for control of virus diseases are therefore widely applied. Despite a considerable amount of research, there are still no chemicals that can be applied routinely to control viruses within the crop: the toxicity of available antiviral chemicals, and the increasing costs and regulatory difficulties of registering new pesticides, make it unlikely that antiviral pesticides will become commercially viable. Control strategies have therefore focused on methods to prevent the occurrence of infection, and natural methods for virus resistance within the crop plant.

Preventing the virus from reaching the host can involve physical, chemical and biological methods. Healthy planting stock, for vegetatively-propagated subjects, and virus-free seeds, can be guaranteed by appropriate indexing and virus-eradication methods. Virus vectors can be controlled by the use of appropriate pesticides, traps and screens. However, the most important crop protection methods involve breeding plants for virus resistance. This can operate indirectly, by effects on the vector, or directly, by preventing virus multiplication or its deleterious effects within the plant.

A number of host factors can affect plant attractiveness to vectors, and thus the efficiency of virus transmission (reviewed by Jones, 1990). They include physical barriers such as leaf hairs or robust leaf surfaces, non-preferred foliage colour, secretion of insect alarm pheromones, and presence in the sap of anti-feedant chemicals which reduce feeding time and thus time for virus acquisition and transmission. These factors are under genetic control and are to varying extents accessible to the plant breeder. But mostly, the genetic basis of the difference between vector resistant and susceptible individuals is not well understood. These mechanisms are not considered further in this review.

In contrast, the genetics of resistance mechanisms operating against viruses within the host plant have been extensively studied, since Holmes (1938) first demonstrated Mendelian inheritance of resistance to tobacco mosaic virus (TMV) in tobacco. I estimate that these types of resistance mechanism have been seriously deployed against at least 100 viruses in a similar number of different crop species, with varying degrees of success (Fraser, 1986, 1990); attempts to discover useful sources of resistance have undoubtedly extended to a much larger number of cultivated species and wild relatives.

My aims in this review are to examine the genetics of the interactions between plants and viruses, and to ask whether knowledge of the genetics and biochemistry of resistance and matching virulence can be applied to enhance the efficiency of resistance breeding.

Some models for plant-virus interactions

The genetics of the interactions between plants and their microbial pathogens have been the subject of extensive theorisation, and there is little doubt that the development of conceptual models such as the gene-for-gene theory (reviewed in Gabriel & Rolfe, 1990), and testing of the predictions made from them, has contributed greatly to our understanding of the genetics and biochemistry of these interactions. For plant-virus interactions, less attention has been given to development of such models, although examination of the phenotype of resistance to viruses in plants suggests that the types of response may well be more varied than in resistance to fungi and bacteria. It is appropriate therefore to consider some of the possible ways in which plants and viruses might interact, and to develop some testable predictions from these models.

Resistance may be separated into three basic types, operating at three different levels of complexity of the host population. An entire species may be resistant to a particular virus – the so-called non-host immunity – with no detectable symptoms or virus multiplication after attempts at inoculation. This type of resistance does not appear to have been successfully exploited in plant breeding. Within a species, certain individuals or populations may contain heritable resistance to a particular virus normally affecting that species: for cultivated species the resistant population equates to a resistant cultivar or landrace. This type of resistance has been widely used in plant breeding. Finally, at the level of the individual plant, various types of non-heritable resistance may be conferred by a variety of treatments such as prior infection or application of chemicals. Most of these have not found use in crop protection, except for cross protection, in which plants are deliberately inoculated with a mild strain of an affecting virus to protect them against subsequent infection by more severe strains. Cross protection has been quite widely used in a small number of crops (reviewed by Urban et al., 1990). Cross protection can be given a heritable basis, by transforming plants with DNA copies of the entire genome of the cross-protecting isolate (Yamaya et al., 1988), or with the gene for coat protein, which is probably the major participant in the mechanism of cross protection (reviewed by Nelson et al., 1990).

Figure 1 shows two models of mechanisms which might be involved in non-host and cultivar resistance. The basic concept is that where a species is susceptible to a particular virus, the host provides specific factors which are essential for particular phases of pathogenesis by that virus: the equipment required for basic compatibility between host and virus. Examples might include host-coded subunits of a virus replicase (Hayes & Buck, 1990), and functions involved in cell-to-cell transport of the virus. Each host-coded entity would be required to recognise a specific virus-coded protein also involved in the particular process. This type of model, together with some of its predictions, was first considered by Bald & Tinsley (1967) and amplified by Fraser (1985).

In this model, a non-host species is such because it lacks the appropriate factors required to support a particular virus; a resistant cultivar of a host species has become resistant because a required factor has been deleted, or altered in function so that it cannot recognise the viral component. Resistance and non-host immunity in this model are therefore 'negative' effects.

Superimposed on the model where a plant is a host because it has basic compatibility with a particular virus is a 'positive' model of resistance: the resistant cultivar contains an inhibitor of that particular virus, or develops one after infection. This implies a recognition event between virus-coded and host-coded components, resulting either in direct inhibition of the virus, or the induction of a separate inhibitor via a signal transduction pathway after the initial recognition event. Strictly speaking, this type of model could be involved in non-host immunity, if all extant members of the species contained a highly effective inhibitor, which has not been overcome by virulent isolates of the virus. However, it seems more likely that 'positive' mechanisms are associated with cultivar resistance, and that non-host immunity may be of the 'negative' type.

Predictions from the resistance models

Some predictions about the genetics and biochemistry of resistance, and about virulence, can be made from the models shown in Fig. 1. In the 'positive' model, resistance alleles will be dominant or incompletely dominant (gene-dosage dependent), depending on the resistance mechanism involved. The mechanism could be either constitutive or induced after a recognition event. Strong dominance is perhaps more likely where a recognition event triggers an induced resistance response: the initial recognition is likely to cause an all-or-nothing response. In contrast, where the resistant plant constitutively contains an inhibitor of virus replication or movement, it is possible that plants homozygous for the resistance allele may express resistance more strongly than those which are heterozygous, and thus contain a lower concentration of the inhibitor. In all forms of the 'positive' model, virulence (resistance breaking behaviour by the virus) is an altered interaction with the inhibitor, or with the host component of the recognition event, such that inhibition or the induction of the resistance mechanism are less effective or completely ineffective. It may be that mutation to virulence against 'positive' resistance mechanisms is a comparatively easy step for the virus; the constraint is that the virus-coded protein must retain its pathogenic function, while losing its ability to be recognized by the host factor.

In the 'negative' model, resistance alleles must be recessive, and the mechanism must be constitutive in homozygous resistant plants. This type of resistance might be expected to confer complete immunity, if the missing host-specified function

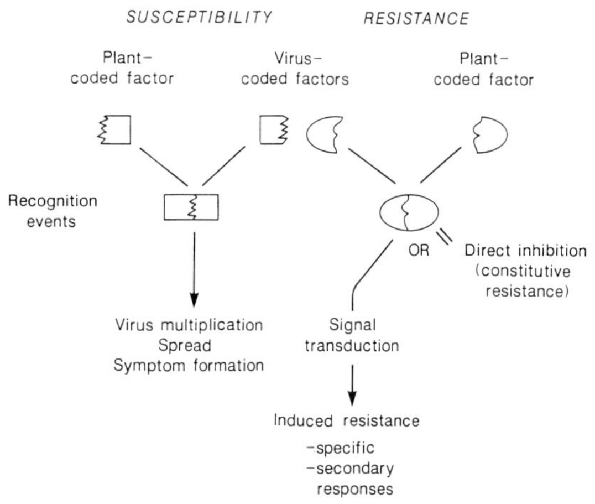

Fig. 1. Two models for plant virus interactions involving recognition events and the consequent responses of susceptibility or resistance. Reproduced from Fraser, 1990.

was involved in an early stage of the virus replicative cycle. Lack of a host function required for cell-to-cell movement of virus could permit virus multiplication in the very small proportion of directly inoculated cells, perhaps analogous to the so-called 'subliminal' infections (Cheo, 1970; Sulzinski & Zaitlin, 1982). It seems likely that it would be difficult for the virus to mutate to a condition where it could replicate or spread without the missing or defective host function: virulence against 'negative' resistance mechanisms might therefore be rarer than against 'positive' mechanisms.

Comparison of the predictions of the models with known examples of resistance

In earlier publications (Fraser, 1986, 1990), I reported the findings of surveys of the genetics of plant resistance and viral virulence. The surveys consisted of literature searches for reports which contained information on the genetics of resistance, the resistance mechanism or phenotype, and whether isolates of the virus that were virulent against the resistance gene in question had been detected. Essentially, the surveys consisted of a random sample of host-virus combinations, although heavily biased towards highly-bred cultivated material rather than weeds or landraces. The findings of these surveys are reported in updated form in Table 1. The primary references are cited in the earlier reviews.

Numbers of loci controlling resistance

It is clear that the majority of resistances investigated were found to be under simple genetic control, with the resistance inherited in a monogenic manner. However, there are numerous reports in the

Table 1. Summary of resistance genetics, mechanisms and occurrence of resistance-breaking isolates. Reproduced from Fraser (1990)

Genetic basis	Number of host-virus combinations
Single dominant gene	38
Incompletely dominant (gene-dosage dependent)	13
Apparently recessive	18
Subtotal: monogenic	69
Possibly oligogenic, or monogenic with modifier genes or effects of host genetic background	18
Total number in sample	87

Resistance phenotype[a]	Immunity or subliminal infection	Local lesions	Partial localization	Systemically effective	Not known
Dominant alleles	5	22	1	3	8
Incompletely	0	0	4	11	0
Recessive dominant	6	0	3	9	4

Virulent isolates reported[a]	Yes	No	Not tested
Dominant alleles	20	4	16
Incompletely dominant	9	3	3
Recessive	9	4	8
Totals	38	11	27

[a] Data are tabulated for all the monogenic resistances, and for those cases of oligogenics or modifiers where the characteristics of individual major genes can be isolated.

literature of resistance to a particular virus which appears to be under more complex control. These may involve several types of effect. The genetic background of the host cultivar in which monogenic resistance is expressed may affect the effectiveness of resistance: in barley, resistance to barley yellow dwarf virus (BYDV), expressed as tolerance, is dependent on the growth rate of the cultivar containing the resistance (Jones & Catherall, 1970). In a very small number of cases, resistance has been shown to be controlled by oligogenic systems involving some interaction between genes at different loci. One of the best documented is that for resistance to bean common mosaic virus (BCMV) in *Phaseolus vulgaris* (Drijfhout, 1978; Day, 1984). The third case is where segregation of resistance after crossing experiments appears to indicate control at a number of loci, operating independently and additively. There can be two problems here: some early reports (reviewed in Fraser, 1986) may not have fully disentangled the effects of genotype and genotype × environment interaction, in that later investigations of the same host-virus systems using more controlled environmental conditions indicated much simpler genetic controls of resistance. Secondly, the different scoring systems used to assess whether plants are resistant or susceptible – involving such disparate parameters as virus multiplication, severity of visible disease symptoms and crop yield – can offer quite different evidence towards the evaluation of genetic complexity. A case in point is resistance to maize dwarf mosaic virus (MDMV) in maize. Rosenkranz & Scott (1984) suggested five genes controlling resistance on the basis of a disease rating scale; subsequent work (Roane et al., 1989) using different methods of evaluation suggested that resistance is controlled by single, allelic, dominant genes. This conclusion is now gaining support as molecular methods of genetic analysis such as restriction fragment length polymorphisms (RFLPs) are applied (McMullen & Louie, 1989; Louie et al., 1991). It is pleasing that modern molecular methods of genetic analysis are bringing clarity to areas where previous attempts to determine the genotype by assessment of the phenotype have suggested more complex solutions.

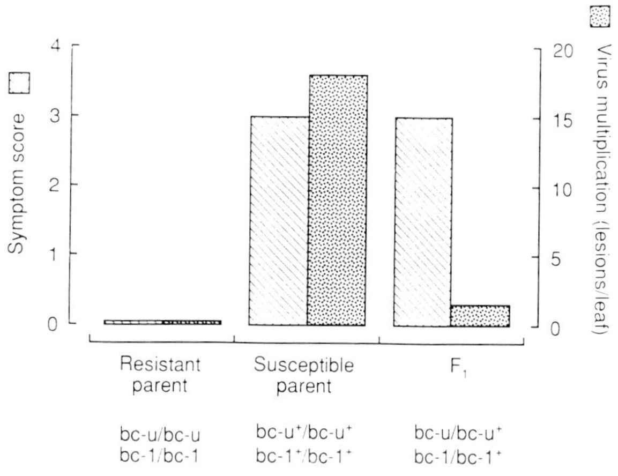

Fig. 2. Resistance to bean common mosaic virus in *Phaseolus vulgaris*. Resistance is expressed as reduction in symptom severity or virus multiplication (measured by infectivity) in systemically infected leaves. Assays were carried out 10 days after inoculation of the primary leaves. Measurements were made in resistant and susceptible pure breeding lines with and without the *bc-u* and *bc-1* resistance genes (Drijfhout, 1978; Day, 1984) and in the F1 hybrid between them. Taken from data in Day (1984).

Dominance relationships between resistance and susceptibility alleles at a single locus

For resistance controlled from a single locus, about half the examples in the survey sample involved resistance as a dominant allele, while smaller proportions indicated resistance which was recessive or incompletely dominant (gene dosage-dependent) (Table 1). However, most judgements of dominance relationships between resistance and susceptibility alleles have been made after examination of symptom expression in F_1 progeny from a cross of pure-breeding resistant and susceptible parents. This may give misleading indications; Figure 2 shows an example. In resistance to BCMV controlled by the complex recessive gene system in *Phaseolus vulgaris* (Drijfhout, 1978), heterozygous plants show just as severe symptoms as the homozygous susceptible parents, and resistance would clearly be classified as recessive (Day, 1984). However, examination of virus multiplication in the heterozygote indicates that it had been quite strongly inhibited compared with the susceptible parent. In this case the resistance allele shows a fair

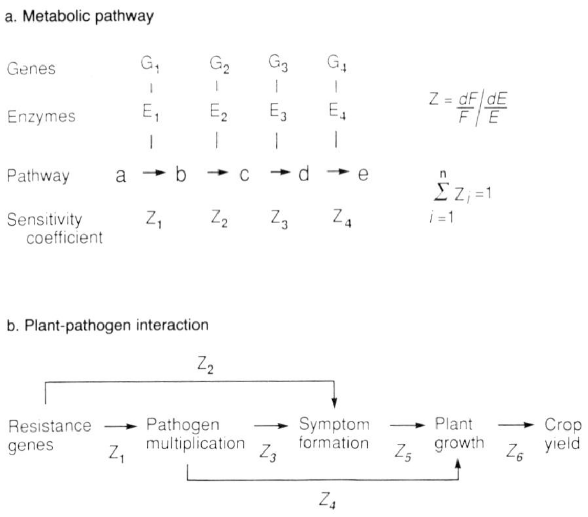

Fig. 3. Relationships between genotype and phenotype. a) The model for genetic control of linear metabolic pathways developed by Kacser & Burns (1981). The sensitivity co-efficient Z is defined as the fractional change in flux (F) through the pathway, divided by the fractional change in enzyme activity (E) for that step. The sum of all the sensitivity co-efficients in the pathway is 1. b) Application of a similar model to plant pathogen interactions. Z_1 and Z_2 involve direct effects of resistance allele dosage on virus multiplication or pathogenesis. Z_3 to Z_6 involve indirect effects which, nevertheless, are responsive to resistance allele dosage.

degree of dominance over the susceptibility allele, and there is a clear effect of resistance allele dosage on virus multiplication.

There are also examples of resistance alleles which give complete suppression of symptom formation in heterozygous plants, thus appearing highly dominant, while showing dosage dependence (incomplete dominance) when virus multiplication is measured (e.g. Fraser & Loughlin, 1980).

Conclusions about the genetic control of resistance – numbers of genes involved, and dominance relationships between resistance and susceptibility alleles – are drawn from observations of the resistance/susceptibility phenotype, which can be assessed in a variety of ways. It is disturbing from the examples given above that such opposing conclusions can be drawn for single host-virus interactions.

A basis for reconciliation and rationalisation may be provided by the theoretical analysis of control of flux in linear metabolic pathways proposed by Kacser & Burns (1981) (Fig. 3a). In their model, each step in the pathway has a sensitivity co-efficient Z. Steps with different values of Z show different relationships between phenotype (pathway flux) and genotype (enzyme activity). By means of the model, it is possible to explain different matches of phenotype to genotype, and dominance relationships, without having to postulate genetic modifiers.

Plant virus multiplication and its pathogenic effects on symptoms, plant growth and crop yield can also be taken to represent a multistep biochemical pathway (Fig. 3b), although the triple interaction between multiplication, symptoms and growth (Fraser et al., 1986) suggests the possible operation of a bypass pathway or shunt, as in several metabolic pathways. Addition of the effects of host resistance genes merely extends the pathway. If the Kacser and Burns model applies to the viral pathway, and each step can have a different value of Z, it follows that when resistance is assessed by any one of virus multiplication, symptom severity, plant growth or crop yield, the conclusions about the dominance relationships of resistance and susceptibility alleles could vary – as is found in practice. By the same token, attempts to establish how many loci control resistance, by assessing some of the more 'distant' parameters in the pathway and inferring segregation ratios, could also be confounded by variations in Z.

A fuller discussion of some aspects of application of the Kacser & Burns (1981) model to plant virus infections is given in Fraser (1986). The model has also been usefully applied to analysis of genetic interactions between plants and fungal pathogens (Crute & Norwood, 1986).

Expression of resistance mechanisms

Table 1 indicates that there are broad associations between types of genetic control of resistance, and the resistance phenotype. Thus dominant resistance alleles are strongly associated with resistance mechanisms which localise the virus, at or around the site of infection, usually with formation of local

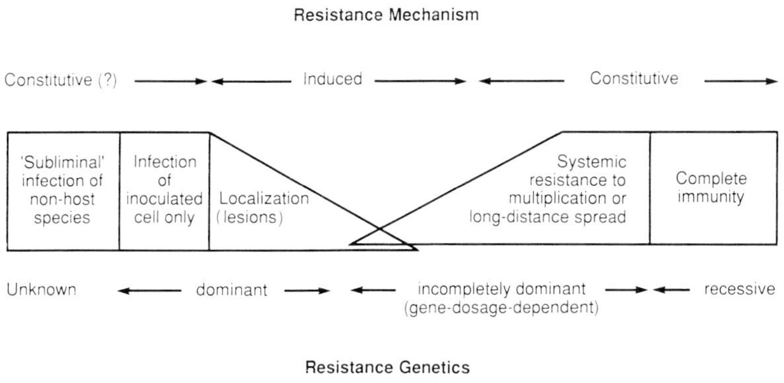

Fig. 4. Some possible relationships between resistance mechanisms and the underlying genetic controls. Modified from Fraser, 1990.

lesions. An extreme case is localisation in the directly inoculated cell only, as happens with the *Tm-2* and *Tm2*^{2} resistances in tomato (Nishiguchi & Motoyoshi, 1987). A similar mechanism may apply in 'subliminal' infections of apparently non-host species (Cheo, 1970; Sulzinski & Zaitlin, 1982). Incompletely dominant resistance alleles are associated with mechanisms giving partial localisation or allowing virus spread throughout the plant while inhibiting multiplication or symptom development. Apparently recessive alleles are either associated with the latter mechanism, or can involve a possible complete immunity, as predicted from Fig. 1.

The question of whether the resistance mechanism is constitutive in genetically resistant plants, or is induced by virus infection, has been addressed for very few examples of resistance. In the hypersensitive resistance to TMV controlled by the *N'* gene in tobacco, the virus localization mechanism appears to be induced, in that an avirulent (lesion forming) strain of the virus was shown to multiply as well as a virulent (systemic) strain during the early part of the infection (Fraser, 1988). Inhibition of multiplication of the avirulent strain commenced shortly before lesion formation. Resistance to TMV in tomato controlled by the *Tm-1* gene (incompletely dominant) is expressed as prevention of symptom formation and inhibition of virus multiplication, although the virus is able to spread systemically (Fraser & Loughlin, 1980). In this case, inhibition of multiplication is demonstrable from the time of inoculation, and the mechanism of resistance is therefore likely to be constitutive. As mentioned earlier, 'negative' resistance mechanisms controlled by recessive alleles are by definition constitutive.

These different types of resistance mechanisms and genetic controls can be summarized as a spectrum of activities as shown in Fig. 4. However, it is emphasized that experimental evidence for some aspects, such as induced versus constitutive resistance, is based on examination of a small number of plant-virus combinations.

The frequency of virulence

Table 1 shows the numbers of the different types of resistance alleles which have been overcome by resistance-breaking isolates. Unfortunately, many reports of resistance genes in the literature make no mention of whether the gene has been overcome by a virulent isolate, or for a newly discovered gene, whether it has been tested against a number of different isolates of the virus. After removing these uncertain cases, it is clear that the majority of dominant and incompletely dominant resistance alleles have virus isolates with matching virulence. For recessive alleles, the sample number is small and it is too early to test the prediction that fully recessive resistance operating by 'negative' mechanisms should prove highly durable. Testing this hypothesis will clearly require a selective approach, rather than the random sampling used for the initial general survey.

Table 1 shows that comparatively low numbers of resistance genes have proven to be highly durable, i.e. have not been overcome by resistance breaking isolates after long exposure to the virus in practical use as well as in the virology or plant breeding laboratory. Examples of highly durable genes include *Ry* against potato virus Y (PVY) in potato (Barker & Harrison, 1984), *bc-3* against BCMV in *Phaseolus vulgaris* (Drijfhout, 1978), *N* against TMV in tobacco (Holmes, 1938) and *Tm-2²* against TMV in tomato (Fraser, 1990). Strictly speaking, the last two examples have been overcome by virulent TMV isolates, but these isolates are somewhat aberrant, and very rare, and have not established themselves in cropping situations (Csillery et al., 1983; Fraser, 1990).

How many virulence 'genes' can a plant virus contain?

The genomes of most plant viruses are large enough to code for between three and about a dozen proteins. For several viruses, the entire genome can now be assigned to specification of particular proteins required for the different phases of virus replication and spread, and for particle structure (Goldbach et al., 1990). It is therefore most unlikely that plant viruses could contain genes solely concerned with determination of virulence/avirulence against a specific resistance gene: the determinant has to be incorporated as a pleiotropic effect within a gene with an overriding function in viral pathogenesis or particle structure. Studies of the location of virulence/avirulence determinants in plant viruses with RNA genomes have recently been made possible by the development of 'reverse genetics' (Culver & Dawson, 1989). cDNA copies of the RNAs of virulent and avirulent isolates can be used to make artificial recombinants, by cutting with appropriate restriction enzymes, then reassembling genomic segments in new combinations. Infectious RNA of the recombinants can be produced using expression vectors, and tested for biological activity, including virulence or avirulence. The precise alteration conferring virulence or avirulence can then be determined by sequencing the small region of the viral genome detected by restriction and recombination.

In TMV RNA, the determinant of virulence/avirulence against the *N'* gene in tobacco is located in the coat protein gene (Culver & Dawson, 1989). The determinants against the *Tm-2* and *Tm-2²* genes in tomato, which prevent cell-to-cell movement of the virus – are located in the gene for the viral 30K protein which is involved in cell-to-cell transport (Meshi et al., 1989; Calder & Palukaitis, 1992). The determinant against the tomato *Tm-1* gene – which inhibits TMV multiplication – is located in the gene for the putative viral replicase (Meshi et al., 1988). It is interesting that in the later cases, the virulence/avirulence determinant is located in a viral function which matches what is known of the host resistance mechanism.

It is clear from studies of a number of mutations to virulence against single resistance genes that the virulence/avirulence determinant is not confined to a single nucleotide position, but can be at a number of nearby sites (Calder & Palukaitis, 1992; Culver & Dawson, 1989; Mundry et al., 1990).

There appears to be no reason why a virus should not accumulate virulence genes against several host resistance genes, if these virulence genes map in different functional regions of the virus genome. However, where virulence against two resistance genes maps in the same viral gene, it is possible that the features required for double virulence might be contradictory or mutually exclusive, making it difficult for the virus to overcome particular pairs of resistance genes. This might explain the lack of isolates virulent against both *Tm-2* and *Tm-2²* in tomato/TMV (Pelham, 1972; Fraser 1990) or against *bc-2* and *bc-2²* in *Phaseolus vulgaris*/BCMV (Drijfhout, 1978).

With that possible exception, it is clear that viruses can contain sufficient virulence/avirulence determinants to build quite complex gene-for-gene relationships, with hosts which have a sufficient number of resistance genes. Well attested examples include tomato/TMV (Pelham, 1972); pepper/TMV (Tobias et al., 1989) and *Phaseolus vulgaris*/BCMV (Drijfhout, 1978; Day, 1984).

Is acquisition of virulence associated with a loss of pathogenic fitness?

Vanderplank (1984) suggested that because of the small genome size of plant viruses, the mutation required to overcome the selection pressure imposed by a resistance gene would be likely to have deleterious side effects on the virus, and that the virus would revert to the parental type if the selection pressure was removed. He cited as evidence the failure of a raspberry ringspot virus (RRV) isolate which overcomes a host resistance to establish itself fully (Murant et al., 1968). It is now clear from wider evidence, such as that cited in Table 1, that virulent strains can establish well and become prevalent in many cases.

In TMV resistance controlled by *Tm-1* in tomato, a number of nitrous-acid induced mutants to virulence did indeed show reduced pathogenic fitness, as measured by their symptom severity and ability to multiply in susceptible hosts (Fraser & Gerwitz, 1987). However, resistance-breaking isolates from commercial nurseries growing *Tm-1* tomatoes were equal to wild type TMV in pathogenic fitness, and had clearly been able to make good any deleterious effects of the initial mutation to virulence in the face of continued selection in *Tm-1* plants.

In contrast, mutation to overcome the $Tm\text{-}2^2$ resistance does seem to be associated with low pathogenic fitness, although this is only expressed in plants containing the 'defeated' resistance gene. A number of separate isolates which overcome $Tm\text{-}2^2$ behave normally in susceptible hosts, but in $Tm\text{-}2^2$ plants they cause very severe stunting, multiply poorly, and have difficulty in establishing infection (R.S.S. Fraser, L. Betti and S.L. Bhattiprolu, unpublished results). This is perhaps more a case of a 'defeated' resistance gene not being fully defeated, than a loss of pathogenic fitness in a fully virulent isolate. However, the failure of the virus fully to overcome the resistance, while obviously being able to go part way, may imply that full virulence is lethal or counterselective.

Consequences and implications for plant breeding

Frequency of virulence

The occurrence of virulent virus isolates overcoming many of the resistance genes in the survey sample obviously poses potential problems for the plant breeder. There is a danger that attempts to incorporate new resistance genes into useful varieties will be confounded if the virus can quickly mutate to resistance-breaking forms. Certainly, any new resistance gene should be screened against the largest practicable number of isolates of the target virus, to check whether virulence against it already exists, before embarking on an expensive full-scale breeding programme. There may also be merit in challenging a new resistance gene with nitrous acid or other artificial mutants of the virus. The ease with which it was possible to produce HNO_2 mutants of TMV overcoming the *Tm-1* gene in tomato (Fraser & Gerwitz, 1987) may, with hindsight, have allowed prediction of the low durability of *Tm-1* when introduced in commercial cultivars (Pelham et al., 1970).

The frequency of virulence against many resistance genes in the survey sample suggests that more effort should be given to 'pyramiding' resistance genes, in an attempt to produce oligogenic resistance systems which may be more difficult for the virus to overcome. To my knowledge, no TMV strains have been reliably shown to overcome both *Tm-2* and $Tm\text{-}2^2$; it may be difficult for the virus to accommodate both virulence determinants within a single viral gene, without them interacting or interfering with the function of the virus movement protein which the gene specifies.

Lack of resistance genes

A major barrier to construction of potentially robust oligogenic resistance systems in many crops, or even to breeding for virus resistance at all, is the scarcity of available resistance genes. J.A. Tomlinson (personal communication) listed 25 viruses affecting 22 horticultural crops in the UK, where no genes for resistance were known. In other crops, the genetic base of available resistance is very narrow, and could easily be eroded by evolution of virulent forms. There is therefore a need to evaluate genetic resource collections, landraces and related wild species for new resistance genes (Lenné & Wood, 1991). Further useful resistances may come from somaclonal variation (Scowcroft & Larkin, 1982) or from spontaneous mutation or chromosome fragment deletion in susceptible cultivars (Toyoda et al., 1989; Pehu et al., 1990). Most promisingly, transgenic plants expressing novel types of resistance derived from portions of the viral genome (Nelson et al., 1990) will offer the opportunity to fill gaps in the available battery of resistance genes.

New challenges in resistance breeding

Discovery of new viruses, and the spread of viruses or their vectors to new geographical areas, creates new challenges for the plant breeder. In the longer term, changes in the global environment, increases in CO_2 concentration, and possible temperature increase, may influence the interactions between plants and viruses. Very little is known about possible effects of increased CO_2 on virus multiplication, crop yield, and operation of resistance mechanisms. An increase in temperature might adversely affect some resistance mechanisms, as many are temperature-sensitive (Fraser 1986; Fraser & Loughlin, 1982). However, a more dramatic effect of increased temperature on plant-virus interactions is likely to be by an indirect effect on insect vectors. Even a change of a few degrees may have major effects on insect population dynamics (Collier et al., 1991).

Acknowledgements

This chapter was written while on a short visit to Sri Venkateswara University, Tirupati, South India. I thank the staff and students of the Virology Department there, and the participants in the First Indian Workshop on Molecular Biology of Plant Viruses, for their hospitality and useful discussions. Financial support was provided by the British Council.

References

Bald, J.G. & T.W. Tinsley, 1967. A quasi-genetic model for plant virus host ranges. II. Differentiation between host ranges. Virology 32: 321–327.

Barker, H. & B.D. Harrison, 1984. Expression of genes for resistance to potato virus Y in potato plants and protoplasts. Ann. Appl. Biol. 105: 539–545.

Calder, V.L. & P. Palukaitis, 1992. Nucleotide sequence analysis of the movement genes of resistance breaking strains of tomato mosaic virus. J. Gen. Virol. 73: 165–168.

Cheo, P.C., 1970. Subliminal infection of cotton by tobacco mosaic virus. Phytopathology 60: 41–46.

Collier, R.H., S. Finch, K. Phelps & A.R. Thompson, 1991. Possible impact of global warming on cabbage root fly (*Delia radicum*) activity in the UK. Ann. Appl. Biol. 118: 261–271.

Crute, I.R. & J.M. Norwood, 1986. Gene-dosage effects on the relationship between *Bremia lactucae* (downy mildew) and *Lactuca sativa* (lettuce): the relevance to a mechanistic understanding of host-parasite specificity. Physiol. Molec. Plant Pathol. 29: 133–145.

Csillery, G., I. Tobias & J. Rusko, 1983. A new pepper strain of tomato mosaic virus. Acta Phytopathol. Acad. Sci. Hung. 18: 195–200.

Culver, J.N. & W.O. Dawson, 1989. Point mutations in the coat protein gene of tobacco mosaic virus induce hypersensitivity in *Nicotiana sylvestris*. Mol. Plant-Microbe Inter. 2: 209–213.

Day, K.L., 1984. Resistance to bean common mosaic virus in *Phaseolus vulgaris* L. PhD Thesis, University of Birmingham, UK.

Drijfhout, E., 1978. Genetic interaction between *Phaseolus vulgaris* and bean common mosaic virus with implications for strain identification and breeding for resistance. Agric. Res. Rep. Wageningen 872: 1–98.

Fraser, R.S.S., 1985. Host range control and non-host immunity to viruses. In: R.S.S. Fraser (Ed.), Mechanisms of Resistance to Plant Diseases, pp. 13–28. Nijhoff/Junk, Dordrecht.

Fraser, R.S.S., 1986. Genes for resistance to plant viruses. CRC Crit. Rev. Plant Sci. 3: 257–294.

Fraser, R.S.S., 1988. Virus recognition and pathogenicity: implications for resistance mechanisms and breeding. Pestic. Sci. 23: 267–275.

Fraser, R.S.S., 1990. The genetics of resistance to plant viruses. Annu. Rev. Phytopathol. 28: 179–200.

Fraser, R.S.S. & A. Gerwitz, 1987. The genetics of resistance and virulence in plant virus disease. In: P.R. Day & G.J. Jellis (Eds.), Genetics and Plant Pathogenesis, pp. 33–44. Blackwell Scientific Publications, Oxford.

Fraser, R.S.S., A. Gerwitz & G.E.L. Morris, 1986. Multiple regression analysis of the relationships between tobacco mosaic virus multiplication, the severity of mosaic symptoms, and the growth of tobacco and tomato. Physiol. Molec. Plant Pathol. 29: 239–249.

Fraser, R.S.S. & S.A.R. Loughlin, 1980. Resistance to tobacco mosaic virus in tomato: effects of the *Tm-1* gene on virus multiplication. J. Gen. Virol. 48: 87–96.

Fraser, R.S.S. & S.A.R. Loughlin, 1982. Effects of temperature on the *Tm-1* gene for resistance in tobacco mosaic virus in tomato. Physiol. Plant Pathol. 20: 109–117.

Gabriel, D.W. & B.G. Rolfe, 1990. Working models of specific recognition in plant-microbe interactions. Annu. Rev. Phytopathol. 28: 365–391.

Goldbach, R., R. Eggen, C. de Jager, A. van Kammen, J. van Lent, G. Rezelman & J. Wellink, 1990. Genetic organization, evolution and expression of plant viral RNA genomes. In: R.S.S. Fraser (Ed.), Recognition and Response in Plant-Virus Interactions, pp. 147–162. Springer-Verlag, Berlin.

Hayes, R.J. & K.W. Buck, 1990. Complete replication of a eukaryotic virus RNA in vitro by a purified RNA-dependent RNA polymerase. Cell 63: 363–368.

Holmes, F.O., 1938. Inheritance of resistance to tobacco mosaic virus in tobacco. Phytopathology 28: 553–561.

Jones, A.T., 1990. Breeding for resistance to virus vectors. Proc. Brighton Crop Prot. Conf. Pests and Diseases 3: 935–938.

Jones, A.T. & P.L. Catherall, 1970. The relationship between growth rate and the expression of tolerance to barley yellow dwarf virus in barley. Ann. Appl. Biol. 65: 137–145.

Kacser, H. & J.A. Burns, 1981. The molecular basis of dominance. Genetics 97: 639–666.

Lenné, J.M. & D. Wood, 1991. Plant diseases and the use of wild germplasm. Annu. Rev. Phytopathol. 29: 35–63.

Louie, R., W.R. Findley, J.K. Knoke & M.D. McMullen, 1991. Genetic basis of resistance to five maize dwarf mosaic virus strains. Crop Sci. 31: 14–18.

McMullen, M.D. & R. Louie, 1989. The linkage of molecular markers to a gene controlling the symptom response in maize to maize dwarf mosaic virus. Molec. Plant-Microbe Interact. 2: 309–314.

Meshi, T., F. Motoyoshi, A. Adachi, Y. Watanabe, N. Takamatsu & Y. Okada, 1988. Two concomitant base substitutions in the putative replicase genes of tobacco mosaic virus confer the ability to overcome the effects of a tomato resistance gene, *Tm-1*. EMBO J. 7: 1575–1522.

Meshi, T., F. Motoyoshi, T. Maeda, S. Yoshiwoka, H. Watanabe & Y. Okada, 1989. Mutations in the tobacco mosaic virus 30-kD protein gene overcome *Tm2* resistance in tomato. Plant Cell 1: 515–522.

Mundry, K.-W., W. Schaible, M. Ellwart-Tschürtz, H. Nitschko & C. Hapke, 1990. Hypersensitivity to tobacco mosaic virus in *N'*-gene hosts: which viral genes are involved? In: R.S.S. Fraser (Ed.), Recognition and Response in Plant-Virus Interactions, pp. 345–359. Springer-Verlag, Berlin.

Murant, A.F., C.E. Taylor & J. Chambers, 1968. Properties, relationships and transmission of a strain of raspberry ringspot virus infecting raspberry cultivars immune to the common Scottish strain. Ann. Appl. Biol. 61: 175–186.

Nelson, R.S., P.A. Powell & R.N. Beachy, 1990. Coat-protein mediated protection against virus infection. In: R.S.S. Fraser (Ed.), Recognition and Response in Plant-Virus Interactions, pp. 427–442. Springer Verlag, Berlin.

Nishiguchi, M. & F. Motoyoshi, 1987. Resistance mechanisms of tobacco mosaic virus strains in tomato and tobacco. In: D. Evered & S. Harnett (Eds.), Plant Resistance to Viruses, pp. 38–56. John Wiley & Sons, Chichester.

Pehu, E., R.W. Gibson, M.G.K. Jones & A. Karp, 1990. Studies on the genetic basis of resistance to potato leaf roll virus, potato virus Y and potato virus X in *Solanum brevidens* using somatic hybrids of *Solanum brevidens* and *Solanum tuberosum*. Plant Science 69: 95–101.

Pelham, J., 1972. Strain-genotype interaction of tobacco mosaic virus in tomato. Ann. Appl. Biol. 71: 219–228.

Pelham, J., J.T. Fletcher & J.H. Hawkins, 1970. The establishment of a new strain of tobacco mosaic virus resulting from the use of resistant varieties of tomato. Ann. Appl. Biol. 65: 293–297.

Roane, C.W., S.A. Tolin & H.S. Aycock, 1989. Genetics of reaction to maize dwarf mosaic virus strain A in several maize inbred lines. Phytopathology 79: 1364–1368.

Rosenkranz, E. & G.E. Scott, 1984. Determination of the number of genes for resistance to maize dwarf mosaic virus strain A in five corn inbred lines. Phytopathology 74: 71–76.

Scowcroft, W.R. & P.J. Larkin, 1982. Somaclonal variation: a new option for plant improvement. In: I.K. Varil, W.R. Scowcroft & K.J. Frey (Eds.), Plant Improvement and Somatic Cell Genetics, pp. 159–178. Academic Press, New York.

Sulzinski, M.A. & M. Zaitlin, 1982. Tobacco mosaic virus replication in resistant and susceptible plants: in some resistant species virus is confined to a small number of initially infected cells. Virology 121: 12–19.

Tobias, I., R.S.S. Fraser & A. Gerwitz, 1989. The gene-for-gene relationship between *Capsicum annuum* L. and tobacco mosaic virus: effects on virus multiplication, ethylene synthesis and accumulation of pathogenesis-related proteins. Physiol. Mol. Plant Pathol. 35: 271–286.

Toyoda, H., K. Chatani, Y. Matsuda & S. Ouchi, 1989. Multiplication of tobacco mosaic virus in tobacco callus tissues and in vitro selection for viral disease resistance. Plant Cell Rep. 8: 433–436.

Urban, L.A., J.L. Sherwood, J.A.M. Rezende & U. Melcher, 1990. Examination of mechanisms of cross protection in non-transgenic plants. In: R.S.S. Fraser (Ed.), Recognition and Response in Plant-Virus Interactions, pp. 415–426. Springer Verlag, Berlin.

Vanderplank, J.E., 1984. Disease Resistance in Plants. Second Edition. Academic Press, New York.

Yamaya, J., M. Yoshioka, T. Meshi, Y. Okada & T. Ohno, 1988. Cross protection in transgenic tobacco plants expressing a mild strain of tobacco mosaic virus. Mol. Gen. Genet. 215: 173–175.

Euphytica **63**: 187–197, 1992.
R. Johnson and G.J. Jellis (eds), Breeding for Disease Resistance

Transgenic potato plants resistant to viruses

Marianne J. Huisman, Ben J.C. Cornelissen & Erik Jongedijk
Mogen International N.V., Einsteinweg 97, 2333 CB Leiden, The Netherlands

Key words: potato breeding, coat protein mediated protection, virus resistance, PVX, *Solanum tuberosum*

Summary

Traditional potato breeding is a laborious process in which, by intercrossing, valuable traits from different parental clones are combined in a progeny genotype. Depending on the availability of genes, molecular techniques can be used to add specific genes to existing cultivars that, although otherwise satisfactory, lack a few commercially important traits. For virus resistance the gene for the coat protein of a given virus transplanted into the genome of the plant renders the plant resistant to that virus. In conferring such resistance to potato varieties it proved to be possible to preserve their intrinsic properties.

Introduction

Conventional potato breeding

In the 16th century the cultivated potato (*Solanum* sp.) was imported into Europe from South and Central America. In the subsequent eras the world accepted the potato as a major food source and vast amounts of agricultural land were used to grow them. Unfortunately, monocultures are renowned for their ability to increase the occurrence of diseases and pests. Multiple introductions of new *Solanum* species from South and Central America have increased the number of diseases carried with the potato. As the main priority of the early breeding programmes was to increase tuber yield, many potato cultivars showed a range of disease susceptibilities. In recent decades, the interest of breeders has shifted towards quality and processing characteristics and disease resistance.

In potato breeding carefully selected parental clones are intercrossed and the best performing F1 hybrids are selected. This process is time-consuming because the cultivated potato is an autotetraploid plant species which is highly heterozygous for many of its valuable traits. Intercrossing for breeding therefore generates enormous variation only some of which is useful. Selection among vast numbers of progeny clones is necessary to establish the presence of valuable traits of the two parental clones in their offspring. Decades of careful crossing and selecting has generated a large number of useful cultivars. However, even the best performing cultivars today contain some undesirable traits (Harris, 1978).

Molecular breeding

Molecular breeding techniques can have an advantage over conventional breeding in that they allow the addition of specific genes encoding desirable traits to a cultivar or advanced breeding line, whilst preserving its intrinsic features. Molecular techniques can thus expedite the generation of new cultivars from suitable parental clones (Jongedijk et al., 1992). The molecular techniques alluded to here can be divided into two sets. Firstly are the cellular techniques such as transformation and regeneration of transformed cells into new plants and

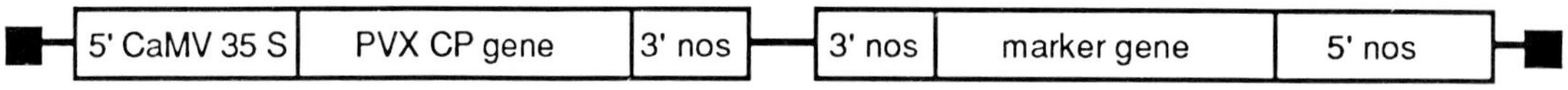

Fig. 1. Schematic representation of the T-DNA region from the binary construct. The region between the solid rectangles represents the piece of T-DNA present in the binary construct, this coincides with the piece of DNA transferred to the genomic DNA of the plant. 5′ CaMV 35S, plant recognizable promoter sequence derived from cauliflower mosaic virus; 5′ nos, plant recognizable promoter sequence derived from nopaline synthase gene from *Agrobacterium tumefaciens;* 3′ nos, plant recognizable terminator sequence derived from the nopaline synthase gene; marker gene, selectable marker gene coding for resistance to the antibiotic kanamycin; PVX CP, coat protein gene of potato virus X.

secondly there are molecular techniques required for identification, isolation and modification of specific genes coding for interesting traits.

Regeneration and transformation

Plant cells are totipotent. Single cells can dedifferentiate, grow out into other kinds of specialized cell types and subsequently develop into a complete plant. The genotype of this new plant may be identical to the genotype of the cell it was regenerated from. Although plant cells are totipotent, some tissue cell types are more readily induced to grow into whole new plants than others. Conditions necessary for efficient regeneration vary among plant cell and tissue types, and also among plant species and plant species cultivars. Some plant species are easily kept in tissue culture, others require intricate media to allow axenic growth. Special plant hormones, cytokinins and auxins, are required to induce shoot and root growth from plant tissue. Formation of shoots or roots depends on the plant type and the tissue type and also on the ratio in which the two hormones are present in the medium (see e.g. Hoekema et al., 1989).

During tissue culture growth, plant cells can accumulate chromosomal aberrations such as variations in chromosome number, chromosome rearrangements and point mutations (Lee & Philips, 1988). Such genotypic deviations from the wild type often give rise to phenotypic changes called somaclonal variation. Somaclonal variation occurs frequently and spontaneously. The frequency may vary with the plant species. Somaclonal variation is well known to occur in potato tissue especially when it is kept in culture for a long time (Potter & Jones, 1991). Moreover, because potato is a highly heterozygous, tetraploid crop species these effects cannot be reduced by self-pollination or backcrossing, without losing the intrinsic genetic constitution of the cultivar.

Upon transformation, a single cell is modified in such a way that it contains an extra piece of DNA, integrated into the genomic DNA of the cell. After regeneration of a transformed cell into a whole plant, this new plant contains one extra piece of genetic information, which will be stably inherited in the offspring. In dicotyledonous plant species this transfer of DNA can be mediated by the plant pathogenic bacterium *Agrobacterium tumefaciens.* The interaction between *A. tumefaciens* and dicotelydonous plants is well established (e.g. Sheerman & Bevan, 1988). During the interaction between *A. tumefaciens* and a plant cell, a specific part of the Ti plasmid from *A. tumefaciens*, the so called T (for transfer)-DNA, is transferred to the plant cell and is integrated into its DNA. The right and left border regions of the T-DNA are essential, all the other sequences of the T-DNA are non-essential for transfer. This permits the insertion of genes into the T-DNA. Thus a selectable marker gene may be included next to the gene of interest, such as a gene conferring resistance to an antibiotic (Fig. 1). In the case of plant transformation the most often used antibiotic is kanamycin and several different genes conferring resistance to kanamycin are available. After transformation with a selectable marker gene as well as a gene of interest, the transformed cells are distinguished from all non-transformed cells by resistance to the antibiotic present in the medium: only transformed cells will grow out into new plants containing the extra genetic information.

For expression of the gene of interest in the transformed plant cell, plant recognizable transcription initiation and stop signals as well as plant recognizable translation initiation and stop signals (all regulatory signals) are essential. The degree to which the gene is expressed depends on the place of integration in the genome of the host cell. With the potato virus X coat protein (PVX CP) gene construct as shown in Fig. 1, selection on kanamycin gave rise to numerous transgenic shoots which after rooting grew into normal plants. Most of the independently obtained transgenic lines produced different amounts of the transgene product. Selection for the level of expression of the chosen gene is an important next step after transgenic plants have been selected on media containing kanamycin.

Gene identification and selection

The availability of suitable genes is a limiting factor in molecular breeding. The transfer of multiple genes present on separate vectors is not only time-consuming because of the separate transformation steps, but is also hampered by lack of enough different selectable marker genes. Another approach would be the simultaneous transfer of multiple genes on one vector with only one selectable marker gene. This strategy is being studied at present, but for the near future, single genes are the most likely candidates amenable to molecular breeding techniques. Although many single genes are available for breeding programmes e.g. in parental lines, a gene is only available for molecular breeding when the coding region of that gene has been identified and isolated and is present in such a form that it can be modified to give rise to the desired phenotype in plants after transformation. Such a coding region must be placed between regulatory sequences necessary for expression in plant cells. The resulting chimaeric gene (Fig. 1) is usually only a few kilobases long.

Coat-protein-mediated virus resistance

When the coat-protein gene of a certain virus is constitutively expressed in a plant, this plant becomes resistant to infection by that specific virus and to a lesser extent resistant also to some closely related viruses (Van Dun & Bol, 1988; Stark & Beachy, 1989; Nejidat & Beachy, 1990; Ling et al., 1991). This phenomenon is called coat-protein-mediated resistance. Coat-protein-mediated resistance is the most frequently used molecular strategy to obtain virus resistance. In 1986 Beachy and co-workers (Powell Abel et al., 1986) showed that transformation of tobacco plants with the coat protein (CP) gene of tobacco mosaic virus (TMV) resulted in resistance against this virus. The addition of the extra gene provided the resistance sought after, without affecting the visible phenotype of that plant species. Later results with alfalfa mosaic virus (AlMV) (Van Dun et al., 1988) and TMV (Powell Abel et al., 1990) indicated that the presence of CP in the transgenic plants and not the expression of the mRNA coding for the protein was responsible for the observed resistance because a truncated AlMV CP gene that could produce mRNA but not CP did not give rise to resistant transgenic plants. Likewise, the TMV CP gene lacking its AUG translational start codon was unable to produce resistance in transgenic plants expressing ample amounts of mRNA transcribed from the transgene. The idea that the protein and not its mRNA is responsible for the resistant phenotype observed, was supported by claims that higher levels of CP gene expression would lead to higher levels of resistance (Hemenway et al., 1988; Powell Abel et al., 1990). However, evidence is accumulating which seems to indicate that the observed resistance in CP transgenic plants need not always be mediated by the viral CP. Resistance to potato leafroll virus (PLRV) could be observed in transgenic potato plants in which the cloned CP gene product could not be detected, although the presence of the CP mRNA could be demonstrated convincingly (Van der Wilk et al., 1991; Kawchuk et al., 1991). Furthermore, in the case of cucumber mosaic virus and potato virus X, RNA expression of CP genes in antisense orientation in transgenic plants was shown to result in protection against the virus, albeit to a much lower extent (Cuozzo et al., 1988; Hemenway et al., 1988). For PLRV antisense CP transgenic plants the protection observed was as high as the sense orientation mediated protection (Van der Wilk et al., 1991; Kawchuk et al., 1991).

In the past 5 years many different plant plus-sense RNA viruses were shown to behave in a similar fashion, i.e. CP gene expression in a transgenic plant rendered the plant resistant to the corresponding virus. Resistance via CP gene expression could be observed in many groups of the plus-sense RNA viruses: tobamo-, tobra-, potex-, poty-, luteo-, cucumo-, and ilarviruses. Also resistance against a plant bunyavirus, an enveloped negative-strand RNA virus (tomato spotted wilt virus, TSWV), was recently obtained after expression of the TSWV CP gene in transgenic tobacco plants (Gielen et al., 1991). The combination of all these data indicates that viral CP genes introduced into plants evoke resistance to the corresponding viruses. Also some evidence is available which shows that different mechanisms might mediate resistance in the ensuing transgenic plants (Van der Wilk et al., 1991). Possibly, differences between viruses from different taxonomic groups in, for example, assembly or uncoating, might be responsible for the differences observed with CP-mediated resistance.

Engineering virus resistant potato plants

The cultivated potato is a crop affected by many diseases. The major virus diseases in potato are caused by very diverse kinds of plus-stranded RNA viruses: potato virus X (PVX), potato virus Y (PVY) and PLRV (Matthews, 1981). PLRV is economically the most important potato virus disease, followed closely by PVY and PVX. PVX infection itself does not give rise to great economic losses, but in conjunction with other potato viruses like PVY serious crop losses can occur because of the synergistic effects when the two viruses are present in one plant (Rochow & Ross, 1955). The biology of the three viruses differs, too. PVX is the type member of the potexviruses, PVY is the type member of the potyviruses and PLRV is the type member of the luteoviruses. Both potex- and potyviruses have their genetic contents in flexuous rods, but luteoviruses have their RNA encapsidated in spherical particals. These capsid structures consist of coat protein molecules. PLRV and PVY are aphid transmissible, the former in a persistent manner the latter non-persistent. PVX is mechanically transmissible only. Spread of potato viruses is controlled by insecticides for PLRV, by roguing of diseased plants and by the use of certified seed potatoes. This way major crop losses can be prevented.

PVX causes a mild mosaic sometimes called the 'healthy potato disease'. Although, the primary symptoms are usually only a slight mottling, secondary infections may have drastic effects (Beemster & de Bokx, 1987). The virus contains one single stranded RNA molecule which putatively encodes five discrete genes. The 5′-proximal gene is translated from the parental RNA and the other four genes from subgenomic mRNAs. The CP gene was shown to be located at the 3′ end of the genome (Huisman et al., 1988). The RNA of PVY contains one large open reading frame for a putative protein of about 200 kD. This large polyprotein is cleaved during translation into smaller functional proteins. As in the case of PVX CP, the PVY CP cistron is located at the 3′ terminus of the RNA but in this case it is processed from a polyprotein (Robaglia et al., 1989). The PLRV CP gene is located near the 3′ end but its leaky UAA translational stop codon is followed immediately by another coding region (Van der Wilk et al., 1989).

Potato virus X

The PVX CP gene with a few extra bases at both its 5′ and its 3′ end was cloned between the necessary plant transcription initiation and termination signals: the cauliflower mosaic virus derived 35S promoter and the nopaline synthase derived terminator sequences (Hoekema et al., 1989), respectively (Fig. 1). The chimaeric gene together with a kanamycin resistance gene as a selectable marker was cloned into a binary vector in *Agrobacterium tumefaciens*. Potato tuber discs of cultivars Escort and Bintje were dipped in the bacterial suspension to obtain transgenic plants (Hoekema et al., 1989). After selection on kanamycin-containing medium numerous transgenic shoots were seen to emerge on the tuber discs. These shoots were cut from the discs and were regenerated into plants.

Analysis of the individual transformants for the

presence of CP was done by the use of the western blot technique: proteins were isolated from leaf samples, size separated on polyacrylamide gels and blotted onto a suitable membrane, followed by detection of the coat protein of PVX with antiserum obtained from a rabbit immunized with PVX purified particles. The amount of coat protein present in the samples was estimated by comparison with standard amounts of purified PVX CP which were included on the same gel. Analysis showed appreciable amounts of PVX CP accumulation: in between 0.05 and 0.3% of the soluble protein fraction (Hoekema et al., 1989). The CP levels were measured in plants grown in pots in a growth room after transfer of the axenically grown plantlets to soil. The CP analysis could not be done on axenically grown plants because virtually no coat protein could be detected in them. In some cases the level of expression of the CP gene appeared to be higher in field grown plants than in those grown in the growth room (Jongedijk et al., 1992).

Since in earlier studies a positive correlation had been suggested between level of CP gene expression and level of resistance, four transgenic plant lines which showed high CP levels were selected for initial indoor resistance tests: axenically grown plantlets were transferred to soil in pots, inoculated with PVX at 1 μg/ml and analysed for PVX accumulation at 2 weeks post inoculation. The experiment showed that the virus titer in transgenic plants at 14 days post inoculation was substantially (100–1000 times) lower than that in regenerated plants or vector transformed plants (Hoekema et al., 1989). This indicated that potato could be engineered for PVX resistance by inserting the PVX CP gene in the potato genome.

Potato leafroll virus

The PLRV CP encoding region with a few bases upstream and a few bases downstream of the cistron was engineered to contain the plant expression signals as with the chimaeric PVX CP gene. Both a sense and an antisense PLRV CP gene construct were transferred to the potato cultivar Désirée. Désirée was selected on the basis of its susceptibility to PLRV infection. Surprisingly, in the transgenic plants sense PLRV CP mRNA could be detected, but its protein could not. These plants showed resistance to PLRV infection both in primary infected plants as well as in the secondary infected plants analyzed in the following growing season (Van der Wilk et al., 1991; Kawchuk et al., 1991). The resistance was measured by ELISA on infected material at various time points after infection. The same level of resistance was obtained with transgenic plants containing antisense PLRV CP. The PLRV CP antisense transgenic plants showed the presence of the antisense CP mRNA but for obvious reasons no protein. The suggestion was made that the RNA itself played a role in resistance similar to that observed with antisense RNA-mediated changes in flower colour in petunia (Van der Krol et al., 1988). Possibly, the sense RNA acted in the same way as the antisense RNA i.e. in preventing replication of RNA molecules by hybrization to the strand that is being replicated.

Potato virus Y

In the literature, the only transformation with PVY CP that has produced resistant plants is reported by Lawson et al. (1990). PVY coat protein does not contain a start codon, because of the polyprotein processing strategy of the virus. Therefore, the construct for the PVY CP chimeric gene was engineered to contain an additional AUG translational start codon next to the required regulatory signals. In transgenic plants the coat protein of PVY was hardly detectable and especially difficult to detect in the plant line showing the best resistance to infection. This makes one wonder about the actual mechanism causing the resistance. Other laboratories, including our own, have tried to obtain transgenic plants expressing PVY CP. However, no other successes have been reported. In most cases a PVY CP gene expression construct with an additional AUG preceding the PVY CP coding region was used. A number of alternative possibilities are feasible to obtain transgenic plants containing PVY CP, like using an AUG start codon present upstream of the CP coding region or the cloning of the specific protease coding region as well in order to let the protease process this small polyprotein.

Resistance by the expression of the potyvirus CP was demonstrated for other potyviruses like soybean mosaic virus (Stark & Beachy, 1989) and papaya ringspot virus (Ling et al., 1991). In those reports the coat protein of one potyvirus was shown to render the transgenic plants resistant to a range of potyviruses. The level of the potyviral CP-mediated resistance is lower than that observed with, for instance, TMV or PVX. It seems more comparable to the resistance observed with PLRV CP-mediated resistance.

Field evaluation of transgenic potato plants

Before placing the PVX CP transgenic plants in the field one extra laboratory test consisted of selection on gross karyotypic changes. Aberrations in karyotype were found in two out of the 41 Bintje lines being tested. One was a polyploid with about 96 chromosomes, the other was an aneuploid with 47 chromosomes (Hoekema et al., 1989). No major karyotype changes could be detected in any of the Escort transgenic plants. After this gross selection the non-aberrant transgenic clones, 39 Bintje and 22 Escort clones, were propagated in the field to test whether the molecular biological techniques described above can be used to transform potato while preserving the intrinsic cultivar properties, whether expression of the transgene is stable under field conditions and whether PVX CP transgenic plants are resistant to PVX in the field.

Stability of the transgene

Stability of transgene expression was analysed by taking leaf samples on a number of occasions during the growing season. These leaf samples were analysed by western blotting as previously described. The conclusion of this analysis was that the expression of the PVX CP gene in transgenic plants showing medium and high coat protein expression was consistently stable over the growing season, i.e. no major fluctuations were observed. Moreover, the expression level in the laboratory and greenhouse tests was appreciably lower than the levels measured in the plants growing in the field (Jongedijk et al., 1992).

Preservation of intrinsic properties

The intrinsic properties of potato cultivars were checked by analysis of yield, grading and the 50 different UPOV (1986) morphological characteristics: 21 plant, 10 flower, 7 tuber and 12 light sprout characteristics (light sprouts are obtained by placing the tubers in continuous low light conditions (Houwing et al., 1986). The UPOV morphological characteristics are currently used for cultivar identification. For the transgenic PVX CP expressing potato clones, analysis of all characteristics was carried out in randomized complete block experiments in two replicates with five plants per plot. For all characteristics the transgenic plants were checked for significant deviations from the control cultivar, Bintje or Escort (Jongedijk et al., 1992). Nine of the characteristics analyzed were found to be unchanged in all Bintje or Escort transgenic lines. All other morphological characteristics showed changes to some degree. Typical changes in morphology included reduced vigour, small and glossy leaves, coalesced leaves, relatively closed foliage (Fig. 2), darker green leaf tissue, stunted inflorescences, changes in light sprout shape and changes in tuber shape as well as reduced yield (Fig. 3). Combining the results for all 52 characteristics, 17.9% of all the Bintje and 81.8% of all the Escort transgenic clones were true-to-type. In a number of true-to-type Bintje transgenic clones some consistent but insignificant effects on plant vigour and tuber yield were observed but for Escort no such variations were observed. Transgenic plants with changes in plant morphology showed pronounced negative effects on plant vigour. Positive effects on vigour were never observed (Jongedijk et al., 1992).

During the analyses it was observed that the morphology of light sprouts of mature tubers harvested from the field and from growth room grown plants were alike. Since aberrant light sprout mor-

Fig. 2. Leaves of transgenic and non-transgenic Escort plants in the field. The upper left panel is taken from Escort, the upper right panel is taken from transgenic Escort line MGE-30 which is true to type. The lower two panels show aberrant phenotypes in MGE-28 and MGE-16: the leaves are darker green and the foliage structure is more closed.

Fig. 3. Tubers harvested from plants grown in the field. The upper left panel is taken from Escort, the upper right panel is taken from transgenic Escort line MGE-38 which is true to type. The lower two panels show aberrant phenotypes of which the lower right MGE-28 shows a clear penalty on yield and the lower left shows an effect on tuber shape.

phology was invariably associated with deviations in plant phenotype in the field this permits the use of light sprout morphology analysis as a tool in the early screening of transgenic clones for aberrations (Jongedijk et al., 1992).

Field resistance

In the literature, wherever field trials are described, the resistance of transgenic plants was assessed by the appearance of symptoms upon challenge inoculation, or more specifically the number of individual plants that showed symptoms after primary infection was recorded. In addition, as an alternative parameter for resistance, enzyme linked immunosorbent assay (ELISA) values for primary infected plants were used as an indication for the amount of virus present (Nelson et al., 1988; Kaniewski et al., 1990; Van der Wilk et al., 1991). However, for vegetatively propagated crops like potato, these tests are inappropriate. A delay in symptom development or virus accumulation after primary infection does not necessarily result in lower frequencies of infection in clonal progeny as is required in seed tuber production. Therefore, in the tests for cultivar entry onto the Dutch Variety List secondary infection rates are measured rather than primary infection rates. For the transgenic Bintje and Escort clones such secondary infection rates were measured to assess field resistance.

The field resistance of six transgenic Bintje and six transgenic Escort clones was assessed by growing them together with eight registered varieties with different ratings for PVX resistance, among which were non-transgenic Bintje and Escort. Tubers were planted in four replicates in randomized complete block experiments with four plants per plot. Each plot was surrounded by a double row of PVX infected plants which were used for the mechanical inoculation of the test plants. When the foliage between rows touched, a harrow was dragged through the field twice to damage the plants and thereby spread the virus. This kind of mechanical damage is sufficient to get a uniform infection over the whole field. Six weeks after infection three tubers were harvested per individual plant and plants emerging from these tubers were analysed for PVX content. In a standard double antibody sandwich ELISA as used for detection of PVX in tuber material, the PVX coat protein in the transgenic plants could only be detected a long time after the substrate has been added. When an infection is established in a potato plant the PVX titer is easily 100 times higher than the amount of PVX CP in the transgenic plants. Therefore, ELISA

Fig. 4. The effect of the dehaulming apparatus stripping the leaves from the tops of the potato plants.

tests read only 1 hour after adding the substrate did not detect this coat protein but only that from virus that had multiplied in the plant. The percentage of plants found to carry PVX is a measure for ranking the resistance of cultivars on the Dutch Variety List. Significantly fewer PVX infected plants were observed among progeny from most transgenic plant lines expressing medium to high levels of the PVX CP. Other transgenic lines expressing low amounts of CP were not significantly more resistant than the starting material. Most of the transgenic plant lines ranked considerably more resistant than non-transgenic Escort and Bintje (Jongedijk et al., 1992).

The 1990 field test described above was done using the standard procedure for entry of new cultivars onto the Dutch Variety List. Because some reports indicate that coat-protein-mediated-resistance may not be stable under extreme infection pressure, another field test was performed in 1991 in which the inoculation was even more rigorous. The trial layout used for this experiment was similar to that in 1990 but this time, instead of a harrow

Table 1. Incidence of PVX infected plants among clonal progeny from previously heavily infected, transgenic Bintje (MGB) and Escort (MGE) clones and untransformed standard cultivars

Cultivar	% Infected progeny		VLS[2]	PVX CP expression		
	Mean[1]	σ		Low	Medium	High[3]
MGB-26	100.0 a	0.0	5	+		
MGB-18	100.0 a	0.0	5			+
MGE-08	100.0 a	0.0	5			+
Escort	100.0 a	0.0	5			
MGB-44	98.4 a	3.1	5			+
MGB-28	98.4 a	3.1	5			+
MGB-13	98.4 a	3.1	5		+	
Bintje	96.9 a	3.6	5			
MGE-04	93.6 ab	8.8	5–7	+		
Amazone	93.8 ab	5.1	7			
Estima	87.5 ab	8.8	6			
Spunta	84.4 bc	13.0	8			
MGB-66	82.4 bcd	9.1	8–9		+	
MGE-21	81.3 cd	13.5	8–9			+
MGE-44	76.0 cd	10.0	8–9		+	
MGE-13	71.9 d	10.8	8–9			+
MGE-32	59.4 e	16.5	8–9			+
Elles	0.0 f	0.0	9			
Bildtstar	0.0 f	0.0	R			
Sante	0.0 f	0.0	R			

[1] Different letters denote a significant difference (Duncan's multiple range test, $\alpha = 0.05$).
[2] VLS, Variety List Score for PVX resistance according to the Dutch Variety List (Parlevliet et al., 1991): 5 to 9 from very susceptible to very resistant; R, resistant in the field (immunity).
[3] Low, 0.0–0.1%; medium, 0.1–0.2%; high, 0.2–0.3% from total soluble protein in axenically grown plants (Hoekema et al., 1989).

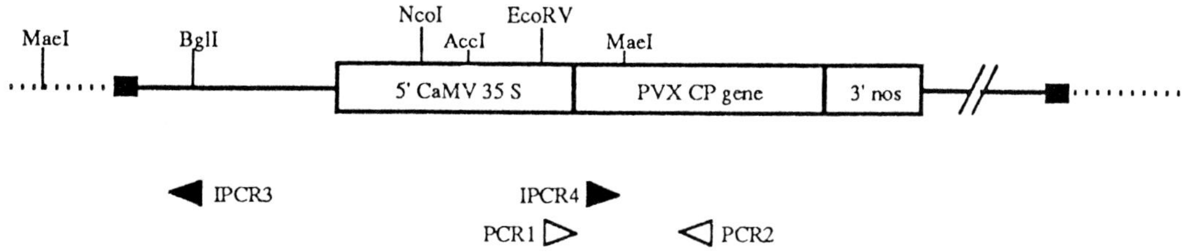

Fig. 5. Schematic representation of the T-DNA region from the binary construct. The region between the solid rectangles represents the piece of T-DNA present in the binary construct, this represents the piece of DNA transferred to the genomic DNA of the plant. Open triangles represent the sites of recognition for the PCR primers PCR 1 and 2, which are used for detection of PVX CP transgenic plants. Solid triangles represent the sites of recognition for the IPCR primers IPCR3 and 4, which are used for the discrimination between transgenic cultivars. The restriction enzyme sites (Mael, Bgll, Ncol, Accl, and EcoRV) denoted above the drawing of the DNA were used in the IPCR procedure (see Does et al., 1991). 5′ CaMV 35S, plant promoter sequence derived from cauliflower mosaic virus; 5′ nos, plant promoter sequence derived from nopaline synthase gene; 3′ nos, plant terminator sequence derived from nopaline synthase gene; marker gene, gene coding for resistance to the antibiotic kanamycin; PVX CP, coat protein gene of potato virus X.

a dehaulming device was used to strip off the leaves from the plants resulting in much more severe damage than that made by a harrow (Fig. 4). The results (Table 1) show that the degree of PVX contamination in clonal progeny was much higher than in the previous experiment (Jongedijk et al., 1992). Moreover, some of the transgenic plant lines that were moderate to highly resistant under standard conditions, like MGE-04, MGE-08, MGB-44, MGB-28, MGB-13 and MGB-18 were not as resistant under the more severe inoculation conditions. However, other plant lines, MGE-04, MGB-66, MGE-21, MGE-44, MGE-13, and MGE-32, clearly demonstrated a higher level of resistance than the original cultivars even under these extreme conditions, comparable to that observed in the best field resistant cultivars, like Spunta and Elles.

Identification of transgenic material

To obtain plant breeders' rights protection for transgenic cultivars, it is necessary to be able to distinguish them unequivocally from the original cultivars. Moreover, it may also be necessary to discriminate between independent transgenic clones. To achieve the former goal a polymerase chain reaction (PCR) was done to discriminate untransformed cultivars from the transgenic clones. Specific primers PCR1 and PCR2 (Fig. 5), which recognized sites in the transgene construct could only amplify a fragment in transgenic clones. Southern blots of gels containing these amplified fragments showed that the PCR amplified fragment present in the transgenic lines corresponded to the PVX CP gene (Jongedijk et al., 1992). To discriminate between independent transgenic plant lines derived from the same cultivar inverted PCR (IPCR) was set up as described by Does et al. (1991). IPCR primers IPCR3 and IPCPR4 (Fig. 5) were used to amplify fragments of DNA corresponding to the site of integration. As expected, no bands could be observed in the non-transgenic cultivars and different patterns of DNA fragments could be seen with different individual transgenic clones. As every transformation is a unique event of integration in the genome and every IPCR of an individual transformant appeared to be unique this showed IPCR to be a useful identification method (Jongedijk et al., 1992).

Conclusions

Although differences may occur in the proportion of true-to-type transgenic plant lines among potato cultivars, transformation can be accomplished with preservation of the intrinsic properties of the cultivar. Since altered light sprout morphology is invariably associated with deviations in overall morphology of the plant line, transgenic potato plant lines can be screened for deviations in the phenotype on the basis of light sprout morphology early in the selection process.

For PVX and PLRV, resistance has been obtained by inserting the corresponding coat protein gene into the genome of potato cultivars. To date, resistance to PLRV infection in PLRV CP transgenic plant lines has been shown only in greenhouse experiments. Field experiments to demonstrate resistance of these plants are planned for the near future. High levels of resistance to PVX were demonstrated in PVX CP transgenic lines both in growth room and in field trials. Under standard infection conditions in the field the majority of the transgenic plant lines tested were ranked considerably higher for resistance than the original material (cultivars Bintje and Escort). Field resistance was shown even under unusually high infection pressure.

In the case of PVX CP transgenic plants, expression of the transgene was shown to be stable over the growing season. With the use of PCR analysis transgenic clones could be distinguished unambiguously from the cultivars from which they were derived and IPCR analysis could be used to distinguish between individual transformants.

Thus it can be concluded that the molecular techniques used to produce transgenic potato plants can indeed be of help to the breeder to enhance the value of new potato cultivars.

References

Beemster, A.B.R. & J.A. de Bokx, 1987. Survey of properties and symptoms, pp. 84–140. In: J.A. de Bokx & J.P.H. van der Want (Eds.), Viruses of potatoes and seed-potato production, 2nd edition. PUDOC, Wageningen, The Netherlands.

Cuozzo, M., K.M. O'Connell, W. Kaniewski, R.-X. Fang, N.-H. Chua & N.E. Tumer, 1988. Viral protection in transgenic plants expressing the cucumber mosaic virus coat protein or its antisense RNA. Bio-Technology 6: 549–557.

Does, M.P., B.M.M. Dekker, M.J.A. de Groot & R. Offringa, 1991. A quick method to estimate the T-DNA copy number in transgenic plants at an early stage after transformation, using inverted PCR. Plant Mol. Biol. 17: 151–153.

Gielen, J.J.L., P. de Haan, A.J. Kool, D. Peters, M.Q.J.M. van Grinsven & R. Goldbach, 1991. Engineered resistance to tomato spotted wilt virus, a negative-strand RNA virus. Bio-Technology 9: 1363–1367.

Harris, P.M. (Ed.), 1978. The Potato Crop: The Scientific Basis for Improvement. Chapman and Hall, London, UK.

Hemenway, C., R.-X. Fang, W. Kaniewski, N.-H. Chua & N.E. Tumer, 1988. Analysis of the mechanism of protection in transgenic plants expressing the potato virus X coat protein or its antisense RNA. EMBO J 7: 1273–1280.

Hoekema, A., M.J. Huisman, L. Molendijk, P.J.M. Van den Elzen & B.J.C. Cornelissen, 1989. The genetic engineering of two commercial potato cultivars for resistance to potato virus X. Bio-Technology 7: 273–278.

Houwing, A., R. Suk & B. Ros, 1986. Generation of light-sprouts suitable for potato variety identification by means of artificial light. Acta Hortic. 182: 359–363.

Huisman, M.J., H.J.M. Linthorst, J.F. Bol & B.J.C. Cornelissen, 1988. The complete nucleotide sequence of potato virus X and its homologies at the amino acid level with various plus-stranded RNA viruses. J. Gen. Virol. 69: 1789–1798.

Jongedijk, E., A.A.J.M. De Schutter, T. Stolte, P.J.M. Van den Elzen & B.J.C. Cornelissen, 1992. Increased resistance to potato virus X and preservation of cultivar properties in transgenic potato under field conditions. Bio-Technology 10: 422–429.

Kaniewski, W., C. Lawson, B. Sammons, L. Haley, J. Hart, X. Delannay & N.E. Tumer, 1990. Field resistance of transgenic Russet Burbank potato to effects of infection by potato virus X and potato virus Y. Bio-Technology 8: 750–754.

Kawchuk, L.M., R.R. Martins & J. McPherson, 1991. Sense and antisense RNA-mediated resistance to potato leafroll virus in Russet Burbank potato plants. Mol. Plant-Microbe Interact. 4: 254–261.

Lawson, C., W. Kaniewski, L. Haley, R. Rozman, C. Newell, P. Sanders & N.E. Tumer, 1990. Engineering resistance to mixed virus infection in a commercial potato cultivar: resistance to potato virus X and potato virus Y in transgenic Russet Burbank. Bio-Technology 8: 127–134.

Lee, M. & R.L. Philips, 1988. The chromosomal basis of somaclonal variation. Annu. Rev. Plant Physiol. Plant Mol. Biol. 39: 413–437.

Ling, K., S. Namba, C. Gonsalves, J.L. Slightom & D. Gonsalves, 1991. Protection against detrimental effects of potyvirus infection in transgenic tobacco plants expressing the papaya ringspot virus coat protein gene. Bio-Technology 9: 752–758.

Matthews, R.E.F., 1981. Plant Virology. Academic Press, New York.

Nejidat, A. & R.N. Beachy, 1990. Transgenic tobacco plants expressing a tobacco mosaic virus coat protein gene are resistant to some tobamoviruses. Mol. Plant-Microbe Interact. 3: 247–251.

Nelson, R.S., S.M. McCormick, X. Delannay, P. Dubé, J. Layton, E.J. Anderson, M. Kaniewska, R.K. Proksch, R.B. Horsch, S.G. Rogers, R.T. Fraley & R.N. Beachy, 1988. Virus tolerance, plant growth, and field performance of transgenic tomato plants expressing coat protein from tobacco mosaic virus. Bio-Technology 6: 403–409.

Potter, R. & M.G.K. Jones, 1991. An assessment of genetic

stability of potato in vitro by molecular and phenotypic analysis. Plant Science 76: 239–248.

Powell Abel, P., R.S. Nelson, B. De, N. Hoffman, S.G. Rogers, R.T. Fraley & R.N. Beachy, 1986. Delay of disease development in transgenic plants that express the tobacco mosaic virus coat protein gene. Science 232: 738–743.

Powell Abel, P., P.R. Sanders, N.E. Tumer, R.T. Fraley & R.N. Beachy, 1990. Protection against tobacco mosaic virus infection in transgenic plants requires accumulation of coat protein rather than coat protein RNA sequences. Virology 175: 124–130.

Robaglia, C., M. Durand-Tardif, M. Tronchet, G. Boudazin, S. Astier-Manifacier, R. & F. Casse-Delbart, 1989. Nucleotide sequence of potato virus Y (N strain) genomic RNA. J. Gen. Virology 70: 935–947.

Rochow, W.F. & F.A. Ross, 1955. Virus multiplication in plants double infected with potato viruses X and Y. Virology 1: 10–27.

Sheerman, S. & M.W. Bevan, 1988. A rapid transformation method for *Solanum tuberosum* using binary *Agrobacterium tumefaciens* vectors. Plant Cell Rep. 7: 13–16.

Stark, D. & R.N. Beachy, 1989. Protection against potyvirus infection in transgenic plants: evidence for broad spectrum resistance. Bio-Technology 7: 1257–1262.

UPOV (International union for the protection of new varieties of plants), 1986. Guidelines for the conduct of tests for distinctness, homogeneity and stability of potato (*Solanum tuberosum*).

Van Dun, C.M.P. & J.F. Bol, 1988. Transgenic tobacco plants accumulating tobacco rattle virus coat protein resist infection with tobacco rattle virus and pea early browning virus. Virology 167: 649–652.

Van Dun, C.M.P., B. Overduin, L. Van Vloten-Doting & J.F. Bol, 1988. Transgenic tobacco expressing tobacco streak virus or mutated alfalfa mosaic virus coat protein does not cross-protect against alfalfa mosaic virus infection. Virology 164: 383–389.

Van der Krol, A.R., P.E. Lenting, J. Veenstra, I.M. van der Meer, R.E. Koes, A.G.M. Gerats, J.N.M. Mol & A.R. Stuitje, 1988. An antisense chalcone synthase gene in transgenic plants inhibits flower pigmentation. Nature 333: 866–869.

Van der Wilk, F., M.J. Huisman, B.J.C. Cornelissen, H. Huttinga & R. Goldbach, 1989. Nucleotide sequence and organization of potato leafroll virus genomic RNA. FEBS Lett. 245: 51–56.

Van der Wilk, F., D. Posthumus-Lutke Willink, M.J. Huisman, H. Huttinga & R. Goldbach, 1991. Expression of the potato leafroll luteovirus coat protein gene in transgenic potato plants inhibits viral infection. Plant Mol. Biol. 17: 431–439.

Euphytica **63**: 199–205, 1992.

Index